Developing Spatial Data Infrastructures: From concept to reality

Developing Spatial Data Infrastructures: From concept to reality

Edited by

Ian Williamson
Abbas Rajabifard
and
Mary-Ellen F. Feeney

CRC Press
Taylor & Francis Group
Boca Raton London New York

CRC Press is an imprint of the
Taylor & Francis Group, an **informa** business

CRC Press
Taylor & Francis Group
6000 Broken Sound Parkway NW, Suite 300
Boca Raton, FL 33487-2742

First issued in paperback 2019

© 2003 by Taylor & Francis Group, LLC
CRC Press is an imprint of Taylor & Francis Group, an Informa business

No claim to original U.S. Government works

ISBN-13: 978-0-415-30265-4 (hbk)
ISBN-13: 978-1-138-37250-4 (pbk)

Library of Congress Cataloging-in-Publication Data

Catalog record is available from the Library of Congress

**Visit the Taylor & Francis Web site at
http://www.taylorandfrancis.com**

**and the CRC Press Web site at
http://www.crcpress.com**

Contents

List of Figures... xv
List of Tables... xvii
Contributors.. xix
Foreword xxiii
Preface xxv
Notes on Editors xxix
Acknowledgments xxxi

PART ONE   Introduction and Background 1

1   SDIs – Setting the Scene...3

    1.1   Introduction...3
    1.2   Building Spatial Data Infrastructures...4
          1.2.1   Spatial Data in Developed and Developing Countries...........5
          1.2.2   SDIs and the "Triple Bottom Line"...............................6
          1.2.3   The Evolving SDI Concept..7
          1.2.4   SDIs and Land Administration....................................7
          1.2.5   SDIs and the Government, Private and Academic Sectors......8
          1.2.6   SDIs and Users...9
          1.2.7   Technology as an SDI Driver 9
    1.3   Challenges Facing SDI Development 10
          1.3.1   Developing an SDI Vision..10
          1.3.2   Raising Community Awareness of Spatial Data..............10
          1.3.3   The Marine Dimension of SDIs.................................11
          1.3.4   SDIs and Privacy 11
          1.3.5   Strengthening Institutional Arrangements for
                  SDI Development 12
          1.3.6   Ensuring Capacity for SDI Development.......................13
          1.3.7   SDI Research and Development 13
    1.4   Conclusion..14
    1.5   References..15

2   Spatial Data Infrastructures: Concept, Nature and SDI
    Hierarchy...17

    2.1   Introduction..17

2.2   The Need for Spatial Data... 17
       2.2.1   Sharing Spatial Data.. 18
       2.2.2   Spatial Data and Decision Support Systems.................... 19
2.3   Spatial Data Infrastructure 20
       2.3.1   SDI Nature and Characteristics.................................. 24
                (a) Partnerships.... 25
                (b) Different Views to Understanding and Development 25
                (c) The Importance of People................................. 26
                (d) Dynamic Nature 27
       2.3.2   Current SDI Initiatives
                (Global, Regional, National, State, Local)..................... 27
2.4   SDI Hierarchy 28
       2.4.1   The Theory of Hierarchy...................................... 29
                (a) Definition of Hierarchy.................................. 29
                (b) Purpose and Levels of a Hierarchical Structure 30
       2.4.2   Hierarchical Reasoning.. 30
                (a) Principles of HSR 30
                (b) Properties of Hierarchies............................... 31
       2.4.3   Different Views on SDI Hierarchy 32
2.5   Applying Hierarchy Theory to SDI 33
       2.5.1   Hierarchy Theory and an SDI Hierarchy......................... 34
2.6   Relationships Among Different SDIs 34
2.7   Conclusion... 36
2.8   References.. 37

PART TWO   **From Global SDI to Local SDI**................................. 41

3   **Global   Initiatives**.. 43

3.1   Introduction... 43
3.2   Global SDI-GSDI.. 43
       3.2.1   GSDI Components and Organizational Model.................... 45
                (a) The GSDI Steering Committee 45
                (b) The GSDI Association 46
                (c) GSDI Program and Business Plan........................... 46
                (d) The SDI Implementation Guide ("Cookbook") 47
                (e) GSDI for Decision Support 47
       3.2.2   GSDI-Current Status of Development.......................... 48
                (a) The Global Level... 48
                (b) The Regional Level 48
                (c) The National Level 49
       3.2.3   Issues and Challenges 49
       3.2.4   Future Plans and Discussion 50

          (a) Implementing the GSDI Association 50
          (b) Main Drivers in GSDI. 51
   3.3   Global Map 51
       3.3.1   Components and Organizational Model 52
          (a) Institutional Arrangements 52
          (b) Technical Standards. 53
          (c) Fundamental Data. 53
       3.3.2   SDI Current Status of Development 53
       3.3.3   Issues and Challenges 55
       3.3.4   Future Plans and Discussion 55
   3.4   Evaluation of Global Initiatives. 56
   3.5   References 57

4   Regional SDIs 59

   4.1   Introduction 59
   4.2   European Umbrella Organization for Geographic Information
       (EUROGI). 59
       4.2.1   Organizational Status. 60
       4.2.2   Issues and Challenges 63
          (a) Strengths 63
          (b) Weaknesses. 63
          (c) Opportunities. 64
          (d) Threats 64
   4.3   Permanent Committee on GIS Infrastructure for Asia and
       the Pacific (PCGIAP). 64
       4.3.1   SDI Components and Organizational Model 65
       4.3.2   Status of Development. 66
          (a) Institutional Framework. 66
          (b) Technical Standards. 67
          (c) Fundamental Datasets 68
          (d) Access Network. 69
       4.3.3   Issues and Challenges 69
       4.3.4   Future Plans and Discussions 70
   4.4   Permanent Committee on SDI for the Americas (PCIDEA) 70
       4.4.1   Organizational Status. 72
       4.4.2   Issues and Challenges 73
   4.5   Comparative Evaluation. 74
   4.6   References 76

**5   SDI Diffusion – A Regional Case Study with
      Relevance to other Levels.** 79

5.1   Introduction..79
5.2   Asia and the Pacific Region and Regional SDI Activities:
      A Case Study...79
      5.2.1   Current Progress of PCGIAP and APSDI Development......81
5.3   SDI Diffusion..82
5.4   Influencing Factors for Regional SDI Diffusion............................85
      5.4.1   Environmental Factors.....................................86
      5.4.2   Capacity Factors...87
      5.4.3   SDI Organization Factors 88
5.5   SDI Development Models 88
5.6   New Strategies and Future Direction.....................................89
      5.6.1   Organizational Restructure................................89
      5.6.2   Redesign Future Strategy Based on Social System............90
      5.6.3   Modify the SDI Conceptual Model.............................90
      5.6.4   Adopting SDI Process-Based Model.............................91
5.7   Conclusion...91
5.8   References...92

6   **National SDI Initiatives** 95

6.1   Introduction..95
6.2   National SDI – The Concept and Nature.....................................95
      6.2.1   Motivation for National SDI Development 96
      6.2.2   Models for National SDI Development...........................97
6.3   Role and Relationships Within SDI Hierarchy...............................98
6.4   Generational Development of SDIs 100
      6.4.1   First Generation of National SDI Initiatives...................101
      6.4.2   Second Generation of National SDI Initiatives.................102
      6.4.3   Comparative Analysis...105
6.5   Recommendations and Future Directions....................................107
6.6   References..108

7   **State SDI Initiatives**..111

7.1   Introduction..111
7.2   The Nature of Spatial Information at a State Level........................112
7.3   State SDI – Organizational Issues...114
7.4   The Operation of State SDI 116
      7.4.1   Range of Datasets to Support State SDI........................116
      7.4.2   The Product 117
      7.4.3   Access Policy...118
      7.4.4   Value Adding Regime..119
      7.4.5   Integration of State SDI Operations...........................119

7.5   Partnerships in State SDI... 121
     7.5.1   Types of Partnerships 121
     7.5.2   Derivative Relationships – Licensing, Royalties and
               Value Adding Resellers 122
     7.5.3   An Evolving SDI Concept Based on Partnerships............ 123
7.6   Evaluation of State SDI Initiatives... 124
7.7   Other Issues and Future Directions of State SDI 125
7.8   References.. 126

PART THREE   **Australian Case Study from National to Local**...... 129

8   Development of the Australian Spatial Data Infrastructure. 131

8.1   Introduction.. 131
8.2   ANZLIC – The Spatial Information Council............................. 132
8.3   ASDI Conceptual Model.. 133
8.4   ASDI Organizational Model.. 133
     8.4.1   ANZLIC Organizational Model.................................. 135
     8.4.2   Public Sector Mapping Agencies 135
     8.4.3   Spatial Information Industry Action Agenda.................. 136
     8.4.4   Australian Spatial Information Business Association 137
     8.4.5   Spatial Sciences Coalition... 137
     8.4.6   Australian Spatial Information Education and
               Research Association.. 138
     8.4.7   Spatial Information Cooperative Research Centre........... 138
8.5   ASDI Implementation 139
     8.5.1   Fundamental Datasets 139
     8.5.2   Australian Spatial Data Directory (ASDD).................... 140
     8.5.3   ASDI Clearinghouse Definition................................. 142
     8.5.4   Pricing Policy... 142
8.6   Current Issues and Future Directions 143
8.7   References.. 145

9   State SDI Development: A Victorian Perspective................ 147

9.1   Introduction.. 147
9.2   State SDI in Victoria... 147
     9.2.1   Victorian Land Administration................................... 147
     9.2.2   Roles and Responsibilities for Developing Victorian SDI. 148
     9.2.3   State Spatial Information Policy.................................. 149
     9.2.4   Victorian Spatial Information Strategy......................... 150
               (a) Framework Data... 151

          (b) Key Business Information...................................... 151
          (c) Custody... 152
          (d) Metadata 152
          (e) Access Infrastructure................................ 152
          (f) Pricing and Licensing 153
          (g) Spatial Accuracy 153
          (h) Awareness... 154
9.3   Linkages to the National Spatial Information Sector................... 154
    9.3.1   Growing the Private Sector..................... 155
    9.3.2   Participation in Standards Development........................ 155
    9.3.3   Jurisdictional Linkages............................... 156
9.4   Key State SDI Initiatives 156
    9.4.1   Victorian Online Title System (VOTS)...................... 156
    9.4.2   Vicmap.. 157
    9.4.3   Property Information Project (PIP)............................ 157
    9.4.4   Online Service Delivery – Land Channel 158
    9.4.5   GPSnet.. 159
9.5   Victoria's SDI: Next Steps...................................... 160
    9.5.1   Land Exchange 160
9.6   Conclusion... 162
9.7   References... 163

10   SDI Development: Roles of Local and Corporate SDIs....... 165

10.1 Introduction... 165
    10.1.1 The Concept of SDI 165
10.2 Local SDI Development at Geelong 167
    10.2.1 History of Geelong SDI Development........................ 167
    10.2.2 Current Geelong SDI (Components and
           Organization Model)..................................... 169
    10.2.3 Linkages of Greater Geelong SDI Through
           Corporate SDI... 172
10.3 Corporate SDI in Multi-Level SDI Development –
    The Case of DNRE...................................... 173
    10.3.1 Catchment and Water............................... 173
    10.3.2 Parks Flora and Fauna............................. 174
    10.3.3 Department Wide Effort............................. 175
10.4 Some Patterns of SDI Development........................... 176
    10.4.1 Some Good Practices............................... 177
    10.4.2 Two Scales of SDI Development 177
10.5 Conclusions 178
10.6 References... 179

PART FOUR Supporting Economic, Environmental and Social Objectives.. 181

11 Sustainable Development, the Place for SDIs, and the Potential of E-Governance................................... 183

11.1 Introduction... 183
11.2 Sustainable Development.. 183
    11.2.1 Information.. 185
    11.2.2 Good Governance..................................... 185
11.3 SDIs for Sustainable Development........................... 186
    (a) Policies.. 187
    (b) Access 188
    (c) Standards 188
    (d) The Role of Partnerships 188
    (e) Data 189
    (f) SDI Design and Equity.. 190
11.4 The Potential of E-Governance............................... 191
11.5 Conclusion.. 192
11.6 References.. 193

12 SDIs and Decision Support.................................... 195

12.1 Introduction... 195
12.2 Decision Support for Sustainable Development.......... 195
    12.2.1 The Role of SDIs...................................... 196
    12.2.2 The Challenges for SDIs 196
12.3 Spatial Decision Support and SDIs 197
12.4 Supporting the Decision Environment....................... 199
    12.4.1 The Data Environment................................ 199
        (a) Availability.. 200
        (b) Accessibility.. 200
        (c) Applicability.. 201
    12.4.2 The Technological Environment 202
    12.4.3 The People Environment............................ 204
12.5 The Developing Decision Support Status of SDIs....... 206
12.6 Decision Support in the Future of SDIs..................... 208
12.7 References.. 208

13 Financing SDI Development: Examining Alternative Funding Models... 211

13.1 Introduction... 211

13.2 The Economic Issues of SDI Implementation............................ 212
13.3 The Concept of SDI Funding Models 213
    13.3.1 Funding Models of the First Generation of SDIs............. 215
13.4 Alternative Funding Models.. 216
    13.4.1 Infrastructure Classification... 217
        (a) Natural Monopoly 218
    13.4.2 Funding Models for SDIs Classified as Classic
        Infrastructures/Natural Monopolies.............................. 219
    13.4.3 Funding Models for SDIs Classified as Network
        Infrastructures.. 221
    13.4.4 Government's Function in this Category....................... 224
13.5 Customizing the Alternative Funding Models for
    Emerging Nations.. 225
13.6 Discussion 227
    13.6.1 Future Directions.. 228
13.7 Conclusion... 229
13.8 References.. 230

14   **Developing Evaluation and Performance Indicators**
       for SDIs 235

14.1 Introduction... 235
14.2 Land Administration Systems and the Role of SDIs 236
14.3 Evaluation and a Framework for Evaluation.............................. 237
14.4 Evaluation of SDIs 241
    (a) Policy Level... 241
    (b) Management Level... 242
    (c) Operational Level.. 242
    (d) Influencing Factors... 242
    (e) Assessment of Performance.. 243
14.5 Conclusions 244
14.6 References.. 245

PART FIVE   **Technical Dimension**.. 247

15   **Administrative Boundary Design in Support of**
       SDI Objectives 249

15.1 Introduction... 249
15.2 A Definition of the Spatial-Hierarchy Problem........................... 250
    15.2.1 The First Solution: Surface Modelling........................... 251
    15.2.2 The Second Solution: Data Interpolation 251

15.2.3   The Third Solution: Derived Boundaries 252
15.2.4   The Fourth Solution: Re-Aggregation......................... 252
15.3 Administrative Boundaries Within SDI 254
15.3.1   Access 254
15.3.2   People 254
15.3.3   Data 255
          (a) Confidentiality.. 255
          (b) The Modifiable-Area-Unit Problem (MAUP) 255
15.3.4   Technical Standards 256
          (a) Hierarchical Spatial Reasoning (HSR) Applied to
          Administrative Boundaries 256
          (b) The Structure of Administrative Boundaries.............. 257
15.3.5   Policy... 258
15.3.6   Summary 258
15.4 Conclusion.. 259
15.5 References... 260

16   **SDI and Location Based Wireless Applications**.................... 263

16.1 Introduction... 263
16.2 SDI as a Foundation for Location Based Services 263
16.3 Augmenting the SDI Model.. 265
16.3.1   Access Network 266
16.3.2   Policies.. 267
16.3.3   Standards 268
16.3.4   Data 269
16.3.5   People 270
16.4 Framework to Facilitate Wireless Applications........................... 271
16.4.1   SDI Requirements.. 271
16.4.2   User Environment.. 272
16.5 Case Study – Public Transport Application 272
16.5.1   Use Case Scenarios 273
16.6 Discussion, Recommendations and Future Directions 276
16.7 References.. 277

17   **Positional Frameworks for SDI**... 281

17.1 Introduction... 281
17.2 SDI and Positional Accuracy... 281
17.3 Opening the Lid on Geodesy 283
17.3.1   Reference Systems, Reference Frames and
          Geodetic Datums.. 283
17.3.2   Dynamic Datums... 284

17.3.3   Height Datums.. 285
17.3.4   Coordinate Systems.. 285
17.3.5   Error.. 285
17.3.6   Positional Accuracy.. 286
17.3.7   Precision... 287
17.4 Spatial Data Referencing.. 287
17.4.1   Adopting a Reference Frame to Support SDI................. 287
17.4.2   Global Reference Frames as a Basis for SDI.................. 288
17.5 Transforming Between Different Reference Systems 289
17.6 Measuring and Recording Positional Accuracy........................... 290
17.7 Case Studies.. 292
17.7.1   Case Study 1: The Geocentric Datum of Australia.......... 292
17.7.2   Case Study 2: PCGIAP... 294
17.8 Conclusions 296
17.9 References... 297

PART SIX   **Future Directions** 299

18   Future Directions for SDI Development............................. 301

18.1 Introduction... 301
18.2 Covering the SDI Landscape... 301
18.3 SDI Development Issues.. 305
18.3.1   Access Networks.. 305
18.3.2   People 306
18.3.3   Data and Standards... 306
18.3.4   Policies.. 307
18.3.5   External Developments 308
18.4 Relationships Between SDI Hierarchy, Issues and Different
       Models of SDI Development... 309
18.5 Conclusion... 310
18.6 References... 311

Index.. 313

List of Figures

Figure 2.1:   Infrastructure and Business Process Modules 24
Figure 2.2:   Nature and Relations Between SDI Components 27
Figure 2.3:   An SDI Hierarchy 28
Figure 2.4:   Hierarchical Structures Represented by Square
                Subdivisions and by a Tree Like Structure 29
Figure 2.5:   A) The Umbrella View of SDI; B) The Building Block
                View of SDI 32
Figure 2.6:   Relationships Between Data Detail, Different Levels of SDIs,
                and Level of Planning 33
Figure 2.7:   The Complex SDI Relationships Within and Between
                Different Levels 34
Figure 2.8.   Countries with Adjacency and Non-Adjacency Areas 35
Figure 3.1:   Global Mapping Project Participation Status as At 21
                November 2002 54
Figure 5.1:   Organizational Innovation Process Model 83
Figure 5.2:   Current Approach Taken by PCGIAP for APSDI
                Development, Based on Organizational Innovation Theory,
                1-Innovation, 2- Communication Channel, 3- Time 84
Figure 5.3:   Extended Conceptual Model, 1- Innovation,
                2- Communication Channel, 3- Time, 4- Social System 85
Figure 5.4:   Factors Influencing the Development of a Regional SDI 86
Figure 6.1:   Relationships Among Different Level of SDIs 98
Figure 6.2:   National SDI- A Particularity 100
Figure 6.3:   From a Techno-Centric Position to a Socio-Technical
                Position 104
Figure 7.1:   Spatial Datasets at a State level 117
Figure 7.2:   Spatial Data Infrastructure – Funding Models 119
Figure 9.1:   Key Strategy Detail Documents of VGIS 2000-2003 151
Figure 9.2:   Land Exchange Inter-Relationships Between Partners
                and Projects 161
Figure 10.1:   Hierarchy of SDI: Umbrella View and Building Block View 166
Figure 10.2:   Productional Perspective of an SDI 116
Figure 10.3:   PLACES, Spatial Information System of City of Greater
                Geelong 169
Figure 10.4:   Multi-level SDI Iteraction Involving the Corporate DNRE
                SDI 174
Figure 11.1:   SDI, Information and Good Governance Provide a Basis to Manage
                the Tensions Between Economic, Social and Environmental
                Imperatives for Sustainable Development 184
Figure 12.1:   Describing the Decision Environment 199
Figure 12.2:   Relative Support for Spatial Data Use in Decision-Making 202

Figure 12.3:   Framework Illustrating the Relationship Between
                Technological SDI Components and Decision Support
                for Accessing Data and Technologies 203
Figure 13.1:   SDI as A Network Model..217
Figure 13.2:   Funding Models for SDIs Viewed as Classic Infrastructures.....221
Figure 13.3:   Funding Models for SDIs Classified as Network
                Infrastructures 224
Figure 13.4:   Funding Models for SDI Implementation in Emerging
                Nations..227
Figure 14.1:   Basic Evaluation Elements and Cycle of Assessment...............238
Figure 14.2:   The Rrelation Between Evaluation Elements and
                Organizational Levels 238
Figure 14.3:   Areas for Evaluating an Administration System 239
Figure 14.4:   Evaluation Areas for SDI..241
Figure 15.1:   An Abstract Illustration of the Various Boundary Layers
                that Exist in Victoria 250
Figure 15.2:   An Illustration of the Difference Between Derived
                Postcode Boundaries and Actual Postcode Boundaries
                in the North West Melbourne Health Division..........................252
Figure 15.3:   Future Hierarchically Organised Administrative Structures 257
Figure 16.1:   SDI Components 264
Figure 16.2:   SDI Requirements for the Diverse User Environment,
                Categorised by People...271
Figure 16.3:   Example Interface Flow for Route Determination 275
Figure 16.4:   Augmented SDI Model for LBS 276
Figure 17.1:   Indicative Relationship Between GDA94, WGS84 and ITRF ...293
Figure 18.1:   Relationships Between SDI Hierarchy and their Similarities
                to the Organizational Structure 310

List of Tables

Table 2.1:   A sample of SDI Definitions...22

Table 3.1:   List of the GSDI Conferences...44

Table 4.1:   The PCGIAP (Permanent Committee) Model.............................74

Table 4.2:   The EUROGI Model...75

Table 6.1:   Behaviour and Inter-Relationships of SDIs................................99

Table 6.2:   Similarities and Differences Between First and Second
Generations of SDI Development...106

Table 8.1:   Extract from ANZLIC's 1996 ASDI Discussion Paper
(initial) Conceptual Model for the Australian Spatial Data
Infrastructure...134

Table 8.2:   Fundamental Datasets Included in the Scoreboard Project........140

Table 8.3:   Directory Audit Findings..141

Table 10.1:   Corporate Spatial Database Council Datasets.............................171

Table 10.2:   System Interfaces for Transaction Updates................................171

Table 12.1:   Potential Mechanisms for Expanding SDI Capabilities
to Support the Decision Environment..205

Table 13.1:   The Application of Financial Instruments to Organize SDI
Implementation...229

Table 14.1:   Evaluation Framework with Possible Aspects, Indicators
and Good Practice for Each Area...240

Table 14.2:   Possible Indicators for Evaluating SDIs....................................243

Table 14.3:   SWOT Matrix Summarizing a General (First-Order)
Evaluation of a State SDI..244

Table 15.1:   The Role of SDI and Recommendations to Facilitate
the Use of Administrative Boundary Data.................................259

Table 16.1:   Public Transport Information System User Requirements.........274

Table 17.1:   Positional Accuracy and its Influences on SDI Objectives........283

Table 17.2:   Local Data Used for Mapping in Kiribati...................................296

Contributors

Neil Brown
Centre for Spatial Data Infrastructures
and Land Administration
Department of Geomatics
The University of Melbourne
Victoria Australia 3010
Tel: +61 3 8344 4431
Fax: +61 3 9347 4128
neb@sunrise.sli.unimelb.edu.au

Santiago Borrero
President
Cartographic Commission
Pan American Institute for
Geography and History
PAIGH
santiago.borrero@alum.mit.edu

Drew Clarke
Executive General Manager
AusIndustry
Department of Industry, Tourism and
Resources
GPO Box 9839
Canberra ACT Australia 2601
Tel: +61 2 6213 7300
Fax: +612 6213 7344
drew.clarke@industry.gov.au

Serryn Eagleson
Centre for Spatial Data Infrastructures
and Land Administration
Department of Geomatics
The University of Melbourne

City of Melbourne
GPO Box 1603M
Melbourne
Vic 3001, Australia
Email: sereag@melbourne.vic.gov.au

Tai On Chan
Manager
Geographic Information Unit
Department of Primary Industries
Level 11, 8 Nicholson Street
PO Box 500, East Melbourne
Victoria Australia 3002
Tel: +61 3 9637 8651
Fax: +61 3 9637 8100
tai.chan@nre.vic.gov.au

Cathy Chipchase
Geospatial Policy Project Leader
Land Information Group,
Land Victoria
13/570 Bourke St, Melbourne
Tel: +61 3 8636 2316
Fax: +61 3 8636 2813
cathy.chipchase@nre.vic.gov.au

David Coleman
Dean, Faculty of Engineering
University of New Brunswick
P.O. Box 4400
Fredericton, N. B., Canada E3B 5A3
Tel: +1 506 453 5003
Fax +1 504 453 4569
dcoleman@unb.ca

Francisco Escobar
Department of Geography
University of Alcalá de Henares
Calle Colegios 2
28801 Alcalá de Henares, Spain
Tel: +34 918 854 429
Fax: +34 918 854 439
Email: Francisco.Escobar@uah.es

Mary-Ellen F. Feeney
Centre for Spatial Data Infrastructures
and Land Administration
Department of Geomatics
The University of Melbourne
Victoria Australia 3010
Tel: +61 3 8344 4431
Fax: +61 3 9347 4128
mef@sunrise.sli.unimelb.edu.au

Don Grant
Professorial Fellow
Department of Geomatics
The University of Melbourne
Victoria Australia 3010
Tel: +61 3 8344 4431
Fax: +61 3 9347 4128
dongrant@ozemail.com.au

Olaf H. Hedberg
P.O. Box 92
Richmond, TAS 1025
Tel: +61 3 6260 4194
Olaf.Hedberg@dpiwe.tas.gov.au

John Manning
Group Leader Geodesy
Geoscience Australia
PO Box 2, Belconnen
ACT Australia 2616
Tel: +61 2 62014352
Fax: +61 2 62014366
johnmanning@auslig.gov.au

Abbas Rajabifard
Deputy Director
Centre for Spatial Data Infrastructures
and Land Administration
Department of Geomatics
The University of Melbourne
Victoria Australia 3010
Tel: +61 3 8344 0234
Fax: +61 3 9347 2916
abbas.r@unimelb.edu.au

Garfield Giff
Department of Geodesy and Geomatics
Engineering
University of New Brunswick
P.O. Box 4400
Fredericton, N. B., Canada E3B 5A3
Tel: +1 506 447 3259/453-5194
Fax: +1 506 453 4943
b17gc@unb.ca

Peter Holland
General Manager
National Mapping Division
Geoscience Australia
PO Box 2, Belconnen
ACT Australia 2616
Tel: +61 2 6201 4262
Fax: +61 2 6201 4368
peter.holland@ga.gov.au

Allison Kealy
Lecturer
Department of Geomatics
The University of Melbourne
Victoria Australia 3010
Tel: +61 3 8344 6804
Fax: +61 3 9347 2916
akealy@unimelb.edu.au

Ian Masser
President GSDI
White Cottage
Priestcliffe Nr Buxton
Derbys England SK17 9TN
Tel/Fax: +44 1298 85232
f.masser@ukonline.co.uk

Jessica Smith
Centre for Spatial Data Infrastructures
and Land Administration
Department of Geomatics
The University of Melbourne
Victoria Australia 3010
Tel: +61 3 8344 4431
Fax: + 61 3 9347 4128
jcsmith@sunrise.sli.unimelb.edu.au

Daniel Steudler
Centre for Spatial Data Infrastructures
and Land Administration
Department of Geomatics
The University of Melbourne
Victoria Australia 3010
 Tel.: +61 3 8344 4431
Fax: +61 3 9347 4128
steudler@sunrise.sli.unimelb.edu.au

Lisa Ting
Fellow
Centre for Spatial Data Infrastructures
and Land Administration
Department of Geomatics
The University of Melbourne
Victoria Australia 3010
Tel: +61 3 9348 1126
ting@sunrise.sli.unimelb.edu.au

Warwick Watkins
Director General and Surveyor General,
and Chair ANZLIC
Department of Information Technology
and Management
Level 3, 1 Prince Albert Road
Queens Square, Sydney Australia 2000
Tel: +61 2 9236 7601
Fax: +61 2 9236 7631
warwick.watkins@ditm.nsw.gov.au

Ian Williamson
Director, Centre for Spatial Data
Infrastructures and Land Administration
Head, Department of Geomatics
The University of Melbourne
Victoria Australia 3010
Tel: +61 3 8344 4431
Fax: +61 3 9347 4128
ianpw@unimelb.edu.au

Bruce Thompson
Acting Director
Land Information Group
Land Victoria
13/570 Bourke St., Melbourne
Tel: +61 3 8636 2316
Fax: +61 3 8636 2813
bruce.thompson@nre.vic.gov.au

Mathew Warnest
Centre for Spatial Data Infrastructures
and Land Administration
Department of Geomatics
The University of Melbourne
Victoria Australia 3010
Tel: +61 3 8344 4431
Fax: +61 3 9347 4128
 mathew@sunrise.sli.unimelb.edu.au

Rick Whitworth
Spatial Systems Administrator
City of Greater Geelong
PO Box 104
Greater Geelong Australia 3220
Tel: +61 3 5227 0269
Fax: +61 3 5227 0922
rwhitworth@geelongcity.vic.gov.au

Acknowledgments

The Editors and authors want to thank the many people who have contributed to and supported the publication of this book. Of particular note are presenters and participants of the International Symposium on Spatial Data Infrastructures which was held at The University of Melbourne, Australia, on 19-20 November 2001.

We particularly want to thank Land Victoria, Department of Sustainability and Environment in the State of Victoria, Australia, which sponsored the Symposium and provided significant financial and in-kind support for the preparation of the book.

The authors of the various chapters wish to thank and acknowledge the following who have assisted them in the research and writing of the respective chapters: Land Victoria, Department of Sustainability and Environment; State Government of Victoria; Land and Property Information NSW, State Government of New South Wales; GeoSciences Australia; Department of Geomatics, The University of Melbourne; Chris Reynolds, Australia Post; Frank Blanchfield, ABS; Neil Williams, Geoscience Australia; Paul Kelly, ANZLIC; Graham Baker, Geoscience Australia; Jürg Kaufmann; Webraska Mobile Technologies; Sensis; Australian Research Council; Allan Stewart; Andrew Wilson; Dora-Ines Rey, PC-IDEA Executive Secretariat; Jim Steed, GeoSciences Australia; the Natural Sciences and Engineering Research Council of Canada; and the GeoConnections Secretariat, Earth Sciences Sector of Natural Resources Canada.

The Editors also want to acknowledge all the members of the Centre for Spatial Data Infrastructures and Land Administration in the Department of Geomatics, The University of Melbourne, Australia, for their encouragement and support. Lastly we want to thank Lillian Cheung for her ongoing support and organisation, without which the book would not have been possible.

Foreword

Individuals and governments alike are recognizing the need to pursue the sustainable development objective of a balance between economic, social and environmental priorities – the 'Triple Bottom Line'. This requires careful management of our natural and built environment, and the development and maintenance of an institutional and regulatory framework that supports planning, good governance and community participation.

Economic and social reforms, together with globalisation, have driven businesses to become more efficient and effective in service delivery. Yet at the same time natural and man-made disasters as well as more subtle impacts from air, noise and water pollution impact our lives. Even as our quality of life is improved by economic growth, there are concomitant environmental and social issues that act to detract from it. Maximising the benefits while minimising the impacts requires well-developed policies and systems that must, in turn, be based on complete, reliable and current information. The key to a sustainable future within this changing world is access to information that leads to better decision-making.

Today information and information and communications technologies (ICT) are the backbone of a modern society. The use of information, and particularly spatial information which provides the "where is" (or location) dimension, is transparent for much of society. Whether it is used for emergency response, for managing our resources, for ordering a taxi, for recreational use or for just finding the way from our home to a restaurant, the use of spatial information is now a central part of our daily lives.

In our information society the "point and click" vision to provide access to both government and private data within a virtual world is almost upon us. Such a vision underpins e-commerce and good governance. However this vision is not possible without a Spatial Data Infrastructure (SDI) in place to support it, in a similar way that an effective road and rail infrastructure supports efficient transport. SDIs underpin the relationship of people to land. They provide the spatial information or the location component to support the systems that allow modern societies to operate and that allow both the natural and built environment to be modelled, understood and managed.

This book is a welcome and timely contribution to the theory and practice of SDIs, and in many respects breaks new ground in improving our understanding of the increasing relevance and value of SDIs. It also explores practical issues and solutions, and provides case studies of successful SDI implementations. I am particularly pleased that the Government of Victoria has been able to play a role in supporting this research and the publication of this book. The book will contribute to the achievement of the Victorian Government's vision, through the Victorian Spatial Information Strategy, of ensuring that all our citizens have access to the spatial information they require to support their everyday lives and more broadly to ensure economic development and environmental sustainability. I hope the book will be equally useful to the many practitioners around the world who wish to better understand, build and use Spatial Data Infrastructures.

The Hon Mary Delahunty, MP
Minister for Planning,
Government of Victoria,
Australia

Preface

The spatial data infrastructure (SDI) concept continues to evolve as it becomes a core infrastructure supporting economic development, environmental management and social stability in developed and developing countries alike. Due to its dynamic and complex nature it is still a fuzzy concept to many, with practitioners, researchers and governments adopting different perspectives depending on their needs and circumstances.

The diversity of perspectives on SDI development offers a rich variety of experiences from which to learn and an extensive resource to the broader SDI community when documented. Contributing to the process of documentation is one of the motivations for this book. The book aims to provide some clarity to the SDI concept by drawing on practitioners and researchers from different backgrounds and jurisdictions to document their understanding of SDI and to share their experiences in building and analysing SDI.

This process began on the 19th and 20th November, 2001 when the Department of Geomatics at the University of Melbourne, convened the International Symposium on Spatial Data Infrastructures. Over 100 delegates from 15 countries attended the Symposium to discuss the issues and challenges facing SDI development. Practitioners presented reports detailing their experiences and achievements from local, state, national, regional and global SDI initiatives. Researchers from the Centre for Spatial Data Infrastructures and Land Administration, the University of Melbourne made a number of presentations on key research areas.

The Symposium enabled contact among, and international comparisons with, those working independently on similar SDI initiatives, who are facing similar problems, and are using similar or alternative approaches. It brought together key researchers and agencies in technology development, management information systems, sociology, and users and producers from the spatial data community. Through this interdisciplinary approach constructive criticism was sought of the methods used and results achieved in SDI development to date;as a result directions were drawn for extending research into priority areas for the future.

The SDI Symposium provided a framework for thinking about the future of SDI development and tackled the following objectives:

- To provide an overview of SDI policies, concepts, standards and practices associated with the implementation and operation of an SDI, with a focus on both developed and developing countries;
- To provide an understanding of the similarities and differences of SDIs operating within and between different jurisdictions;
- To explore the institutional and technical issues influencing the development of SDIs;

- To identify forces affecting the future of SDIs in both developed and developing countries based on assessment of current SDI initiatives.

Each Symposium session culminated in detailed discussion of the issues raised by presenters, which were then revisited at the end of the Symposium in a panel discussion on the 'Future directions for SDI development'. Issues facing the development of a globally compatible and workable SDI that extends from local government, through state, national and regional jurisdictions were examined. The forum considered key steps toward the facilitation and promotion of a spatially enabled society over the next five to ten years. Of particular relevance is our professional and institutional capacity for developing integrated and operational SDIs. In this respect the ideas raised and discussed during the course of the Symposium, as well as those for a longer-term education agenda, have contributed to the development of this book.

The aim of the book is to provide an introduction to the concepts, organizational models and progress made on SDI developments and the cross-jurisdictional relationships of these developments, for those participating in and managing SDI implementation. The book is designed to be an educational and professional resource to help build information resource management capacity for the spatial industry in the context of SDI. Although directed at spatial scientists, technologists, professionals, managers, policy makers, students and researchers, it will have broader applications for other disciplines as the concept of SDI continues to adapt in response to user needs. As summarised below, the book is divided into six parts with each comprising a number of chapters.

Part One – Introduction and Background

The first part of the book addresses the need for a broad understanding of the complexity and nature of SDIs and explores some of the key drivers influencing SDI development. One of the challenging questions is the definition of SDI or what constitutes SDI. The difficulty is that SDI is an evolving concept that sustains various perspectives or views depending on the user's interests and its role within the broader SDI hierarchy. There is general recognition that SDIs are not just about data (or maps) but include standards, institutional arrangements, delivery or access mechanisms, as well as people. This section also endeavours to provide a review of SDI policies, standards and practices associated with the implementation and operation of an SDI.

Part Two – From Global to Local SDI

The second part examines the role of SDIs at all levels, recognising that SDIs are hierarchical, ranging from corporate through global initiatives, and are based on dynamic relationships, which are both inter-jurisdictional and intra-jurisdictional. The interaction between spatial data users, suppliers and value-adding agents drive the development of any SDI and present significant influences on the changing spatial data relationships within the context of SDI jurisdictions. This challenges us to understand the similarities and differences of SDIs operating within and between different jurisdictions. Part Two looks at a variety of studies for each jurisdiction, from local/state, national, regional and global SDI initiatives. It is complemented by an examination of diffusion theory and the role of diffusion in the development of SDIs.

Part Three – Australian Case Study from National to Local

Discussions at the Symposium emphasised the need for in-depth operational examples of existing initiatives. This has been pursued in the third part of the book with a multi-levelled examination of SDI development occurring across Australia as a case study. Australia was one of the first countries to establish an operational concept of SDI, which has now manifested at many different levels from local through state and national initiatives, all at different stages of development and maturity.

The Australian SDI has gone through several phases of revision and modification and is contemporarily concentrating on activities consolidating integrated spatial information at a national level. Key initiatives have been the launching of the Spatial Information Industry Action Agenda, incorporation of the Public Sector Mapping Agencies of Australia, establishment of the Australian Spatial Information Business Association, and development of a Spatial Sciences Coalition - an amalgam of the key professional and learned societies representing spatial information interests in Australia.

All these changes are healthy for the development of a strong spatial information industry in Australia. In particular they are an example of the utility of partnerships to align objectives and motivate the voluntary participation of a variety of agencies and organizations, which is necessary in a federated system of government. It is hoped that this case study will provide useful principles and strategies for others who are building SDIs.

Part Four – Supporting Economic, Environmental and Social Objectives

Part Four reviews a variety of the institutional, economic and organizational questions that need to be addressed if efficient, effective and appropriate SDIs are to be built and operated. Global drivers (globalization, sustainable development, urbanization, economic reform and technology) are resulting in an increasing complexity of rights, restrictions and responsibilities pertaining to land, natural resources and the marine environment at private, public, state and corporate levels. In developing SDIs to support the human relationship to these issues it is not just a matter of understanding who the users are. Increasingly complex and accountable decision-support is required for the spatial dimension of social, environmental and economic decision-making. This requires the development of a diverse range of institutional mechanisms to support governance with appropriate technical tools, as well as an understanding of the economic models supporting SDI development. It also requires methods for monitoring and developing SDIs within a quality-assured environment, which introduces the need for evaluation and performance indicators.

Part Five – Technical Dimensions

A selection of the many technical issues associated with the development of SDIs are the focus for Part Five. Technology impacts on all the dimensions of the human relationship to land and the marine environment – from boundary delineation, to administration, mapping and visualization of rights, restrictions and responsibilities. These processes rely on an accurate and up-to-date geodetic framework, intelligent design of spatial databases, interoperable software and hardware, access to data through the Internet and the use of information and communication technologies (for example wireless applications). In this positioning and communication environment technologies will link with spatial databases to form SDIs with functionality not yet conceived. As a result Part Five will look at some of the prominent technical issues in regard to spatial data quality, administration and dissemination.

Part Six – Future Directions

The book concludes with a discussion on future directions for SDI development, which in particular recognises the breadth and diversity of contributors to the SDI Symposium and the continuing need to bring the private, public and academic sectors together. This collaboration is fundamental to stimulate continuing debate and education about the dynamic nature of SDIs.

The SDI community is multi-disciplinary and draws on a wide range of experiences from the geographic information systems, computer science, land administration, geography, surveying and mapping, legal and public administration disciplines. The editors are very grateful for the cooperation and input of authors from these disciplines to the individual chapters as well as to the overall concept of the book. It is hoped that the book achieves its objective of providing an introduction to the evolving SDI concept, recognising that it will be only one of many written in the years ahead.

The Editors
Ian Williamson
Abbas Rajabifard
Mary-Ellen F. Feeney

February 15, 2003

Notes on Editors

Professor Ian P. Williamson *AM FTSE* is Director, Centre for Spatial Data Infrastructures and Land Administration and Head, Department of Geomatics at the University of Melbourne. His teaching and research is concerned with land administration and spatial data infrastructures. He was appointed a Member of the Order of Australia 2003, and is a Fellow of the Academy of Technological Sciences and Engineering, Australia, a Fellow of the Institution of Surveyors Australia Inc., a Fellow of the Institution of Engineers Australia, an Honorary Fellow of The Mapping Sciences Institute, Australia and an Honorary Member of the International Federation of Surveyors. He has undertaken research or consultancies world-wide including for AusAID, the United Nations and the World Bank. He was Chair of Commission 7 (Cadastre and Land Management) 1994-98 and is currently Director, United Nations Liaison of the International Federation of Surveyors and member of the Executive and Chair WG3 (Cadastre) of the UN sponsored Permanent Committee for GIS Infrastructure for Asia and the Pacific.

Dr Abbas Rajabifard is Deputy Director for Centre for Spatial Data Infrastructures and Land Administration, and a Research Fellow in the Department of Geomatics, The University of Melbourne. He holds PhD (Melb), MSc (ITC), Postgrad-Dipl (ITC), and BSurv (Tehran). He worked for the National Cartographic Centre (NCC), Iran (1990-98), where he was Head of the GIS Department, managing the National Topographic Database and National GIS in Iran. From 1994-1999 he has been an Executive Board member and National representative to the UN sponsored Permanent Committee on GIS Infrastructure for Asia and the Pacific (PCGIAP). Since then he has been an active member of PCGIAP-WG2 (Regional Fundamental Data) and is now Research coordinator in PCGIAP-WG3 (Cadastre). He has also been a member of the International Steering Committee for Global Mapping 1997-2001. His research interests include SDI Hierarchy, SDI modelling, organisational management and partnerships.

Ms Mary-Ellen F. Feeney is completing her PhD at the Centre for Spatial Data Infrastructures and Land Administration, at the Department of Geomatics, The University of Melbourne. She has research interests in the development of SDI to support decision-making at local through global levels through institutional and technical capacity building. She is a member of the Global SDI and Decision Support System (DSS) Working Group. She holds BSc App.Geog.(Hons) (NSW), completing her research with the Commonwealth Science and Industry Research Organisation (CSIRO) and University of NSW in 1997. She has held positions as CSIRO Research Assistant (1996-1997), Australian Hydrographic Service Cartographer (1998) and Technical Development Officer (1998-1999), and has been involved in the development of the Australian Digital Hydrographic Database, GIS infrastructure and capacity building for the Department of Defence. Since 2000 she has been involved in lecturing, tutoring and course development in a number of subjects at the Department of Geomatics, The University of Melbourne.

PART ONE

Introduction and Background

CHAPTER ONE

SDIs – Setting the Scene

Ian Williamson

1.1 INTRODUCTION

We live in an age of information with this information being essential to tackle the issues of today's society. Spatial information in particular is one of the most critical elements underpinning decision-making for many disciplines. In the past we used maps to show where people and objects were located. Today this has evolved into a complex digital environment with sophisticated spatial and related textual databases, satellite positioning, communication networks like the Internet as well as wireless applications.

It seems for decades that the spatial information industry has been endeavouring to raise the profile and importance of spatial information within the wider society, yet it is only in the last few years that the message is starting to be heard. The reasons are many and varied, but the result is the same – governments are starting to listen. Two recent examples in Australia are firstly the Prime Minister (Howard, 2002), in announcing the national research priorities, identified geo-informatics as an example of a frontier technology for building and transforming Australian industries, along with bio-technology and nano-technology. Secondly, the Minister for Science (McGauran, 2002) announced the awarding of a major research grant to support a joint industry, government and university Cooperative Research Centre for Spatial Information. This attention is placing the management of spatial information and Spatial Data Infrastructures (SDIs) under the spotlight. It follows earlier statements such as in the USA by President Bill Clinton (Executive Order, 1994) and Vice President Al Gore (1998).

SDIs, which are the focus of this book, are allowing spatial information to be integrated and accessible within a complex digital environment and in so doing are increasingly underpinning the relationship of humankind to land, and the management of natural resources and the marine environment by enabling spatial information to support planning and decision-making. SDI is an initiative intended to create an environment in which all stakeholders can co-operate with each other and interact with technology, to better achieve their objectives at different political and administrative levels. In simple terms SDIs facilitate the sharing of data. By avoiding duplication associated with generation and maintenance of data and integration with other datasets, and facilitating integration and development of innovative business applications, SDIs can produce significant human and resource savings and returns.

SDIs have thus become important in determining the way in which spatial data are used throughout an organization, a state, a nation, different regions and the world resulting in the consequent development of the SDI concept at different political and administrative levels and the development of the SDI hierarchy.

This book builds on a number of previous initiatives which have focussed on the concept and development of SDIs and the sharing of development experiences. The latter has included an increasing number of conferences on SDIs, with the leading forums being the Global Spatial Data Infrastructure (GSDI) conferences, United Nations Regional Cartographic Conferences, the meetings of the UN-sponsored Permanent Committee on GIS Infrastructure for Asia and the Pacific (PCGIAP) and such initiatives as the Digital Earth conference and the International Symposium on SDI which gave the initial impetus for this book. These forums form the basis of professional SDI development networks facilitating the exchange of experiences and the sharing of problems as well as opportunities, to look for solutions in the experiences of others in different jurisdictions.

To support the evolving concept and practitioner development, the SDI Cookbook (GSDI, 2000) was produced by the GSDI as an online resource drawing examples from a variety of national level developments, to give practitioners access to SDI development experiences. At the same time there have been a number of books written which deal with the evolving concept, such as "Framework for the World" by Rhind (1997), "Governments and geographic information" by Masser (1998), "Geospatial Data Infrastructure – concepts, cases and good practice" by Groot and McLaughlin (2000) and the forthcoming book "World Spatial Metadata Standards" edited by Moellering *et al.,* (2003) to be published by the International Cartographic Association (ICA).

All these and many conference papers, journal articles and book chapters such as the chapter by Rajabifard and Williamson (2002) titled "Spatial Data Infrastructures: an Initiative to facilitate Spatial Data Sharing" in "Global Environmental Databases – Present Situation; Future Directions" published by the International Society of Photogrammetry and Remote Sensing (ISPRS), are contributing to our understanding of the evolving SDI concept.

1.2 BUILDING SPATIAL DATA INFRASTRUCTURES

Our understanding of the evolving SDI concept is essential to the process of building SDIs. Building SDIs is a complex task not just because of the evolving nature of the SDI concept, but as much because of the social, political, cultural and technological context to which such development must respond. The latter is shaped by the need for spatial information to meet and support the objectives of sustainable development, land administration, a broadening spectrum of users and developers, not to mention the different challenges of developed and developing nations. This section will highlight some of these issues influencing the development of SDIs.

1.2.1 Spatial Data in Developed and Developing Countries

Spatial data and spatial data products are becoming a consumer good in countries throughout the world, especially when connected to positioning technologies such as Global Positioning Systems (GPS) and when utilising a range of communications networks including the Internet, as well as wireless applications. The key driver in the commercialization and public use of spatial data is the ready availability of spatial datasets at medium to large scales, showing road networks, the land parcel framework, street addresses and topography. The use of these datasets for government, the private sector and recreational use is expanding rapidly as societies are becoming more spatially enabled - that is technology is available to assist in the use of spatial information to support decision-making and people are able to use the technology and access data to support their decisions.

Spatial data is considered a force multiplier in the military and provides a strategic advantage to many businesses. This use of spatial data and the rapid expansion of spatial data applications are driving the need for jurisdiction-wide policies surrounding the creation, maintenance, custodianship and use of spatial data – in other words the development of the SDI concept. As fast as information technology is changing, the SDI concept is evolving and adapting to capitalise on the new technologies and to meet the ever changing needs of society. One of the objectives of this book is to identify these changing needs and to explore the evolving SDI concept.

The reality is that every country is at a different point in the 'SDI development continuum'. The situation described above is most common in highly developed countries which may be considered to be positioned at the front end of the SDI development continuum. The situation however is very different in many developing countries, which are just starting SDI development, or those countries where adoption of the SDI concept is not even under consideration.

In developed countries the push towards a spatially enabled society and the need for jurisdiction-wide spatial data policies has resulted in major institutional changes. This is reflected by a world trend for spatial data and land related information activities to come together in one organization (as seen in Chapters 7, 8 and 9). This re-engineering of the institutional arrangements in developed countries has not been easy and has resulted in many political, administrative and professional challenges.

On the other hand, many developed countries still have fragmented institutional arrangements in the spatial data and land information area. It is not uncommon to find a range of different government departments, often in different ministries, responsible for different aspects of the management of spatial data. The result is that spatial data is held in independent silos with often little contact between them. Examples of different government agencies include those for land registration, cadastral surveying and cadastral mapping, planning, land valuation, administration of state lands, geodetic control and national mapping. The problem is often exacerbated by some functions being under the control of state governments, national governments and defence organizations. It is not surprising that moves to establish SDIs under these circumstances are problematic at best or non existent at worst. The difficulties this places on developing countries moving to establish an information society, which is sensitive to sustainable development objectives, are huge.

Between these two extremes are the majority of countries which are positioned at different points on the SDI development continuum. While this book focuses on SDI developments in a spatially enabled information society, the book recognises that there are many countries which are not ready or in a position to build an SDI in this technological environment. However even the poorest and least developed country can still adopt SDI principles and implement strategies which can lead them to develop SDI in the future, such as the creation of a common base map and reforming institutional arrangements for spatial data.

1.2.2 SDIs and the "Triple Bottom Line"

While the growth of spatial datasets and the use of spatial data rapidly expanded in the latter part of the 21st Century, it was not until the creation of jurisdiction-wide spatial databases that Geographic Information System (GIS) functionality could be incorporated into mainstream decision-making and government administration, and could support mainstream private sector activities. The resulting government administrative structures supporting the establishment and maintenance of these jurisdiction-wide spatial databases promoted the growth of and need for the SDI concept.

In the meantime no country today can ignore the economic, social and environmental dimensions of sustainable development. In most cases the developed world has embraced the concepts more than most developing countries with the "Triple Bottom Line" (economic, social and environmental considerations) now driving a great deal of government policy.

The growing inter-dependency between sustainable development and SDIs has evolved over the last decade (see Chapter 11 for more detailed discussion). The economic, social and environmental dimensions of sustainable development and the use of GIS initially had a project focus in the natural resources sector but have been gradually tied to the evolution of land administration systems and in turn the evolution of the SDI concept. This has been emphasised and highlighted in such international statements as the joint UN-FIG Bogor Declaration on Cadastral Reform (FIG, 1996), the UN-FIG Bathurst Declaration on Land Administration for Sustainable Development (FIG, 1999), FIG Agenda 21 (FIG, 2001), The Nairobi Statement on Spatial Information for Sustainable Development (FIG, 2002a), Land Information Management for Sustainable Development (FIG, 2002b) and resolutions of UN Regional Cartographic Conferences for Asia and the Pacific (UN, 2000) and the Americas (UN, 2001).

It was logical that governments recognised the important link and inter-dependency between achieving sustainable development objectives and establishing and maintaining SDIs as described above. In many countries SDIs are now being seen as a key component in delivering a jurisdiction's "Triple Bottom Line".

1.2.3 The Evolving SDI Concept

The evolution of the SDI concept has paralleled the development of complete digital spatial datasets in jurisdictions. As a result of the availability of jurisdictional-wide spatial data and the many opportunities this presents to both the government and private sectors, the institutional, legal and technical arrangements to support and facilitate the use of this spatial data have been re-engineered. One result highlighted over the last decade has been the evolution of the SDI concept.

Originally the focus of SDIs was on the type, the development of and access to the various spatial datasets. As the concept has evolved it has expanded to include a focus on people, access, policies and standards in relation to the data, as well as adopting a shift in emphasis from what can be called a "product-based" approach to a "process-based" approach in SDI development (Rajabifard *et al,* 2002). This will be discuss further in Chapter 5.

Discussion of the SDI concept also initially focussed on nations as an entity, while the last few years have seen more attention given to understanding the SDI hierarchy, from local level, through to state, national, regional and global levels. In general the various levels in the SDI hierarchy are a function of scale with the local government and state level SDIs usually concerned with large (1:5,000) and medium (1:25,000) scale data, whereas National SDIs tend to be small scale (1:25,000-1:100,000) with regional and Global SDIs adopting the scale of the global map of the world (1:1,000,000).

With an improved understanding of the SDI hierarchy has come the challenge to improve the relationships between SDIs in different jurisdictions as well as between different spatial data initiatives. As identified throughout this book the key to building successful SDIs is in the establishment of these relationships, especially through mutually beneficial partnerships, which are both inter- and intra-jurisdictional within the SDI hierarchy.

What is certain is that in the foreseeable future the SDI concept will continue to develop to a large degree in parallel with the evolution of information technology and the mainstreaming of sustainable development objectives in government polices.

1.2.4 SDIs and Land Administration

While the early focus of conferences and forums on SDIs was at a national level, and then in parallel with the growth of Regional SDIs such as the PCGIAP, the European initiatives and of late the Permanent Committee on GIS Infrastructure for the Americas (PC IDEA), one area of SDI development which has equivalent impact on society has not considered in depth. This is SDIs based on large scale data, such as those closely linked to land administration activities. These forms of SDI, which are linked to land administration activities such as the operation of land markets (land registration, cadastral surveying, land use planning, valuation, local government administration, administration of utilities and services) or natural resources management, have not received the same amount of attention internationally, yet could be considered to be more complex.

The key aspect about the form of SDI which is linked to land administration is that it is fuelled by people-relevant data – this data provides a richness to the SDI which distinguishes it from the typically small scale data in national, regional or Global SDIs. While the last decade has seen the evolving SDI concept focus on National SDIs, there is an expectation that the next decade will focus much more on large scale SDIs and particularly those related to land administration activities (see Chapter 7 for more discussion).

The very close relationship between land administration and SDIs in a large-scale context was the driving force behind the establishment of the Centre for Spatial Data Infrastructures and Land Administration in the Department of Geomatics at the University of Melbourne (University of Melbourne, 2003). This Centre receives significant funding from the State Government of Victoria, due to the need to better understand the complex issues surrounding the role of spatial data in an information society with a focus on the role of SDIs and land administration.

1.2.5 SDIs and the Government, Private and Academic Sectors

Another early focus on SDIs has been the emphasis on the role of government as having the primary or even sole responsibility for their development. As more jurisdiction-wide spatial datasets are completed, and increasingly in larger scales, so the role of the private sector has increased in the collection, maintenance and provision of spatial data within the context of SDIs. The private sector is now playing a greater role in both developed and developing countries in supporting the establishment and maintenance of SDIs.

The sector which has not had the same attention as the government and private sectors, is the academic sector, which is responsible for education, training and research in SDIs. While there are some excellent examples of commitment from the academic sector to SDI education and research, particularly in Western Europe, North America and countries like Australia, the attention is minor and in its infancy compared to the more traditional disciplines within the spatial data sector, such as GIS, positioning technologies (GPS), data collection and technologies such as remote sensing. What is certain is that the continued development of the SDI concept, like other areas within the spatial information discipline, is dependent on a close working and mutually beneficial relationship between the government, private and academic sectors.

It is hoped that this book will encourage a greater awareness of the need to pursue education, training and research in the development and maintenance of SDIs and to ensure that the academic sector plays an equivalent role to the government and private sectors by providing the capacity, and the research and development to ensure that the SDI concept will continue to evolve and be relevant to the users of spatial data.

1.2.6 SDIs and Users

The move into the digital environment has seen a very significant institutional change to the position where users are now the dominant driver in the establishment of spatial databases and the underpinning SDIs. While there is still a great deal of interest in international forums on investigating and better understanding SDIs, without a use or a business application an infrastructure such as SDI has no justification for its existence. A simple analogy is the road network, which is a classic example of an infrastructure. If there were no vehicles to travel along the roads then there is little justification for today's sophisticated road network.

Another reason why users are having far greater say than in the past on what spatial data is required and how it should be accessed is due to the rapid increase in the different forms of uses being found for spatial data. A good example is the increasing interest over the last couple of years on the need for improved emergency response, which can range from the provision of road networks and satellite positioning devices to support for police, fire and ambulance in an emergency, to a better understanding of the complex spatial requirements in mapping bushfires and responding to bushfire threats or other environmental disasters. As the wider community is learning to better use spatial data for making better decisions, then we are seeing the decision makers increasingly demanding more in the way of content, quality and ease of access to spatial data.

As a result, in very simple terms the future development of SDIs and the use of spatial data will be driven by the users. In many countries this will require a change in focus in the development of SDIs, away from government directives to listening to the needs of the wider community and non-government organizations to a far greater extent.

1.2.7 Technology as an SDI Driver

Technology has clearly been one of the most important, if not the most important, driver in influencing the evolving SDI concept. Ongoing changes and improvements in technology will ensure that the SDI concept continues to evolve for many years to come. To a large extent the SDI concept is tied to the development of jurisdiction-wide digital spatial databases, which in turn have promoted the need for SDIs. From a theoretical point of view, SDIs are not technology dependent; however, all the components of an SDI are influenced by technology with all the spatial technologies having an influence in one way or another on SDI development.

There is currently great optimism about the potential of information and communication technologies in revolutionising most applications that are dependent on spatial data. Communication technologies such as the Internet and wireless applications are revolutionising methods of maintaining, disseminating and accessing spatial data. The convergence of wireless communications, positioning technology and network computing is now capable of providing new facilities and new applications and as a result, new challenges for spatial data providers and users. To fully utilise these technologies there must be a clear understanding of how they impact on and assist in the development of an SDI.

1.3 CHALLENGES FACING SDI DEVELOPMENT

As set out in the Preface, this book focuses on the nature of the SDI concept and particularly the SDI hierarchy. With regard to the SDI hierarchy, the book has chapters that discuss global, regional, national and state SDI initiatives from a general perspective. The book also looks at an Australian case study from the national to the local level and in the latter part of the book, looks at social, technical and economic dimensions, and concludes by identifying some future directions for SDI development. However there are some over-arching challenges in developing SDIs that are discussed below.

1.3.1 Developing an SDI Vision

Even though good management practices demand the development of a vision prior to the determination of specific objectives and implementation strategies for managing projects, it is surprising how few jurisdictions have developed a simple vision for their SDI. Such a vision provides a road map and a way forward for the development of more detailed objectives and implementation strategies. Typically such a vision is part of a three to five year implementation strategy, although three years is preferable due to the continued evolution of the SDI concept and fast changing technology.

1.3.2 Raising Community Awareness of Spatial Data

The last couple of years have seen spatial data and associated technologies becoming mass consumer goods (such as the availability of low cost hand held GPS receivers having national street address datasets), a scenario that was hard to imagine a decade ago. Today in most developed countries spatial data has permeated the wider community albeit in a transparent manner.

Unfortunately, one of the draw backs of such transparency is that the term "spatial" is still misunderstood by the wider community with many people believing it relates to activities or associated technologies in Space. While professionals in the discipline may refer to a community becoming 'spatially enabled', that community more often than not is not aware how spatial data and the associated technologies are supporting activities which identify street address, location and other everyday activities.

There is an argument that it is not necessary for the wider public to understand the role of spatial data and associated technologies in delivering emergency services or something as simple as ordering a taxicab, just in the same way that they do not have to know anything about information technology to be able to use a computer or the technology in a car in order to drive one. In the same way that much of the wider community still has difficulty in reading maps, it is likely there will continue to be a lack of understanding about spatial data and associated technologies in the years ahead.

On the other hand, just as some understanding of the operation of a computer or the operation of a motor vehicle allows the user or driver to gain greater benefit from the technology, so does a greater understanding of spatial data and associated technologies provide users with a greater ability to use the data to better serve their needs. Therefore, while it is unreasonable to expect the wider community to appreciate the detailed nature of spatial data and associated science and technologies, there is no doubt that appropriate education and training of the use of spatial data and associated technologies will improve both the spatial information industry and benefit the community at large.

Therefore, while governments and the private sector will continue to develop SDIs they will need to invest significant resources into explaining the use of spatial data and technologies to the wider community if the full potential of SDIs is to be realised.

1.3.3 The Marine Dimension of SDIs

In the same way that land administration systems have traditionally only focussed on land and have stopped at High-Water Mark, SDIs have also typically only related to land. With the increasing focus on large to medium scale data in SDIs, the closer relationship with land administration systems, and the trend for rights, restrictions and responsibilities relating to land to move from the land into the marine environment, there is a trend for SDIs to cover both the land and marine environments depending on the responsibility of the specific jurisdiction. This is particularly relevant in the coastal zone.

If an SDI is seen as an infrastructure for a jurisdiction or nation, then that infrastructure should cover those areas which are the responsibility of the jurisdiction - for most countries this includes both land and marine environments – so it is inevitable that SDIs will include marine areas.

1.3.4 SDIs and Privacy

There have always been issues in many countries with regard to restricted access to maps as a result of security and defence considerations. Unfortunately the laws and policies of these countries have not kept pace with technology whereby medium scale maps can now be produced of any country from high resolution satellite imagery. As a result, the historic restrictions on access to spatial data, at medium to small scales, is no longer valid.

This is not usually the case for large scale spatial data when it concerns land parcel data resulting from land administration systems, and where it is possible to identify the owner and/or occupier of a particular property. The use of spatial technologies for these purposes does raise significant issues of privacy which should be taken into consideration when designing those components dealing with large to medium scale data in SDIs. Similar arguments relate to spatial technologies used to identify the location of mobile phone users as another example.

Since the trend is for more large scale spatial data becoming available in SDIs, the issues surrounding privacy will increasingly require attention in the development and administration of future SDIs.

1.3.5 Strengthening Institutional Arrangements for SDI Development

The emergence of a spatially-enabled information society as a result of the development of jurisdiction-wide digital spatial datasets has had a major effect on those institutions, both in government and in the professions, which support spatial information. For the past fifty or possibly a hundred years or so, the institutional arrangements for organizations in government responsible for surveying, mapping and land administration were relatively stable.

These government institutions were often led by a Surveyor-General or Chief Surveyor, supported by a surveying profession. They have had a long and well established history of over a hundred years in many countries. Historically the surveying profession was heavily influenced by the cadastral surveying or land boundary surveying segment of the profession, which was the largest sector of the profession.

The latter half of the 20th Century saw the growth of other professional bodies, such as cartography, and international bodies to represent the scientific interests of specialists in photogrammetry and geodesy. The last half of the 20th Century also saw the growth of associations concerned with the use of GIS technology such as the Urban & Regional Information Systems Association in North America (URISA).

While different countries and different regions of the world have had different interpretations or slightly different structures in accommodating these discipline areas, in general there was a similarity in the institutional arrangements, both in government and in the professions, supporting the spatial information industry (typically the surveying and mapping industry). Within this context the academic sector was represented primarily by schools of surveying which supported the surveying profession. Again there was great similarity between these academic programs around the world.

The impact of the spatial information revolution has significantly altered the historic institutional landscape in those countries which have moved heavily into information technology with their communities becoming spatially-enabled. This trend has also impacted on the manner in which the private sector organises itself to take advantage of the new opportunities presented by the new spatial information environment. Other chapters discuss this in greater depth but the result has been a continued re-evaluation of both government institutional structures and the professional organizations which support the industry.

What is clear from this brief overview is that if countries or jurisdictions are to take advantage of the spatial information revolution and the SDI concept, then major institutional changes are most probably required in government, in professional bodies, in higher education and in the private sector.

1.3.6 Ensuring Capacity for SDI Development

An educated and trained workforce is an important component in building SDIs, a position which is usually not questioned. What is unclear is what form of education and training is required to produce this work force to support the growth of SDIs. For example the question could be asked where the professionals will be educated or trained who will have the educational background as reflected by the chapters of this and other SDI books? The same question can also be asked for many other sections within the spatial information discipline, ranging from positioning technologies such as GPS to GIS, to data collection including photogrammetry and remote sensing, to measurement science through to land administration.

It is proposed in the medium term that a similar approach is taken as for the other sections of the spatial information discipline outlined above, and that is that specific courses on SDI be included in professional degrees in geographic or spatial information science, geomatic engineering, survey engineering, surveying and related disciplines. The SDI work force will be sourced from these programs and the graduates will have the knowledge and skills in geographic information science and technology as well as organizational, legal, institutional and sustainable development factors. In the short to medium term these formal courses can be supplemented by short courses run by organizations such as PCGIAP, GSDI, professional bodies or by universities which have the requisite expertise.

To complement this form of education it is expected there will be a continued expansion of educational initiatives concerned with SDI such as professional and research forums, text books, workshops and on-line facilities which are designed to enhance our understanding and development of the SDI concept and the potential of spatial information.

Just as the SDI concept will continue to evolve, so SDI education and training will need to evolve, recognising it will need to be a partnership between the educational institutions, the professional bodies and the government and private sectors.

1.3.7 SDI Research and Development

As in all technical disciplines, research and development is essential to the ongoing evolution of the technology and associated concepts. Research into SDIs is similarly essential if the concept is to grow and reach its full potential. At this point in time, international research in SDIs is in its infancy with only a handful of universities around the world actively pursuing SDI research. While much of the research which supports the development of SDIs can be considered as being undertaken under many related discipline areas in the spatial information area, such as in data collection, positioning, geographic information science etc., specific SDI related research could include:

- understanding, identifying and promoting the nature of SDI
- developing conceptual models of SDI within the SDI hierarchy
- comparing SDI initiatives to identify best practices
- investigating differences between the various levels in the SDI hierarchy

- investigating technical issues in support of SDI development and implementation including testing and evaluating prototypes
- technical issues concerned with interoperability and access
- data issues of privacy, intellectual property and security
- pricing policies and funding models
- statutory control of spatial data
- cultural and indigenous issues concerned with the establishment and maintenance of SDI
- establishment and integration of marine SDI within the SDI concept

While this is by no means a comprehensive coverage of the diverse range of challenges facing SDI development, it simply demonstrates some of the areas for research ranging from social and cultural dimensions, legal, policy and institutional considerations, through to technical issues and their intersection with the former. Nevertheless there is a whole range of issues which impact on the development of SDIs which need to be researched if the SDI concept is going to deliver its potential.

1.4 CONCLUSION

If we are going to design relevant SDIs we have to understand the spatial needs of society, the social system in which the SDI will operate, and the technical environment which the SDI will be required to support. We need a Global SDI vision which can facilitate the concept of Digital Earth and related initiatives. This latter concept is important for the future requirements of SDI and extends the concept of SDI further to incorporate a political, institutional and social dimension. It clearly shows the role that SDIs play in supporting good governance.

The concept being discussed in some jurisdictions is that government will enter into a contract with the community that all government decisions and policies would be based on data which is part of "The Virtual State" and is freely available to all citizens of the jurisdiction over the Internet. This promotes transparency in government decision-making. It is based on freely available data with the result that the wider community can review government policy decisions thereby promoting good governance and civil society. The importance of such a strategy is that all government spatial datasets need to be compliant with "The Virtual State" standards.

A vision such as "The Virtual State" raises issues of interoperability including data identification, data storage, data integration and accessibility of data. It also raises issues concerned with intellectual property and privacy of data. However, it is initiatives such as "The Virtual State" which will be one of the key driving forces for the development of future SDIs, recognising this will only be one of the many drivers influencing the development of SDIs. It is only a matter of time before the "The Virtual State" concept manifests itself from local government all the way to a national level in many jurisdictions.

"The Virtual State" concept discussed above highlights that SDIs cannot be developed in isolation. They must be user-driven; otherwise they have no justification or purpose. The challenge for governments, the private sector, educators and researchers is to develop appropriate SDIs that can support the complex demands of society as described above and as explored throughout the course of this book.

1.5 REFERENCES

Executive Order, 1994, Coordinating geographic data acquisition and access, the National Spatial Data Infrastructure. *Executive Order 12906*, Federal Register 59, 1767117674, Executive Office of the President, USA.

FIG, 2002a, FIG Publication No. 30. The Nairobi Statement on Spatial Information for Sustainable Development, Online. <http://www.ddl.org/figtree/pub/figpub/pub30/figpub30.htm > (Accessed February 2003).

FIG, 2002b, FIG Publication No. 31. Land Information Management for Sustainable Development, Online. <http://www.ddl.org/figtree/pub/figpub/pub31/figpub31.htm > (Accessed February 2003).

FIG, 2001, FIG Publication No. 23. FIG Agenda 21, Online. <http://www.ddl.org/figtree/pub/figpub/pub23/figpub23.htm> (Accessed February 2003).

FIG, 1999, FIG Publication No. 21 The Bathurst Declaration on Land Administration for Sustainable Development, Online. <http://www.ddl.org/figtree/pub/figpub/pub21/figpub21.htm> (Accessed January 2003).

FIG, 1996, UN-FIG Bogor Declaration on Cadastral Reform, Online. <http://www.geom.unimelb.edu.au/fig7/Bogor/BogorDeclaration.html> (Accessed January 2003).

Gore, A., 1998, The Digital Earth: understanding our planet in the 21st century, *The Australian Surveyor* 43(2): 89-91.

Groot, R. and McLaughlin, J. D., (Eds) 2000, *Geospatial Data Infrastructure: Concepts, Cases and Good Practice*, (Oxford, UK: Oxford University Press).

GSDI, 2002, SDI Cookbook v1, Online. <http://www.gsdi.org> (Accessed October 2002).

Howard, J., 2002, Research priorities for Australia's future prosperity. Media Release, Prime Minister of Australia, 5 December, 2002, Online. <Http://www.pm.gov.au> (Accessed January 2003).

Masser, I., 1998, *Governments and Geographic Information*. Taylor & Francis, London, UK.

McGauran, 2002, Media Release (MIN 108/02) from Minister for Science/Deputy Leader of the House, on Record Funding for Cooperative Research Centres, Online. <http://www.crc.gov.au/whats_new.htm> (Accessed January 2003).

Moellering, H., Aalders, H.J.G.L. and Crane, A. (Eds), 2003, World Spatial Metadata Standards, International Cartographic Association, ISBN-008439497, Forthcoming August 2003.

Rajabifard, A. and Williamson, I. P., 2002, Spatial Data Infrastructures: an initiative to facilitate spatial data sharing, In *Global Environmental Databases-Present Situation and Future Directions*, Volume 2 Edited by R. Tateishi and D.

Hastings (108-136). International Society for Photogrammetry and Remote Sensing (ISPRS-WG IV/8), (Hong Kong: GeoCarto International Centre).

Rajabifard, A. Feeney, M. and Williamson, I.P., 2002, Future Directions for SDI Development. *International Journal of Applied Earth Observation and Geoinformation*, Vol. 4, No. 1, pp. 11-22, The Netherlands.

Rhind, D.(Ed.), 1997, *Framework for the World.* (Cambridge: GeoInformation International).

UN, 2000, Resolutions of 15th UN Regional Cartographic Conference for Asia and the Pacific, Kuala Lumpur, Malaysia, 11-14 April 2000, United Nations, E/CONF.92/1.

UN, 2001, Resolutions of 7th UN Regional Cartographic Conference for the Americas, New York, USA, 22-26 January 2001, United Nations, E/CONF.93/3.

University of Melbourne, 2003, Centre for Spatial Data Infrastructures and Land Administration website, Online. <http://www.geom.unimelb.edu.au/research/SDI_research/index.html.> (Accessed January 2003).

CHAPTER TWO

Spatial Data Infrastructures: Concept, Nature and SDI Hierarchy

Abbas Rajabifard, Mary-Ellen F. Feeney and Ian Williamson

2.1 INTRODUCTION

Defining the role of Spatial Data Infrastructure (SDI) in society is important for acceptance of the concept and its alignment with spatial industry objectives. Much has been done to describe and understand the components and operation of different aspects of SDIs and their integration into the spatial data community. However, what is often misunderstood is that the role SDI plays is by necessity greater than the sum of individual components of SDI and stakeholder groups.

This chapter first aims to discuss the nature and concept of SDI, including the components which have helped to build the current understanding about the importance of an infrastructure to support the interactions of the spatial data community and their partnerships. Several examples of how SDIs have been described to date are offered to aid understanding of their complexity. The need for descriptions to represent the discrepancies between the role and deliverables of an SDI and thus contribute to a simpler, but dynamic, understanding of the complexity of the SDI concept, are postulated. Then, based on the characteristics and nature of SDIs, the chapter demonstrates the fitness and applicability of Hierarchical Spatial Theory as a theoretical framework to model the multi-dimensional nature of SDIs. The chapter introduces the concept of an SDI hierarchy followed by a review of hierarchical reasoning and its properties. The chapter then proposes that through understanding and demonstrating the nature of an SDI hierarchy, any SDI development can gain support from a wider community of both government and non-government data users and providers.

2.2 THE NEED FOR SPATIAL DATA

Spatial data are items of information related to a location on the Earth, particularly information on natural phenomena, cultural and human resources. Examples are topography, including geographic features, place names, height data, land cover, hydrography; cadastre (property-boundary information); administrative boundaries; resources and environment; socio-economic information, including demographics (OSDM, 2002). Spatial data are critical to promote economic development, improve our stewardship of natural resources and to protect the environment (Executive Order, 1994). People need spatial data and its derived information to establish the position of identified features on the surface of the Earth. But why is position important? This question can be viewed from different perspectives. First, knowledge of the location of an activity allows it to be linked to other activities or features that occur in the same or nearby locations. Second, locations allow distances to be calculated, maps to be made, directions to be given and decisions to be made about complex, inter-related issues (Mapping Science Committee, 1995).

It has been estimated that over 80% of governmental data has a locational basis (Budic and Pinto, 1999; Rhind, 1999; Lemmens, 2001). Examples range from local to national, regional and global scales and address issues such as land-use planning and zoning, new schools or shopping centres, environmental regulation, emergency relief and economic developments - the potential list of uses is enormous (Masser, 1998; Mapping Science Committee, 1997; GI2000, 1995).

There are two major forces driving the development of spatial data. The first is a growing need for governments and businesses to improve their decision-making and increase their efficiency with the help of proper spatial analysis (Gore, 1998). In most of the developed countries it is widely acknowledged that spatial data is part of the national infrastructure and extensive efforts are being expended on this (Clarke, 2000). In the last two decades nations have made unprecedented investments in information and the means to assemble, store, process, analyse and disseminate it. Many organizations, agencies and departments in all levels of government, private and non-profit sectors and academia throughout the world spend billions of dollars each year producing and using spatial information (FGDC, 1997).

The second force is the advent of cheap, powerful information and communications technology, which facilitate the more effective handling of large quantities of spatial data (Openshaw, 1993). The rapid advancement in spatial data capture technologies has made the capture of digital spatial data a relatively quick and easy process, such as satellite imagery with digital image processing techniques as well as using global positioning systems (Openshaw, 1993). Moreover, apart from rapid advances in information and communication technologies, there have been five principal drivers for most of the technological changes that have occurred over the past three decades. These drivers are technological developments, environmental awareness, political unrest and war, urbanization and peacetime economy.

2.2.1 Sharing Spatial Data

People need to share spatial data to avoid duplication of expenses, associated with generation and maintenance of data and their integration with other data. Also, it is apparent that spatial data constitutes much of the data required for physical disaster planning, management and recovery work. Given that natural and human disasters will continue to occur, a major issue is the ability of various users to share and access necessary data and information to prepare for the effects and to minimise loss of life.

Geographic Information systems (GIS) are common tools used to store, manage and utilise digital spatial data. GIS benefits are increased by data sharing among organizations. Often the spatial data produced for one application can be applied in others, thus saving money by sharing data. For many organizations, building and using a GIS requires large quantities of current and accurate digital data. They can save significant time, money and effort when they share the burden of data collection and maintenance. This is important, not only to the organizations looking for the data, but also for the organizations with the data. The more partners there are, the more the savings and the greater the efficiency.

Sharing data can also improve data quality by increasing the number of individuals who find and correct errors. Savings realised on the production of common data can be used for other areas, such as application development. In addition, resources that would be used to collect repetitive data can be diverted into quality control, data management and collection of other necessary data.

Working together in a geographic area can also provide data coverage in a common form over a wider area. This aids cross-jurisdictional or cross-organizational analyses and decision-making. For example, adjoining jurisdictions may have a common interest in an environmental issue. A transit operator may serve a region, rather than stopping at country boundaries (FGDC, 1997). Sharing geographic data of common interest enables countries to defray some of the costs of producing and maintaining the data. Mechanisms to facilitate the use and exchange of spatial data are a major justification for developing and expanding any type of SDI.

The importance of spatial data and its sharing to the economy goes far beyond the potential development of the industry itself. It has the potential to impact widely on society due to its ability to represent a host of important characteristics spatially and thus provide support in areas as diverse as town planning, oil exploration and environmental monitoring. Spatial information has long been used in the military field as an aid to strategy and many existing structures for spatial data and information have their roots in the military. However, spatial information can help governments to make informed decisions in a wide range of other areas, from environmental management to crime prevention.

In the private sector it can aid companies in their investment and marketing decisions and help individuals to better understand the world in which they live (Mennecke, 1997). Thus this tool can improve the ability of many societal actors to make informed choices. The impact of this intangible aspect is difficult to measure. The economic advantages of a company choosing the best location for their factory, or of emergency services more effectively controlling a forest fire, cannot always be readily quantified. However, these benefits can be considerable.

2.2.2 Spatial Data and Decision Support Systems

Decision-making may be broadly defined to include any choice or selection of alternative course of action (Malczewski, 1999). A preliminary step toward achieving decision-making for complex problems has been increasing recognition of the role of spatial information to generate knowledge, provide added value to identify problems, assist in proposing alternatives and defining a course of action, information discovery, access and use. The importance of spatial information to support decision-making and management of growing national, regional, and global issues, such as deforestation and pollution, was specifically cited in the 1992 Rio

Declaration on Environment and Development and has been made one of the key themes in subsequent meetings of the Commission on Sustainable Development (CSD, 2001).

All decisions require data, yet the rights, restrictions and responsibilities influencing the relationship of people to land as well as to data become increasingly complex, through compelling and often competing issues of social, environmental and economic management. Since the Rio Declaration' call to develop strategies to guarantee the existence of all life, not just human life, there has been acknowledgement of the need to integrate environmental and developmental aspirations at all levels of decision-making.

Decision problems that involve spatial data and information are referred to as spatial decision problems. Spatial decision problems often require that a large number of feasible alternatives be evaluated on the basis of multiple criteria, thus spatial decisions are multi-criteria in nature (Massam, 1980). Multi-criteria decision-making is more complex than that based on a single criterion, because of the difficulty finding an alternative that dominates all others with respect to all criteria. For instance, the need for sustainable development to link social, economic and environmental issues, and examine the use of land in an integrated manner to minimise conflicts, requires multi-criteria decision-making .

The number of people involved in the decision-making process also influences the complexity of spatial decision problems (Massam, 1988; Malczewski, 1996). Spatial decision problems may be characterised by different preferences with respect to the decision consequences and the relative importance of the evaluation criteria. The incorporation of values and preferences into decision-making models is an important function of multi-criteria analysis in complex decision problems, and often requires the aid of sophisticated technologies to structure the decision process and the outcomes.

Decision-makers requiring technological support for complex forms of spatial decision-making face challenges to establishing linkages between data, information and decision support systems (DSS) without SDI (Birk, 2000). DSS are "an interactive, computer-based tool or collection of tools that uses information and models to improve both the process and the outcomes of decision-making" (Lessard and Gunther, 1999). DSS are developed to access and utilise knowledge bases (of expertise or experience) to support decision-making by the generation of alternative solution scenarios between multiple criteria, and often spatial representations of these through maps and cartographic tools. SDIs have the potential to promote widespread use of the available spatial datasets, which are essential to optimise spatial technology support for decision-making processes, such as DSS.

2.3 SPATIAL DATA INFRASTRUCTURE

Spatial Data Infrastructure (SDI) is an initiative which is defined in many different ways, however its common intent is to create an environment in which all stakeholders can cooperate with each other and interact with technology to better achieve their objectives at different political/administrative levels. SDI initiatives have evolved in response to the need for cooperation between users and producers of spatial data to nurture the means and environment for spatial data sharing and development (McLaughlin and Nichols, 1992; Coleman and McLaughlin, 1998; Rajabifard *et al.* 1999; Rajabifard *et al.* 2000b). The ultimate objectives of these initiatives, as summarised by Masser (1998), are to promote economic development, to stimulate better government and to foster environmental sustainability.

SDI is fundamentally about facilitation and coordination of the exchange and sharing of spatial data between stakeholders in the spatial data community. SDI constitutes dynamic partnerships between inter- and intra-jurisdictional stakeholders. The principal objective for developing SDI for any political and administrative level, as highlighted by Rajabifard (2002), is to achieve better outcomes for the level through improved economic, social and environmental decision–making. SDIs have become very important in determining the way in which spatial data are used throughout an organization, a state or province, a nation, different regions and the world. In this regard, as suggested in the SDI Cookbook, without a coherent and consistent SDI in place, there are inefficiencies and lost opportunities in the use of geographic information to solve problems (SDI Cookbook, 2000). In principle, SDIs allow the sharing of data, which is extremely useful, as it enables users to save resources, time and effort when trying to acquire new datasets by avoiding duplication of expenses associated with generation and maintenance of data and their integration with other datasets. By reducing duplication and facilitating integration and development of new and innovative business applications, SDIs can produce significant human and resource savings and returns.

The design and implementation of an SDI is not only a matter of technology but also one of designing institutions, the legislative and regulatory frameworks and acquiring new types of skills (Remkes, 2000). Balancing these elements to develop an SDI enables intra- and inter-jurisdictional dynamics of spatial data sharing (Feeney & Williamson, 2000; Rajabifard *et al.* 2002). Moreover, SDI development requires new relationships and partnerships among different levels of government and between public and private sector entities to be established as raised in Chapter 1. These partnerships allow and require organizations to assume responsibilities that may differ to those of the past (Tosta, 1997). With this arrangement, an effective SDI allows all cooperating bodies to access accurate and consistent spatial databases used to inform local and inter-jurisdictional decisions and to support implementation of the resulting initiatives. Therefore, an SDI has to ensure the jurisdictional consistency of content to meet user needs. In particular the needs of cooperating members must be met with the additional provision for other non-participating members to join.

Within this framework, fundamental datasets can be collected and maintained through partnerships (Jacoby *et al.* 2002). These datasets include all data necessary to understand the jurisdiction, both spatially and aspatially. To maximise the benefits from investment in data collection and maintenance from both a jurisdictional perspective and that of the individual members, it is important that SDIs are focused and coordinated. However, current progress of SDI initiatives shows that SDI is understood differently by stakeholders from different disciplines or backgrounds. In this regard, researchers and various agencies have attempted to capture the nature of SDI in definitions produced in various contexts (Table 2.1).

Table 2.1: A Sample of SDI Definitions (Adapted from Chan *et. al* 2001)

Source (reference)	Definition of SDI
Australia New Zealand Land Information Council (ANZLIC, 1996)	A national spatial data infrastructure comprises four core components - institutional framework, technical standards, fundamental datasets, and clearing house networks
Dutch Council for Real Estate Information (Ravi) (Masser, 1998)	The National Geographic Information Infrastructure is a collection of policy, datasets, standards, technology (hardware, software and electronic communications) and knowledge providing a user with the geographic information needed to carry out a task
European Commission (European Commission, 1995)	The European Geographic Information Infrastructure (EGII) is the European policy framework creating the necessary conditions for achieving the objectives. It thus encompasses all policies, regulations, incentives and structures set up by the EU Institutions and the Member States.
Executive Order of US President (Executive Order, 1994)	National Spatial Data Infrastructure (NSDI) means the technology, policies, standards, and human resources necessary to acquire, process, store, distribute, and improve utilization of geospatial data
Federal Geographic Data Committee (FGDC, 1997)	National SDI is an umbrella of policies, standards, and procedures under which organizations and technologies interact to foster more efficient use, management, and production of geospatial data.
Global Spatial Data Infrastructure Conference 1997 (GSDI, 1997)	Global Spatial Data Infrastructure (GSDI) should generally encompass the policies, organizational remits, data, technologies, standards, delivery mechanisms, and financial and human resources necessary to ensure that those working at the global and regional scale are not impeded in meeting their objectives
McLaughlin and Nichols (1992)	The components of a spatial data infrastructure should include sources of spatial data, databases and metadata, data networks, technology (dealing with data collection, management and representation), institutional arrangements, policies and standards and end-users
Queensland Spatial Information Infrastructure Council (Department of Natural Resources, 1999)	The Queensland Spatial Information Infrastructure comprises the datasets, institutional arrangements, technical standards, products and services required to meet the needs of government, industry and the community
Victoria's Geospatial Information Strategic Plan of the State Government of Victoria, Australia (Land Victoria, 1999)	The concept of a spatial data infrastructure is extended to include more than just the data itself – it now encompasses all organizations and customers involved in the entire process, from data capture to data access, including the geodetic framework

Whilst these existing definitions provide a useful base for the understanding different aspects of SDI, or an SDI at a snapshot in time, the variety of descriptions have resulted in a fragmentation of the identities and nature of SDI, derived for the varied purposes of promotion, funding and support. Lack of a more holistic representation and understanding of SDI has limited the ability to adapt to its evolution in response to the technical and user environment.

Existing definitions have been slow to incorporate the concept of an integrated, multi-leveled SDI. Current investigation indicates that SDI is multi-leveled in nature, formed from a hierarchy of inter-connected SDIs at corporate, local, state or provincial, national, regional (multi-national) and global levels (Rajabifard *et al.* 1999, 2000b). SDI development at a state level also suggests that an SDI is a dynamic entity; its identity and functionality change and become more complex over time (Chan and Williamson, 1999). Failing to acknowledge these characteristics of SDI the multi-dimensionality and dynamic mechanistic and functional roles of the SDI, have rendered many descriptions of SDI inadequate to describe the complexity and the dynamics of SDI as it develops and thus ultimately constrain SDI achieving developmental potential in the future.

With this in mind, in order to understand an SDI, as suggested by Coleman and McLaughlin (1998) a first approximation of its term can be achieved by defining its components:

- McKee (1996) defined 'geographic' data as those data describing phenomena directly or indirectly associated with a location and time relative to the surface of the Earth.
- The Webster Dictionary defines "data" as "factual information (as measurements or statistics) used as a basis for reasoning, discussion, or calculation". The word "infrastructure" is defined as "...the underlying foundation or framework of a system or organization."

The challenge is to come up with a definition which is not too restrictive and does not artificially limit thinking. This is especially critical in an SDI for wider areas such as national, regional and global, which reflect the convergence of telecommunications, information services and information technology sectors, but yet is more than just the physical facilities used to transmit, store, process and display voice, spatial data and images.

In a broader context, Pepper of the U.S. Federal Communications Commission, as cited by Coleman and McLaughlin (1998), expresses this challenge in the following manner:

> "When we talk about infrastructure, we tend to think about wires-hardware. Infrastructure is far more than that. It is people, it is laws, it is the education to be able to use systems. If you think about the highway system, we tend to think about bridges and interstates, but the infrastructure also includes the highway laws, drivers' licenses, gas stations, the people who cut the grass along the highways, and all of those support systems. You cannot talk about infrastructure in the telecom-information sector without also talking about the human support systems."

Beyond these components, Kelley (1993) describes "infrastructure" as sharing the following characteristics with data and information:

- It exists to support other economic or social activities, not as an end in itself;
- It incurs a relatively high initial capital cost; and
- It has a relatively long life. So, it requires long-term management and commitment of funds.

In summary, while most people will accept that spatial data (and in an historical context maps) are component of SDIs, however SDI is much more than data and goes far beyond surveying and mapping - it is an evolving concept. It provides an environment within which organizations and/or nations interact with technologies to foster activities for using, managing and producing geographic data. With the rapid improvement in spatial data collection and communications technologies, SDIs have become very important in the way spatial data are used throughout a company, a governmental agency, a state or province, nation, throughout regions and even the world. SDIs facilitate the sharing of data, which is extremely useful, as it enables spatial data users and producers to conserve efforts when trying to acquire new datasets. Importantly it must be users or business systems which drive the development of SDIs. In turn the business systems which rely on the infrastructure in turn become infrastructure for successive business systems. Along this line, Chan and Williamson (1999) suggested that an SDI does not exist as a single entity but as a hierarchy of modules of infrastructure linked by business processes (Figure 2.1). As a result, a complex arrangement of partnerships develop as the SDI develops.

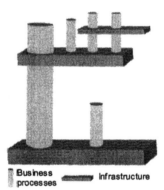

Figure 2.1: Infrastructure and Business Process Modules
(Adapted from Chan and Williamson, 1999)

2.3.1 SDI Nature and Characteristics

It is the needs of the user community that drive SDI development. These present significant influences on the changing spatial data relationships within the context of SDI jurisdictions. Reliable information infrastructures are needed to record environmental, social and economic rights, restrictions and responsibilities as well as provide spatial data to facilitate appropriate decision-making and support conflict resolution (Ting and Williamson, 2000). These drivers in turn affect the resulting spatial data industry environment and SDI vision, in particular partnership concepts. Partnerships, social system, the dynamic nature and stakeholders' different views on SDI development according to the needs of users determine the SDI nature and its characteristics.

(a) Partnerships

There has been a trend for countries to expand their efforts in developing SDIs through partnerships. In the 1990s National SDI development took a broad-base approach to encourage cooperation among stakeholders to pool data assets. Based on this approach, an ideal SDI should have all datasets in the corporate SDI fully integrated. Constrained by existing technical and institutional arrangements, SDI developing agencies have focused on promoting adoption of common standards, as well as fast-tracking integration among certain strategic datasets through partnership arrangements (ANZLIC, 1996; Jacoby *et al.* 2002). Partnerships are formed to create business consortia to develop specific data products or services for strategic users, by adopting a focussed approach to SDI development.

Coleman and McLaughlin (1998) identify four different perspectives of SDI, which provide an insight to the spatial data environment. These perspectives were developed to represent the varied directions of SDI initiatives, as shaped by the participant stakeholders, namely, spatial data supplier, technology supplier, spatial data and technology users and the collection of all three. Coleman and McLaughlin (1998) also point out that these groups interact widely with one another, suggesting that the SDI environment be made up of these interacting stakeholder groups.

In a similar line, Chan et al. (2001) suggested a system of classification to organise the many definitions and various aspects of the nature of SDI in which to better understand the multi-dimensional nature of SDIs. The definition classification system groups the definitions of SDIs into four perspectives: identificational, technological, organizational and productional perspectives. Based on this classification, they argue that the definitions fall within the first three perspectives with the organizational perspective being the most popular approach adopted by government, Regional and Global SDI developing agencies. However, it is the fourth perspective, the productional perspective of SDI, that is potentially most useful in facilitating SDI development and diffusion.

(b) Different Views to Understanding and Development

As was summarised in Table 2.1, different views of SDI can also be derived from different countries' approach to the understanding and development of SDIs. The Federal Geographic Data Committee (FGDC, 1997), defines the United States' National SDI as an umbrella of policies, standards and procedures under which organizations and technologies interact to foster more efficient use, management and production of geospatial data. It further explains that SDIs consist of organizations and individuals that generate or use geospatial data and the technologies that facilitate use and transfer of geospatial data.

The Australian and New Zealand Land Information Council (ANZLIC, 1998) define a National SDI as comprising four core components: an institutional framework, technical standards, fundamental datasets and clearinghouse networks. The institutional framework defines the policy, standards and administrative arrangements for building, maintaining, accessing and applying the datasets. The technical standards define the technical characteristics of the fundamental datasets. The fundamental datasets are produced within the institutional framework and fully comply with the technical standards. The clearinghouse network is the means by which the fundamental datasets are made accessible to the community, in accordance with policy determined within the institutional framework and to agreed technical standards.

According to the Canadian Geospatial Data Infrastructure (CGDI) vision, the CGDI initiative aims to facilitate the sharing of geographic databases, provide mechanisms which transcend the copyright and licensing restrictions, permits data exchange among agencies, and includes funding mechanisms and defines the databases (Turnbull and Loukes, 1997). This initiative has five inter-related technical components, namely data access, geospatial framework, standards, partnerships and supportive policy environment (Labonte *et al.* 1998).

After reviewing the varied histories and values underlying the vision of SDIs, including those cited, Coleman and McLaughlin (1998) defined the Global SDI as encompassing 'the policies, technologies, standards and human resources necessary for the effective collection, management, access, delivery and utilization of geospatial data in a global community'. The principal objective of developing an SDI is to provide a proper environment in which all stakeholders, both users and producers, of spatial information can cooperate with each other in a cost-efficient and cost-effective way to better achieve their targets. In this context, Coleman and McLaughlin regard the ANZLIC definition of SDI as data-centric; not taking into consideration the interactions between the suppliers and users of spatial data which is a key driver in SDI development.

(c) The Importance of People

Based on these selected samples of definitions of an SDI, it is proposed that an SDI comprises not only the four basic components identified for the Australian SDI, but also an important additional component, namely, people. People are the key to transaction processing and decision-making. All decisions require data and as data becomes more sensitive, human issues of data sharing, security, accuracy and access emphasise the need for more defined relationships between people and data. The rights, restrictions and responsibilities influencing the relationship of people to data become increasingly complex, through compelling and often competing issues of social, environmental and economic management. Facilitating the role of people and data in governance that appropriately supports decision-making and sustainable development objectives is central to the concept of SDI (this is discussed in detail in Chapter 11). This relationship involves the spatial data users and suppliers and any value-adding agents in between, who interact to drive the development of the SDI. For this reason, the formation of cross-jurisdictional partnerships have been the foundation of SDI initiatives to date and emphasise the important enabling capacity of the people component.

(d) Dynamic Nature

Viewing the core components of SDI, different categories can be formed based on the different nature of their interactions within the SDI framework. Considering the important and fundamental interaction between people and data as one category, the second can be considered the access network, policy and standards – the main technological components. The nature of both categories is dynamic due to the change of communities (people) and their needs, which in return require different sets of data, and due to the rapidity with which technology develops, so the need for mediation of rights, restrictions and responsibilities between people and data may change (Figure 2.2). This suggests an integrated SDI cannot be composed of spatial data, value-added services and end-users alone, but instead involves other important issues regarding interoperability, policies and networks. This in turn reflects the dynamic nature of the whole SDI concept (Rajabifard, 2002). This is an issue which is also highlighted by Groot and McLaughlin (2000). According to Figure 2.2, anyone (data users through producers) wishing to access datasets must utilise the technological components as described by Rajabifard *et al.* (2002). The influence of the level of SDI and the focus for the technical components have an important influence on the approach taken for aligning components towards the development of SDIs.

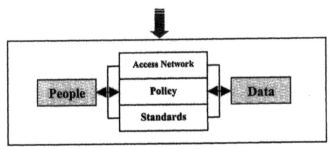

Figure 2.2: Nature and Relations Between SDI Components

2.3.2 Current SDI Initiatives (Global, Regional, National, State, Local)

Different groups of people, organizations and agencies can participate in development and implementation of an SDI. Although different organizations have characteristic data use patterns, all organizations need different resolutions of data at different times, particularly when they are working together.

Local governments typically create and use a great deal of detailed information covering small areas that fall within their jurisdictional boundaries. They need the framework datasets of the respective countries as a base for their applications and they frequently integrate such data when they build GIS. Local governments may use data at smaller scales over wider areas, when they are working on regional issues but their focus is on large scale datasets.

State governments also are characterised by the use of large scale and particularly land parcel data. However they also use less detailed data covering large regions and pertaining to a particular layer.

At the national level, government agencies are also characterised by use of lower resolution or small scale data, frequently producing and using data that have a low level of detail and cover broad areas. They also tend to produce and use individual data themes related to their operations. While their focus is on small scale data, national agencies often need and produce higher resolution data, particularly in managing national owned lands or facilities, or working on specific projects. Depending on the organization's activities, data use may range from higher resolution data over small areas, as in facility management, to low resolution data over wide areas, as in national environmental studies.

At the regional and the global levels, nations are interested to cooperate with each other in different fields, such as business and economic development, global mapping, environmental management and social purposes, as well as other issues which need lower resolution data. In these levels, there are many issues, such as atmospheric pollution, global warming and water catchment management, which do not follow national boundaries and transcend the national interest. These issues require spatial information at the regional and global level (usually at scales such as 1:1 million). To make decisions on global issues requires spatial information appropriate for these purposes. This information must be shared and integrated across national boundaries.

As a result of developing SDIs at different political and administrative levels a model of SDI hierarchy that includes SDIs developed at different political-administrative levels can be developed. The next section introduces and discusses this SDI hierarchy in detail, and the next part of the book discusses different SDI levels.

2.4 SDI HIERARCHY

As discussed above, many countries are developing SDI at different levels ranging from local to state/provincial, national and regional levels, to a global level, to better manage and utilise spatial data assets. The most important objectives of these initiatives, as summarised by Masser (1998), are to promote economic development, to stimulate better government and to foster environmental sustainability. As a result of developing SDIs at different political and administrative levels, a model of SDI hierarchy that includes SDIs developed at different political-administrative levels has been developed (Rajabifard, *et al.* 1999; Chan and Williamson, 1999, Rajabifard, *et al.* 2000a, 2000b). Figure 2.3 illustrates this model in which an SDI hierarchy is made up of inter-connected SDIs at corporate, local, state/provincial, national, regional (multi-national) and global levels.

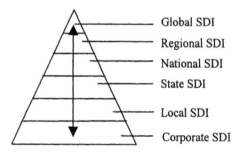

Global SDI

Regional SDI

National SDI

State SDI

Local SDI

Corporate SDI

Figure 2.3: AN SDI Hierarchy

In the above model, a corporate SDI is deemed to be an SDI at the corporate level - the base level of the hierarchy (Chan and Williamson, 1999). Each SDI, at the local level or above, is primarily formed by the integration of spatial datasets originally developed for use in corporations operating at that level and below.

The next sections will review the concept of hierarchy theory and its properties, and will then discuss the main reasons why the hierarchy concept is applied to the SDI concept.

2.4.1 The Theory of Hierarchy

In the past much research has been conducted toward maximising the efficiency of computational processes by using hierarchies to break complex tasks into smaller, simpler tasks (Car, 1997; Timpf, 1998). Hierarchical principles are used in many different disciplines to break complex problems down to sub problems that can be solved in an effective manner. Examples of hierarchical applications include classification of road networks (Car, 1997) and development of political subdivisions and land-use classification (Volta and Egenhofer, 1993). The complexity of the spatial field as highlighted by Timpf (1998) is primarily due to space being continuous and viewed from an infinite number of perspectives at a range of scales.

(a) Definition of Hierarchy

Koestler (1968), as cited by Car (1997), used the term hierarchy for a tree-like structure of a system which can be subdivided into smaller sub-systems, which in turn can be further subdivided into smaller sub-systems, and so on. In Figure 2.4, an example of a hierarchical structure is given, where each new square can be divided into a set of four smaller squares. Z consists of four sub-squares. This can be recursively subdivided as long as subdivision makes sense. This hierarchical arrangement can also be represented as a tree.

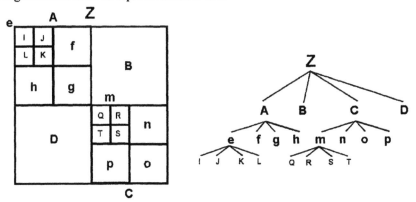

Figure 2.4: Hierarchical Structures Represented by Square Sub Divisions and by a Tree Like Structure (Adapted from Car, 1997)

Hierarchies are usually distinguished by their functions, which produce different types of hierarchies. Timpf (1998) recognised aggregation, generalization and filtering as the three most important functions to produce three different types of hierarchies. The aggregation hierarchy is built by aggregating objects. The generalization hierarchy defines how classes are related to more generic super, or higher order, classes. The filter hierarchy filters objects according to a criterion.

(b) Purpose and Levels of a Hierarchical Structure

There are good reasons why hierarchies develop and persist. Decreasing the processing time of a system (Pattee, 1973; Car, 1997) is one reason to introduce a hierarchy; a process being a sequence of actions performed in a particular way and leading to some result, and the processing time thought of as the time needed either for development or evolution of the system. A hierarchically structured system evolves much faster than a non-hierarchical system containing the same number of elements (Simon, 1973). Increasing the stability of any system is another reason to form hierarchies (Pattee, 1973). Also, hierarchies break down the task into manageable portions and enhance the potential for parallel processing (Timpf *et al.* 1992). The hierarchical approach was especially adopted in the description of complex dynamic systems (Mesarovic *et al.* 1970 as cited by Timpf, 1998), which Simon (1981) states have several advantages to a hierarchical structure.

With regard to the levels in a hierarchical structure, a set is divided into subsets or levels. A level is described by criteria determining which elements of the initial set belong to this level, and in turn, how this level is related to other levels in a hierarchy. The number of levels determines the depth of the hierarchy. The number of elements on each level determines its span and in turn the span of the tree.

2.4.2 Hierarchical Reasoning

Hierarchical reasoning is any reasoning process that applies hierarchy either to sub-divide the task, problem, process or space. Hierarchical reasoning adopts the principle of using the least detailed representation to answer a question. All data are inherently imprecise, but decisions do not require perfect information, instead information that is sufficiently precise (Timpf and Frank, 1997).

Hierarchical Spatial Reasoning (HSR) is defined by Car (1997) as part of the spatial information theory that utilises the hierarchical structuring of space for efficient reasoning. It is only recently, through the works of Car (1997) for way-finding, Glasgow (1995) for spatial planning and Frank and Timpf (1994) devising the intelligent zoom, that this theory has started to be applied in the spatial industry.

(a) Principles of HSR

The framework supporting HSR has three important components - representation, properties and applications. Hierarchies have been represented using alternative methods: Coffey (1981) devised triangles to represent a hierarchical structure; and Car (1997) illustrates how triangles can also be represented as a tree-like structure.

Although there are different representations of hierarchically organised systems, all provide the same function to breakdown the complexity of problems into smaller sub systems that can be efficiently handled and modelled.

In the past HSR research has focused on zero and one-dimensional structures to model urban systems (as points), road and drainage networks (as lines), and to a certain extent, to model simple bi-dimensional objects such as square polygons in quad-trees. However, HSR has focused now on three-dimensional structures to break down the complexities of polygons in the case of Australian administrative boundary design (Eagleson *et al.* 1999). From this research it has become evident that the properties required to model polygon hierarchy are more complex than those utilised for the modelling of points or networks.

(b) Properties of Hierarchies

Hierarchies in various phenomena, both natural and artificial, have properties specific to a particular context, but they also have common properties. These common properties are general relationships among structure, movement and function that are independent of their specific context (Car, 1997). Some of the properties of a hierarchical structure that are relevant to the understanding of hierarchies in general and spatial hierarchies in particular, are as follows:

- Part-Whole Property
In a hierarchy, an element on a higher level consists of one or more elements on the lower level. In view of a part-whole relationship, a higher level is a whole and a lower element is its part (Car, 1997). For example, in Figure 2.4, quadrangle A is a whole made up of quadrangles e, f, g and h. Similarly, A is also part of quadrangle Z.

- Janus-Effect
An element at a hierarchical level has two different faces, one looking toward wholes in a higher level and the other looking toward parts in a lower level. This property was introduced by Koestler (1968, cited by Car, 1997) as a fundamental property of all types of hierarchy. In Figure 2.4, each quadrangle is directly related to both above and below level quadrangles. Thus, e faces A but also I, J, K and L.

- Near Decomposability
The third fundamental property of hierarchy is called near decomposability (Simon, 1973). It is related to the nesting of systems within larger sub-systems and is based on the fact that interactions between various kinds of systems decrease in strength with distance. Components that are closer to each other interact more strongly than components that are far apart, many of them being at the same level. The definition of this property does not refer to whether elements on the same level should or should not be closer and have more interaction than elements in other levels. In Figure 3.2, elements such as J or K are closer to A than to other elements on the same level such as T or Q. In the tree structure part of the same diagram, it is clear how elements within the same level do not necessarily interact with themselves. It is believed, and it will be discussed later, that elements within the same level in the hierarchy should have a way to

communicate or interact in a better way than what is already present amongst levels.

Other than properties, hierarchies may also have special functional features such as uniqueness in particular roles. A feature such as this uniqueness may distinguish one level of the hierarchy due to its inter-relatedness with the other levels of the hierarchy. This feature is known as particularity in the system of hierarchies.

2.4.3 Different Views on SDI Hierarchy

The existence of hierarchical capability for SDIs will enable utilization of the advantages of this concept. Rajabifard *el al.* (2000b) introduced two views on the nature of this SDI hierarchy to better describe the concept and nature of this hierarchy (Figure 2.5). However, there is no major difference between these two views, as they both contribute to a better understanding of the concept of this form of hierarchy.

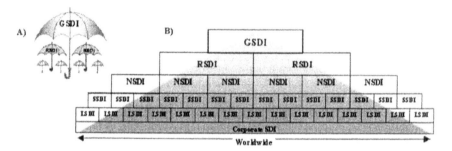

Figure 2.5: A) The Umbrella View of SDI; B) The Building Block View of SDI

The first view is the umbrella view (Figure 2.5A) in which the SDI at a higher level, say the Global level, encompasses all the components of SDIs at levels below. This suggests that ideally at a Global level, the necessary institutional framework, technical standards, access network and people are in place to support sharing of fundamental spatial datasets kept at lower levels, such as the Regional and National levels. The second view is called the building block view (Figure 2.5B). According to this view, any level of SDI, for example the State level, serves as the building blocks supporting the provision of spatial data needed by SDIs at higher levels in the hierarchy, such as the National or Regional levels.

Based on these two views, the SDI hierarchy creates an environment, in which decision-makers working at any level can draw on data from other levels, depending on the themes, scales, currency and coverage of the data needed (Figure 2.6). The double-ended arrow in Figure 2.6 represents the continuum of the relationship between different levels of detail for the data to be used at the different levels of planning corresponding to the hierarchy of SDIs.

As this figure illustrates, users at different levels of planning can have access and need to have access to a certain level of detail to take full advantage of using the SDI. However, it is quite difficult to define a boundary for data detail that can satisfy all user needs at a specific level. Sometimes due to the lack of availability or inaccessibility of the preferred level of data, different users may be required to compromise and use the available data to satisfy their needs.

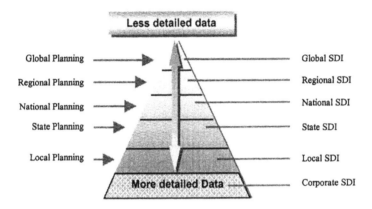

Figure 2.6: Relationships Between Data Detail, Different Levels of SDIs, and Level of Planning

2.5 APPLYING HIERARCHY THEORY TO SDI

The main reason that a hierarchy concept is applied is that all common properties and reasons for developing a hierarchical structure are also applicable to SDI concepts. For example, according to the part-whole property, an SDI at a high level, like a global level, consists of one or more SDIs from the lower level, such as different Regional SDIs like the APSDI in the Asia-Pacific and the EGII in Europe.

Moreover, a Regional SDI is a "whole" for a regional level and is a "part" of the global level. This is also applicable to the individual components of an SDI. Alternatively, according to the Janus-Effect, any element at a hierarchical level, say a National SDI, in the SDI hierarchy has two different faces, one looking toward "wholes" in a higher level, in this case regional and the global levels, and the other looking toward "parts" in lower levels of SDIs such as State and Local levels. This is illustrated by a double-ended arrow in Figure 2.3. According to Timpf (1998), the most common function to build a hierarchy is the aggregation function. Classes of individuals are aggregated because they share a common property or attribute. This is the other reason that a hierarchical concept can be applied to SDIs since different SDI initiatives at a certain political/administrative level can aggregate together to form the next higher level of hierarchy. This is the most common type of construction of hierarchy as introduced by Timpf (1998).

2.5.1 Hierarchy Theory and an SDI Hierarchy

As was discussed above, Hierarchy theory provides an expandable framework to demonstrate the concept of SDI. However, the existing properties of Hierarchy theory have been particularly well adapted to describe the vertical relationships between political or administrative levels of SDIs.

Additional to these vertical relationships (inter-jurisdictional) there are also complex relationships between SDIs within a political/administrative level, at an 'horizontal' level, of an SDI hierarchy. Figure 2.7 is a concept diagram published by Rajabifard *et al.* (2000b) that represents the complex vertical relationships between SDIs at levels in an SDI hierarchy (↕) as well as the complex horizontal relationships (intra-jurisdictional) between SDIs in any one level of such a hierarchy (↔). These 'horizontal' relationships have been less well explored within current HSR theory in respect to SDIs.

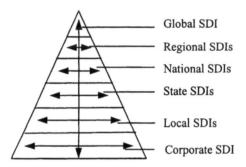

Global SDI

Regional SDIs

National SDIs

State SDIs

Local SDIs

Corporate SDI

Figure 2.7: The Complex SDI Relationships Within and Between
Different Levels

2.6 RELATIONSHIPS AMONG DIFFERENT SDIs

The relationships among different levels of SDIs are complex. This complexity is due to the dynamic, inter- and intra-jurisdictional nature of SDIs as identified above. One way to observe and map these relationships in the context of an SDI hierarchy can be to assess the impact and relationships of each component of any level of SDI on the same component of an SDI at a different level. Rajabifard *et al.* (2000a) observed the behaviour and inter-relationships between any level of SDI on the other levels through each of the components, and demonstrated a general pattern of direct and indirect potential impacts and relationships between them.

According to the pattern, a National SDI initiative in a non-federated system for example, has a full impact and relationship on the other levels of the SDI hierarchy through its components. In terms of policy, National SDI have an important effect on the upper and lower levels. However, policy at a global level has only a direct impact on and relationship with Regional and National SDIs. In terms of fundamental datasets, a National SDI has an important role in forming this component of the upper levels, and its datasets are created based on the datasets from the lower levels of SDIs. But the fundamental datasets at a national level can have an indirect impact on the fundamental datasets at a state level. Users at a state level might need to use national fundamental datasets (such as a geodetic framework) for their applications before using state datasets that are in more detail. In terms of technical standards, a National SDI has a direct influence on the State and Local SDIs, and its position is important for the upper levels to decide on their strategies and standards.

An SDI at the national level therefore, has stronger relationships as well as a more important role, in building the other levels of SDI. The role of a National SDI in an SDI hierarchy displays a particularity not present in the other levels of the SDI hierarchy. This particularity is that bottom levels of an SDI hierarchy, such as local and state, have no strong links to the upper levels of the hierarchy, like to the GSDI. So, there is a crucial level to the lower and higher links, which is the National SDI. Similar situations may exist when the first three levels (local, state and national) of an SDI hierarchy are to be considered, especially within the federated nations. In this case a State SDI is a crucial level to the local and national levels.

The horizontal relationships between individual SDI initiatives within any level of an SDI hierarchy become more important when the respective jurisdictions are spatially adjacent and proximate. SDIs belonging to adjacent jurisdictions play more important roles and have more influence and impact on each other than on SDIs of non-adjacent jurisdictions. For example, at a regional level, the policies and standards used on preparation of fundamental datasets of country A and country B, in Figure 2.8, have more impact on each other than country A with country C or D, when they are supposed to be integrated together forming datasets of the region. Using a global example, the policies and standards of SDIs of the European countries have more impact on each other than they do on the policies and standards adopted for SDIs by countries from the Asia and Pacific region as an example, or Africa. This is a result of the principles of adjacency and proximity.

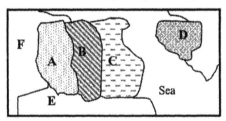

Figure 2.8: Countries with Adjacency and Non-Adjacency Areas

Based on the above discussion, it is proposed that a new property must exist when applying HSR principles to SDI (Rajabifard *et al.* 2000b). This is a horizontal property which defines the levelled nature of SDI within a hierarchically organised system. This property states that within each level of the SDI hierarchy, any SDI is interconnected with another in the same level and has horizontal relationships with them in which they impact on each other. The horizontal property encompasses the relationship between SDIs that are proximate as well as those that are distant.

2.7 CONCLUSION

This chapter begins with a brief review of the need for spatial data and data sharing and introduces major forces driving the development of such data. It then introduced and discussed the nature and concept of SDIs, including the components, by reviewing a number of the more current definitions of SDI. These reviews have helped to build the current understanding about the importance of an infrastructure to support the interactions of the spatial data community.

According to these reviews, SDI is understood and described differently by stakeholders from different disciplines and different political and administrative levels. It is argued that while they provide a useful base for the understanding of SDI, individually on their own they are inadequate for SDI development in the future. Further, it is argued that current SDI definitions are individually insufficient to describe the dynamic and multi-dimensional nature of SDI. Despite the international interest and activities toward SDI development, SDI remains very much an innovation even among practitioners. There are still doubts regarding the nature and identities of SDI, particularly in connection with how they evolve over time to meet user needs. With this in mind, this chapter discussed the concept of SDIs in such a way as to better clarify their nature to facilitate their development and progressive uptake and utilization among members of a community (diffusion). Based on this discussion, it is proposed that an SDI comprises not only the four basic components of institutional framework, technical standards, fundamental datasets and access networks, but also an important additional component, namely, people (human resources). This component of the SDI includes the spatial data users and suppliers and any value-adding agents in between, which interact to drive the development of the SDI.

The chapter then discussed the needs of spatial data for different levels of jurisdictions, followed with an overview of current SDI initiatives worldwide. According to this overview, many countries are developing SDI at different levels to better manage and utilise spatial data assets.

The theme of this chapter is that SDIs are a much-needed tool to better facilitate data sharing as well as jurisdictional cooperation and partnerships. However, an understanding of key SDI principles, such as the hierarchy of SDIs in a jurisdiction and the dynamic nature of SDIs, are also important but not fully understood.

The chapter proposes that the realignment of SDIs, based on HSR, will resolve many of the present issues constraining understanding of the nature and concept of SDIs. Based on the concept, properties and reasons for using a hierarchical structure, a model of SDI hierarchy is discussed, and found suitable to apply to the concept of SDI development. HSR provides an expandable framework to demonstrate the concept of SDI and represent the complexities of the different levels of SDI based on hierarchical principles.

Based on the relationships among different SDIs, a new property, namely horizontal property, is proposed. According to this property, any SDI is interconnected with the other SDIs in the same level and has horizontal relationships with them in which they impact on each other. Within interconnectivity, there is also an impact influenced by the adjacency of two areas.

Particularity of the SDI hierarchy is also identified. This suggests that an SDI at a national level has a crucial role in the development and implementation of the other SDI levels. Therefore, countries that are able to develop an efficient National SDI will be well placed to contribute to the development of Regional and Global SDI initiatives.

In summary this chapter investigated the hierarchical relationships between different political or administrative levels of SDIs and the applicability of hierarchical spatial theory as a theoretical framework to describe the multi-dimensional nature of SDIs.

2.8 REFERENCES

ANZLIC, 1998, Discussion paper, Spatial Data Infrastructure for Australia and New Zealand. Online. <http://www.anzlic.org.au/anzdiscu.htm> (Accessed November 1998).

ANZLIC, 1996, National spatial data infrastructure for Australia and New Zealand. ANZLIC Discussion Paper, Commonwealth of Australia. Online. <http://www.anzlic.org.au/anzdiscu.htm> (Accessed November 1998).

Birk, R. J., 2000, Decision Support. *Space Imaging* May/June 15(3), Online. <www.imagingnotes.com> (Accessed July 2000).

Budic, Z. D. and Pinto, J.K., 1999, Interorganizational GIS: Issues and prospects. *The Annals of Regional Science* (Springer-Verlag).

Car, A., 1997, 'Hierarchical Spatial Reasoning: Theoretical Consideration and its Application to Modeling Wayfinding'. *PhD thesis*, Department of Geoinformation, Technical University Vienna.

Chan, T. O., Feeney, M., Rajabifard, A. and Williamson I.P., 2001, The Dynamic Nature of Spatial Data Infrastructures: A Method of Descriptive Classification. *Geomatica Journal* 55(1): 65-72.

Chan, T. O. and Williamson, I. P., 1999, Spatial Data Infrastructure Management: Lessons from corporate GIS development. *Proceedings of AURISA '99*, Blue Mountains, NSW, AURISA'99: CD-ROM.

Clarke, D., 2000, The Global SDI and Emerging Nations - Challenges and Opportunities for Global Cooperation. *15th UNRCC Conference and 6th PCGIAP meeting*, 11-14 April 2000, Kuala Lumpur, Malaysia, United Nations.

Coffey, W.J., 1981, *Geography: Towards a General Spatial Systems Approach.* (London, New York: Methuen Co).

Coleman, D.J. and McLaughlin, J., 1998, Defining global geospatial data infrastructure (GGDI): components, stakeholders and interfaces. *Geomatics Journal*, Canadian Institute of Geomatics, Vol. 52, No. 2, pp. 129-144.

CSD, 2001, Commission on Sustainable Development Global Issues, environment. Australian Department of Foreign Affairs and Trade. Online. <http://www.dfat.gov.au/environment/csd.html > (Accessed March 2001).

Department of Natural Resources, 1999, Home page of Queensland Spatial Information Infrastructure Strategy, Queensland Spatial Information Infrastructure Council. Online. <http://www.qsiis.qld.gov.au/> (Accessed December 1999).

Eagleson, S., Escobar, F. and Williamson, I.P., 1999, Spatial Hierarchical Reasoning Applied to Administrative Boundary Design Using GIS. *Proceeding of the Australian National Surveying Congress*, Perth, Australia.

Executive Order, 1994, Coordinating geographic data acquisition and access, the National Spatial Data Infrastructure. *Executive Order 12906*, Federal Register 59, 1767117674, Executive Office of the President, USA.

European Commission, 1995, GI2000-Towards a European Geographic Information Infrastructure (EGII). A discussion document for consultation with the European GI community, European Commission. Online. <http://tempus1.utc.sk/gis/txts/gi2000xz.htm> (Accessed December 1999).

Feeney, M. and Williamson, I.P., 2000, Researching Frameworks for Evolving Spatial Data Infrastructures. *Proceedings of SIRC 2000*, The 12th Annual Colloquium of the Spatial Information Research Centre, University of Otago, Dunedin, New Zealand, 10-13 December 2000. p 93-105.

FGDC, 1997, *Framework, introduction and guide*. (Washington: Federal Geographic Data Committee), pp 106.

Frank, A. U. and Timpf, S., 1994, Multiple Representations for cartographic objects in a multi-scale tree - an intelligent graphical zoom. *Computers and Graphics Special Issue on Modelling and Visualization of Spatial Data in GIS*.

GI2000, 1995, Towards a European Geographic Information Infrastructure (EGII). A document of GI-2000 homepage. <http://www.echo.lu/gi/en/gi2000/egii.html> (Accessed January 1999).

Glasgow, J., 1995, A Formalism for Model-Based Spatial Planning, *In Spatial Information Theory - A Theoretical Basis for GIS (International Conference COSIT'95)*, edited by Frank, A.U. and Kuhn, W. 988. (Berlin-Heidelberg: Springer-Verlag)

Gore, A., 1998, The Digital Earth: understanding our planet in the 21st century. *The Australian Surveyor* 43(2): 89-91.

Groot, R. and McLaughlin, J., 2000, *Geospatial Data Infrastructure: concepts, cases and good practice*. (New York: Oxford University Press), 286 p.

GSDI, 1997, Global Spatial Data Infrastructure Conference findings and resolutions, Chapel Hill, North Carolina, 21 October 1997.

Jacoby, S., Smith, J., Ting, L. and Williamson, I.P., 2002, Developing a common spatial data infrastructure between State and Local Government – an Australian case study. *Journal of International of GIS*, Vol 16, No 4, 305-322.

Kelley, P.C., 1993, A National Spatial Information Infrastructure. *Proceedings of the 1993 Conference of the Australian Urban and Regional Information Systems Association* (AURISA), Adelaide, South Australia, Australia.

Koestler, A., 1968, *Das Gespesnt in der Maschine* (The Ghost in the Machine), (Wien-Munchen-Zurich: Verlag Fritz Molden).

Labonte, J., Corey, M. and Evangelatos, T., 1998, Canadian Geospatial Data Infrastructure (CGDI) - geospatial information for the knowledge economy. *Geomatica 1998*, Canadian Geoconnections, Online. <http://cgdi.gc.ca/english/publications/index.html> (Accessed February 2001).

Land Victoria, 1999, Home page of GI Connections. Online. <http://www.giconnections.vic.gov.au/> (Accessed December 1999).

Lemmens, M. J. P. M., 2001, An European Perspective on Geo-Information Infrastructure (GII) Issues. *GIS Development.net. 1*.

Lessard, G. and Gunther, T., 1999, Introduction. In Report of Decision Support Systems Workshop, Denver, Colorado, February 18-20, 1998. U.S. Department of the Interior and U.S. Geological Survey. USGS Open-File Report 99-351, 1999, pp. 1-5.

Malczewski, J., 1999, *GIS and Multicriteria Decision Analysis*. (New York: John Wiley and Sons), 392pp.

Malczewski, J., 1996, A GIS-based approach to multiple criteria group decision making. *International Journal of Geographical Information Systems* 10(8): 955-971.

Mapping Science Committee, 1997, *The Future of Spatial Data and Society*. (National Academy Press), Washington DC.

Mapping Science Committee, 1995, *A Data Foundation for the National Spatial Data Infrastructure*. (National Academy Press), Washington DC.

Massam, B.H., 1980, *Spatial search*. (Oxford: Pergamon Press).

Massam, B.H., 1988, Multi-criteria decision making (MCDM) techniques in planning. *Progress in Planning* 30(1): 1-84.

Masser, I., 1998, The first generation of national geographic information strategies. *Selected Conference Papers of the 3rd Global Spatial Data Infrastructure Conference*, 17-19 November 1998,Canberra, Australia.

McKee, L., 1996, Building the GSDI-discussion paper. *Proceedings of the 1996 Conference on Emerging Global Spatial Data Infrastructure*, September 1996, Konigswinter, Bundesrepublik, Deutschland (Germany), European Umbrella Organization for Geographical Information (EUROGI): 19 pages.

McLaughlin, J.D. and Nichols S.E., 1992, Building a national spatial data infrastructure. *Computing Canada*, 6th January: 24

Mennecke, B.E., 1997, Understanding the Role of Geographical Information Technologies in Business: Applications and Research Directions. *Journal of Geographical Information and Decision Analysis* 1(1): 44-68.

Mesarovic, M.D., Macko, D., and Takahara, Y., 1970, Theory of hierarchical, multilevel, systems. In *Mathematics in science and engineering*, ed. Bellmann, Richard. (New York: Academic Press).

Openshaw, S., 1993, Over twenty years of data handling and computing in Environment and Planning. *Environment and Planning A*, Anniversary Issue: 69-78.

OSDM, 2002, Office of Spatial Data Management Glossary, http://www.osdm.gov.au/osdm/glossary.html, (Accessed May 2002).

Pattee, H. H., 1973, *Hierarchy Theory – The Challenge of Complex Systems*. (New York: Braziller).

Rajabifard, A., 2002, 'Diffusion for Regional Spatial Data Infrastructures: particular reference to Asia and the Pacific'. *PhD Thesis*, The University of Melbourne, Melbourne, Australia, Online.
<http://www.geom.unimelb.edu.au/research/publications/Rajabifard_thesis.pdf>.

Rajabifard, A., Feeney, M. and Williamson, I.P., 2002, Future Directions for SDI Development. *International Journal of Applied Earth Observation and Geoinformation*, Vol. 4, No. 1, pp. 11-22, The Netherlands.

Rajabifard, A., Williamson, I.P., Holland, P. and Johnstone, G., 2000a, From Local to Global SDI initiatives: a pyramid building blocks. *Proceedings of the 4th*

Global Spatial Data Infrastructures Conferences, 13-15 March 2000, Cape Town, South Africa.

Rajabifard, A., Escobar, F. and Williamson, I.P., 2000b, Hierarchical Spatial Reasoning Applied to Spatial Data Infrastructures. *Cartography Journal 29(2)*, Australia.

Rajabifard, A., Chan, T. O., and Williamson, I. P., 1999, The Nature of Regional Spatial Data Infrastructure. *Proceedings of the AURISA 99*, 22-26 November 1999, Blue Mountains, NSW, Australia.

Remkes, J.W., 2000, Foreword In *Geospatial Data Infrastructure- Cases, concepts and good practice,* Edited by Groot, R. and J. McLaughlin, (New York: Oxford University Press).

Rhind, D., 1999, *Key Economic Characteristics of Information.* Ordnance Survey, U.K.

SDI Cookbook, 2000, Developing Spatial Data Infrastructures: The SDI Cookbook, Version 1.0, Prepared and released by the GSDI-Technical Working Group. Online. <http://www.gsdi.org/pubs/cookbook/cookbook0515.pdf> (Accessed July 2001).

Simon, H.A., 1981, *The Sciences of the Artificial,* (Cambridge, MA: MIT Press).

Simon, H. A., 1973, The Organization of Complex Systems, In *Hierarchy Theory- the challenge of complex systems,* Edited by Pattee H., pp. 1-27 (New York: Braziller).

Timpf, S., 1998, 'Hierarchical Structures in Map Series'. *PhD thesis*, Department of Geoinformation, Technical University Vienna.

Timpf, S. and Frank, A. U., 1997, Using hierarchical spatial data structures for hierarchical spatial reasoning. In *Spatial Information Theory - A Theoretical Basis for GIS (International Conference COSIT'97)*, Edited by Hirtle, Stephen C. and Frank, Andrew U. Lecture Notes in Computer Science 1329, (Berlin-Heidelberg: Springer-Verlag).

Timpf, S., Volta, G.S., Pollock, D.W., and Egenhofer, M.J., 1992, A Conceptual Model of Wayfinding Using Multiple Levels of Abstractions. In *Theories and Methods of Spatio-Temporal Reasoning in Geographic Space*, edited by Frank, A.U., Campari, I., and Formentini, U. 639, (Heidelberg-Berlin: Springer Verlag).

Ting, L. and Williamson, I. P., 2000, Spatial Data Infrastructures and Good Governance: Frameworks for Land Administration Reform to Support Sustainable Development. *Proceedings of the 4th Global Spatial Data Infrastructures Conferences*, 13-15 March 2000, Cape Town, South Africa.

Tosta, N., 1997, Building national spatial data infrastructures: Roles and Responsibilities. Online. <http://www.gisqatar.org.qa/conf97/links/g1.html> (Accessed February 1999).

Turnbull, J. and Loukes, D., 1997, Development of an integrated Canadian spatial data model and implementation concept. *Proceedings, URISA'97 Conference*, 19-23 July 1997, Toronto, Canada. 12 p.

Volta, G., and Egenhofer, M., 1993, Interaction with GIS Attribute Data Based on Categorical Coverages. In *European Conference on Spatial Information Theory*, Edited by A. Frank and I. Campari, (Italy: Marciana Marina).

PART TWO

From Global SDI to Local SDI

CHAPTER THREE

Global Initiatives

Peter Holland and Santiago Borrero

3.1 INTRODUCTION

Spatial Data Infrastructure at the global level (Global SDI) is promoting a global and open process for coordinating the organization, management and use of geospatial data and related activities, primarily by linking national and regional initiatives. The chapter will explore the growing role of the regional SDI level, the 'pivot' between the State and National SDI's and the possibilities to attain the Global SDI vision. The chapter demonstrates current global initiatives such as Global SDI and Global Map and their relationships with each other as well as with other SDI initiatives. It traces the history and the background of both initiatives, analyses both in terms of components and organizational models, documents their current stage of development and presents some issues and challenges confronting Global SDI and Global Map and their likely future direction. The chapter concludes with a comparison of both activities in terms of key features and related strengths-weaknesses-opportunities-threats.

3.2 GLOBAL SDI-GSDI

The emergence of the GSDI concept, similar to the other SDI levels, can be traced through the mapping and land information system developments of the 1960's and 70's; the acceptance of the notion of information as a corporate resource in the 1980's; rapid improvements in computing, communications and positioning technologies through the 1980's and 90's; and finally, implementation of national SDI's in the 1990's. In the final few years of the millennium the concept of a GSDI, and its potential realization, has captured the imagination and attention of policy-makers, administrators, industry, and the professions. Although not widely known in the general community or commonly understood by its proponents, the GSDI is seen by many as a central element in the global response to the challenge of sustainable development (Holland, 1999).

The background to the emerging interest in a GSDI is well described in Coleman and McLaughlin's theme paper for the GSDI 2 Conference:

> By the early 1990's, the concept of spatial data infrastructure (SDI) development was being proposed in support of accelerating geographic information exchange standards efforts, selected national mapping programs and the establishment of nation-wide spatial information networks ... Finally, the Santa Barbara Statement prepared from the Interregional Seminar on

Global Mapping for Implementation of Multi-National Environmental Agreements ... made a strong plea for the accelerated collection, promotion and use of the output from national and global mapping programs and the coordinated development of a global spatial data infrastructure.

(Coleman and McLaughlin, 1997)

A GSDI is currently advancing through the leadership of many nations and organizations, represented initially on the GSDI Steering Committee, and subsequently on the board of the new GSDI Association, which serves as the guiding body for the initiative (Borrero, 2002). The definition and vision for the GSDI are agreed as

The Global Spatial Data Infrastructure supports ready global access to geographic information. This is achieved through the coordinated actions of nations and organizations that promote awareness and implementation of complementary policies, common standards and effective mechanisms for the development and availability of interoperable digital geographic data and technologies to support decision-making at all scales for multiple purposes. These actions encompass the policies, organizational remits, data, technologies, standards, delivery mechanisms, and financial and human resources necessary to ensure that those working at the global and regional scale are not impeded in meeting their objectives.

(GSDI 5, 2001)

Conferences have been integral to the development of the GSDI initiative and its vision. Six successful GSDI Conferences have been held, with a seventh planned for early 2004 in Bangalore, India. A list of the GSDI Conferences is shown in Table 3.1.

Table 3.1: List of the GSDI Conferences

Conferences	Year	Location	Conference Theme
GSDI 1	1996	Bonn, Germany	"The Emerging GSDI"
GSDI 2	1997	Chapel Hill, United States	"Towards Sustainable Development Worldwide"
GSDI 3	1998	Canberra, Australia	"Policy and Organizational Framework for GSDI"
GSDI 4	2000	Cape Town, South Africa	"Engaging Emerging Economies"
GSDI 5	2001	Cartagena, Colombia	"Sustainable Development: GSDI for Improved Decision-making"
GSDI 6	2002	Budapest, Hungary	"GSDI: From global to local"
GSDI 7	2004	Bangalore, India (forthcoming)	"SDI-Empowerment and good governance" (provisional)

GSDI 1 identified the critical opportunities and threats inherent in creating a Global SDI. GSDI 2 established the GSDI Steering Committee, grounded GSDI in the context of worldwide sustainable development and encouraged the creation, development and linkage of local, national, regional and global geospatial data infrastructures. GSDI 3 contributed to the formation of global policy, addressed specific initiatives to harmonize existing data by encouraging the development, implementation and maintenance of standards, and proposed an organizational framework for a GSDI. GSDI 4 focused on the needs of developing nations, financial support and capacity building, by national governments and international agencies, and established the GSDI Secretariat. GSDI 5 recognised the importance of the SDI roles in decision support and set the path for the GSDI Association. Lastly, GSDI 6 advanced the transition of the GSDI organization to a permanent and sustainable Association status, reaffirmed its goals and gave specific attention to data interoperability.

3.2.1 GSDI Components and Organizational Model

In essence, by linking national and regional SDIs, GSDI is promoting a global and open process for coordinating the organization, management and use of geospatial data and related activities. GSDI offers the prospect of better decision-making and is backed by international standards, guidelines and policies on access to data, as needed to support global economic growth and its social and environmental objectives.

(a) The GSDI Steering Committee

The GSDI Steering Committee was formed at GSDI 3. It was tasked with establishing a permanent global umbrella organization to take the GSDI into the future, bringing together national and regional committees and other relevant international institutions, in the context of principles of flexibility, inclusivity, simplicity and subsidiarity. The characteristics of this global umbrella organization model were described in a Conference theme paper (Brand, 1998). The goals of the Steering Committee were to help advance awareness, acceptance and implementation of globally compatible SDIs at the local, national and regional levels.

The Steering Committee was comprised of representatives from the regions of the world and a cross section of the GSDI community, basically, government, academia and the private sector. It included the Executive Committee, an Advisory Group, standing Working Groups, a Secretariat for program coordination and an Executive Group in charge of GSDI Conference organization.

The Chair of the Steering Committee was a representative of the region or nation that hosted the most recent GSDI conference and the Vice Chairperson was selected by the host organization of the next conference and was responsible for conference planning.

Each region, National SDI and stakeholder organization named a representative to the Advisory Group. The Chair and Vice Chair proposed regional representatives and stakeholder community representatives from participants at GSDI Conferences and other appropriate individuals as needed to ensure balanced representation of interests from around the world (Holland, 1999).

Two Working Groups implemented the Steering Committee work plan, a Technical Working Group to advise on technical aspects of GSDI, and a Legal and Economic Working Group to advise on economic, legal and funding mechanisms underpinning the GSDI. A Permanent Secretariat was established in 2000 to oversee implementation of the umbrella organization structure, the administration of GSDI and to inform the broad community about the GSDI concept and the value of spatial data.

(b) The GSDI Association

At GSDI 5 it was resolved to form a not-for-profit corporation (a GSDI Association) to replace the GSDI Steering Committee. The Steering Committee would continue as the GSDI guiding authority during the intervening period. In early 2002 the general bylaws, statutes and organizational structure of the GSDI Association were approved at a meeting in Washington DC, USA.

The mission of the Association is, mainly, to serve as the focal point for the community involved in advancing SDI at the global level to support sustainable social, economic and environmental development. In essence the goal of the Association is to support the establishment and expansion of globally compatible SDIs. As well, it promotes better public policy and scientific decision-making.

The GSDI Association provides a unique opportunity for all those interested in facilitating access and applicability of spatial data to promote new development alternatives and democracy, in particular, for developing nations. Being this inclusive, the GSDI Association should be the answer to the complexity and overlap created by the growing number of global spatial data initiatives (Borrero, 2002).

(c) GSDI Program and Business Plan

In 1998, a GSDI Program Plan was prepared to identify and prioritize key programs and activities deemed necessary to promote the objectives of the GSDI organization. The Plan was updated by the Steering Committee in 2002. On the technical side, the Plan included implementation of SDI case studies; an SDI survey; geospatial information and services metadata profiles; the GSDI geospatial search engine; and the development of universal data product identifiers. On the awareness and outreach side, the Plan included the promotion of the GSDI website, brochure and newsletter; coordinated efforts to raise industry, government and donor participation; and focused training and capacity building activities.

Core to the work of the Steering Committee was the development of a business case for investment in an SDI. Two major studies were commissioned, one by the Australian Surveying and Land and Information Group (Centre for International Economics, 2000) and the other by the GSDI Secretariat through the USGS/FGDC (RAND Science and Technology Policy Institute, 2001).

The first of these studies scoped the SDI business case. The study suggested a methodology for undertaking the business case; identified areas of risk; and, defined the terms of reference, timetable for completion, and budget for a project manager to undertake a full business case analysis. Although the Steering Committee saw merit in this comprehensive proposal, it decided to proceed initially in one particular recommended area - by developing a series of SDI case studies.

The second study focused on lessons to be learned from related international collaborative activity, and organizations. The study concentrated on organizations that were successfully promoting global science or technology infrastructures relevant to the GSDI. The study identified the key characteristics of, and the key challenges and decisions facing, the GSDI. Ten examples of successful international and regional collaboration related to the GSDI were documented. Six organizations were analyzed to understand what was important for each organization's past, current, and future success, and what relevant lessons there were for GSDI. Based on this analysis the study presented suggestions and ideas for GSDI development and implementation.

(d) The SDI Implementation Guide ("Cookbook")

The rapid pace of development in technology creates a complex landscape of standards and policies. This has prompted the Technical Working Group to develop an SDI implementation guide, commonly known as the Cookbook. The Cookbook is a useful and dynamic resource, providing advice and context for the adoption of SDI practices. The Guide identifies existing and emerging standards, supportive organizational strategies and policies, free- or low-cost software solutions based on these standards, and best practices associated with building a globally compatible SDI.

Contributions to the Cookbook have been made by many organizations. This effort reflects a commitment to a non-duplicative approach to SDI development. Contributors provide source material, engage in discussions on this information, and provide updates for periodic re-distribution of the Cookbook as technologies and practices evolve. The Guide is available in multiple languages, is accessible on the Internet and is also being distributed on CD ROM and in paper format (Nebert, 2001).

(e) GSDI for Decision Support

One of the key purposes of the GSDI is to improve the availability, accessibility, and applicability of spatial information for decision-making. This has resulted in the establishment, at GSDI 5, of a Working Group to conduct a study on the relationship of decision support systems to GSDI supporting distributed spatial decision-making. The Working Group has described three environments that GSDI decision support must address - the problem environment, the technical environment, and the people environment - and has identified a role for the GSDI in each environment (Feeney, 2002).

3.2.2 GSDI-Current Status of Development

GSDI has made significant progress since GSDI 1 in 1996. Today, many nations are either planning, or are in the process of developing, a National SDI. Equally important, the regions of the world are developing regional coordination mechanisms to bring about linkages amongst nations, regions and sub-regions. Thus, the prospect of a GSDI built upon common elements of national and regional SDI seems very real.

Generally, GSDI is: contributing to the creation of permanent committees in charge of GIS/SDI regional issues; inducing and supporting the formation of National SDIs; increasing awareness on the need "to sustain" SDI's in the context of sustainable development, in the context of Agenda 21 and the Johannesburg World Summit on Sustainable Development; impacting the production of geographic information, for instance, by developing global metadata quality standards, terminology and global geospatial search engine capabilities; promoting national recognition of the geographic sector as strategic for development; and, stimulating the development of Decision Support Systems (DSS).

More specifically, GSDI is related to activities at the global, regional and national level.

(a) The Global Level

There are at least 63 existing documented global SDI initiatives and the number of new initiatives is growing rapidly. Moreover, there are at least 160 initiatives at the global and regional levels based on the use of spatial data, evidencing the increasing importance of geographic information for decision-making and the wide range of applications in need of new mechanisms to improve current levels of access, sharing and applicability of data and interoperability (Borrero, 2002).

Implementation of a GSDI is in its initial stages. As a result the implementation process is sometimes inefficient and not well articulated. A number of implementation initiatives overlap, and could be more efficiently undertaken if they were better linked. This is causing confusion amongst some practitioners.

As a consequence, there is a rising need for cooperation, coordination and integration actions amongst global initiatives. This is one of the main roles of the GSDI Association, to support the establishment and expansion of Local, National, and Regional (multi-nation) SDIs that are globally compatible, and to provide an organization to foster international communication and collaborative efforts for advancing SDI innovations (GSDI 6, 2002).

(b) The Regional Level

At the regional level, there are three main permanent committees established and active: the European Umbrella Organization for Geographical Information – EUROGI; the Permanent Committee on GIS Infrastructure for Asia and the Pacific – PCGIAP; and, the Permanent Committee on SDI for the Americas – PC IDEA. There is also activity at the sub-regional level, such as the Central American Geographic Information Project - PROCIG, and the Caribbean Islands SDI protocol.

The situation in Africa is evolving positively, taking into consideration the complexity of the continent. As has happened in other regions, Africa is moving ahead at its own pace. At the 2nd United Nations Committee on Development Information (CODI) meeting held in Addis Ababa in 2001, there was greater awareness about SDI (based on the number of National SDI initiatives underway in Africa) with a number of national inter-institutional committees already in place and in a few cases, executive orders or laws for geo-information management already enacted. The meeting reached an in-principle agreement to form a Permanent Committee for Spatial Data Infrastructure for Africa before the next CODI meeting.

(c) The National Level

There is an intense and growing degree of activity at the national level with at least 55 countries reporting SDI activities at different stages and pace. Some are at the ideas stage, many are already formally established, and a significant group, representing all regions of the world, are officially approved at high levels in government and very active and productive.

There is strong SDI activity at the state-level in many nations with federal governments, for example, in Australia, the United States and Germany. Also, many local environmental and development authorities, metropolitan areas and urban centers in developed and developing nations show increasing levels of SDI activity. This is likely to be one of fastest growing areas of SDI implementation, given that the majority of spatial data is produced and applied at the local level.

3.2.3 Issues and Challenges

A GSDI backed by international standards, guidelines and policies on access to the data is needed to support global economic growth, and its social and environmental projects (Petersohn, 1997).

The concept of an SDI, and the process by which SDI are being built today, is an essential part of global geography. The way data is being produced, organized and analyzed, from local to global levels (and vice versa), is causing a broad and dynamic exchange of ideas, leading to a better understanding of natural and cultural diversity, and impacting decision-making, territorial planning and sustainable development. In terms of benefits and beneficiaries, it is clear that the GSDI is of vital importance to the implementation of Agenda 21 and to every multi-national environmental convention.

In every nation, but in particular in developing nations, information infrastructures are being recognized as a key to progress and democracy.

However, the accomplishment of GSDI objectives is not an easy task, particularly given the time it is taking some countries well advanced in SDI implementation to reach a level where basic goals are being attained and the infrastructures are performing properly. One of the key questions is what are the costs and the level of resources required? The investment in information infrastructures is one of the largest investments being made by developed nations today. There is a tremendous difference in the productivity of knowledge between countries, industries, and organizations who have or have not made this investment. In all cases data must be accessible, documented, structured and reliable or, otherwise, is practically non-existent.

In looking today at SDI in the developing world , one can observe that most nations have not formally adopted a national policy concerning the use of digital geo-information, especially pertaining to the ways in which it should be used to promote wealth and development according to local needs. As for production, a great deal of old data is still being digitized. There is a kind of paradox in this situation: Although there is more geo-referenced information available, quality applications are limited. This situation is also showing the need for education. For instance, case studies are indicating the existence of a significant gap between technology tools available and poor levels of data use due to low availability of specialized human resources and, to some extent, the quality of technical assistance. Consequently, the lesson learnt seems to be that data production and SDI building, to be successful, demands a capacity building strategy, supported at the highest level in government, where a minimum level of financial, technical and institutional sustainability is required (Borrero, 1999).

Therefore, the challenges confronting the GSDI are many - raising the level of awareness, acceptance and support; recognizing and complementing related initiatives; including all stakeholders and engaging the less developed economies of the world; maintaining enthusiasm and momentum; and finally, delivering beneficial outcomes. Notwithstanding these significant challenges the potential of the GSDI to contribute to a better world in the future drives those who seek its realization (Holland, 1999).

3.2.4 Future Plans and Discussion

The GSDI is an ambitious initiative dedicated to international cooperation and collaboration in support of local, national and international SDI developments that would allow nations to better address social, economic, and environmental issues of pressing importance. GSDI is serving as a point of contact and effective voice for those in the global community involved in developing, implementing and advancing SDI concepts, fostering SDIs that support sustainable social, economic, and environmental systems integrated from local to global scales, and promoting the informed and responsible use of geographic information and spatial technologies for the benefit of society.

(a) Implementing the GSDI Association

In order to better implement the GSDI program and business plan and to attend to the need for global cooperation, coordination and integration amongst the many related global SDI activities, the initiative is in the process of evolving into a not-for-profit Association to replace the Steering Committee and to oversee the implementation of the GSDI in the future. The GSDI Association, formally introduced at GSDI 6 in 2002, will allow many stakeholders to become members of GSDI, bringing in talent to contribute directly to the goals of GSDI. In the long run the idea is to have as many members as possible from throughout the globe implementing and experimenting with SDI concepts at all levels within government and the private sector.

(b) Main Drivers in GSDI

As expressed in its definition, the purpose of a GSDI originates in the necessity to ensure that those working with spatial information are not impeded in meeting their objectives. The goal of the GSDI is to improve the availability, accessibility, and applicability of spatial information. In order to reach this goal the following strategies will need to be pursued:

- Development of mechanisms to enable digital geographic data and technologies to support decision-making at all scales and for multiple purposes.
- Promotion of common implementation practices that will help establish an interoperable GSDI framework and derived products and services. This will encourage maximum consistency of adoption and interpretation and would enhance the commitment to relevant standards by multiple participant sectors.
- Increment the dissemination of proven practices, as emerging from case studies, surveys and other types of research.
- Growing interest about the non-technical aspects of SDI, showing the relevance of multiculturalism, the economics of SDI and the need for improved coordination mechanisms.
- Continue with the outreach activities in specific regions. In this context, there is the need to solve the funding and the capacity building components of SDI.
- Include all stakeholders.
- Complement and coordinate with related initiatives. The GSDI is but one of many initiatives aimed at improving access to geographic information. It is essential that these initiatives are identified, recognised and appropriately supported so that the maximum synergy can be obtained from their collective outcomes.

In summary, the GSDI Association will foster SDI developments in support of important worldwide needs such as: improving local to national economic competitiveness; addressing local to global environmental quality and change; increasing efficiency, effectiveness, and equity in all levels of government, and advancing the health, safety and social wellbeing of humankind in all nations.

3.3 GLOBAL MAP

The Global Mapping project is not an SDI in the strict sense of the term. It does, however, exhibit many elements of an SDI in terms of its processes and its outputs. The Global Mapping project operates within an environment of institutional arrangements and technical standards. Most importantly the Global Map is an important fundamental dataset in the Global SDI.

The Ministry of Construction of Japan first proposed the concept of a Global Map, and an international body to manage its creation, in 1992. The concept was to build a global mapping database through international cooperation.

A major factor currently hindering the understanding of global environmental issues is inadequate mapping of many areas of the world, in terms of map scale, content, and timeliness. This is equally true for both the developed and the developing world. At present a map of the entire globe must be compiled from material from different sources, material that has been produced to different standards of accuracy and content, material that is generally out of date and incomplete, and material that is in many cases inaccessible. The Global Mapping project is designed to assist in addressing this problem. Since 1992 significant progress has been made in building and disseminating the Global Map and many countries are now involved in the initiative.

3.3.1 Components and Organizational Model

(a) Institutional Arrangements

The Global Mapping project is managed by the International Steering Committee for Global Mapping (ISCGM). The purpose of the ISCGM is to facilitate the creation, maintenance, dissemination and use of the Global Map. In order to achieve this outcome it relies heavily on the cooperation and contribution of participating countries, and the active involvement of committee members.

The ISCGM consists of the heads of national mapping agencies (NMA), and representatives of international organizations and academic institutions. Ad-hoc working groups, and a permanent secretariat located in the Geographical Survey Institute of Japan, support the committee. Professor John E. Estes (deceased), of the University of California Santa Barbara, chaired the committee from its inception until 2001. The ISCGM operates under an agreed set of procedural rules.

The ISCGM has met on 9 occasions, in Tsukuba Japan (1996), Santa Barbara USA (1996), Gifu Japan (1997), Sioux Falls USA (1998), Canberra Australia (1998), Cambridge UK (1999), Cape Town South Africa (2000), Cartagena Columbia (2001) and Budapest Hungary (2002). The committee has sponsored several Global Mapping forums as a means of facilitating communication among participating countries.

The secretariat of the ISCGM maintains a comprehensive Internet site. All matters dealt with by the committee are publicly accessible. A Global Map newsletter is produced 4 times a year by the secretariat, the latest and 25th edition released early in 2002.

The ISCGM has established relationships with kindred organizations. It holds liaison status with ISO TC-211 (Technical Committee on spatial data) and is represented on the GSDI steering committee. It works with and through regional SDI groups in Asia and the Pacific (PCGIAP), the Americas (PC-IDEA), Europe (EUROGI), Africa (UN-CODI) and Antarctica (SCAR). It has working arrangements with National Mapping Agencies (NMAs), groups of NMAs (Eurogeographics), academic and research institutions.

Participating countries are classified in terms of their level of participation. Level A countries have committed to contribute data of their own and other countries; level B countries have committed to contribute data of their own countries; and, level C countries require assistance in the production and contribution of data over their own country. It is envisaged that level C countries will collaborate with level A countries in making their contribution.

(b) Technical Standards

The Global Map is being created to a common set of specifications. These are based on the standards work of ISO TC-211 and the Open GIS Consortium, and on globally accepted map specifications produced by major government agencies. Version 1.1 of the Global Map specification is held on the Internet site.

(c) Fundamental Data

The Global Map contains elevation, vegetation, land-use, drainage systems, transportation and administrative boundaries data. The data are represented in a consistent manner and are referenced to a common geodetic datum, thus allowing data layers to be overlaid and integrated with other datasets. The data has a nominal ground resolution of 1 kilometre (equivalent to conventional maps at scales of 1:1,000,000).

Global Map data is presently available in 2 versions. Global Map version 0 data comprises full global coverage of data from several international programs. Global Map version 0 data is out of date in many cases. Global Map version 1 data, initially released in 2000, comprises up to date data of particular countries supplied by their NMAs. Only a small portion of the globe is presently covered with Global Map version 1 data. Both version 0 and version 1 data is constructed to version 1.1 of the Global Map specification.

Global Map data comprises elevation in the form of a digital elevation model; land cover, land use and vegetation in the form of a gridded dataset; and, drainage system (rivers, streams, lakes), transportation networks (roads, railways, airports), political boundaries (including coastlines) and populated places in the form of a vector dataset. Version 0 data has been sourced from the Global 30 Arc Second Elevation Dataset (GTOPO30) of the United States Geological Survey (USGS), EROS Data Centre; the Global Land Cover Characteristics Database of the USGS, the University of Nebraska-Lincoln and the European Commission Joint Research Centre; and, the VMAP Level 0 product of the National Imagery and Mapping Agency USA.

3.3.2 SDI Current Status of Development

A decade into the Global Map initiative 127 countries have committed to participate in the Global Mapping project. A further 17 countries are considering participation. Global Map version 1 data is available over 12 countries - Australia, Bangladesh, Colombia, Japan, Kenya, Lao PDR, Sri Lanka, Mongolia, Nepal, Panama, the Philippines and Thailand. These commitments are shown in Figure 3.1.

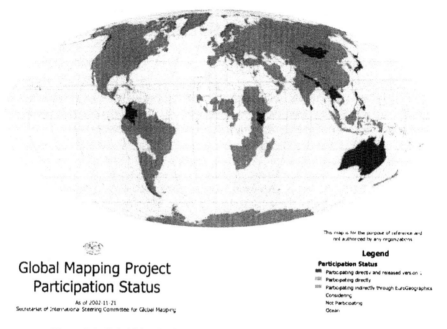

Global Mapping Project
Participation Status

As of 2002-11-21
Secretariat of International Steering Committee for Global Mapping

This map is for the purpose of reference and
not authorized by any organizations

Legend

Participation Status
- Participating directly and released version .
- Participating directly
- Participating indirectly through EuroGeographics
- Considering
- Not Participating
- Ocean

Figure 3.1: Global Mapping Project Participation Status as at 21 November 2002
(ISCGM web site: http://www.iscgm.org/html4/index_c3_s3_ss1.html#doc11_3432)

Global Map data is presently archived and distributed by the ISCGM secretariat. Data may be downloaded from the Global Map Internet site and several mirror sites. Data is distributed at no cost for non-commercial use via the Internet. Version 0 data carries no copyright and is in the public domain. Copyright in version 1 data is held by the contributing country and must be acknowledged by the user. Commercial users are obliged to deal directly with the copyright holder.

ESRI announced a GSDI-Global Map grant program, in honour of the inaugural chairman of ISCGM, in 2001. The program aims to assist countries in developing data by providing software, support and training. 85 countries have currently received a grant through this program.

Several initiatives have been taken with the United Nations. The ISCGM is a participant on the United Nations Geographic Information Working Group. Global Map data has been recognized as an important element of the planned United Nations Geographic Database. ISCGM holds non-government organization status at the World Summit on Sustainable Development (WSSD) in 2002. Several ISCGM members have been actively involved in preparatory meeting for the WSSD. As a result of these efforts Global Map is specifically mentioned as an important enabling technology in the draft plan of implementation of the WSSD.

3.3.3 Issues and Challenges

The immediate challenges facing the Global Mapping project relate to issues of production, innovation and distribution. In the longer term the challenges relate more to issues of commitment, relationships, marketing and coordination.

Notwithstanding the present high level of participation in the Global Mapping project, only a few countries have provided updates to version 0 data of their areas of interest. The slow development of the version 1 product is not a major issue while demand levels are relatively low, once demand picks up, and this is ultimately the objective of the project, it will focus customer attention on deficiencies in product quality. Although it is easy to point to the lack of version 1 data at present it is another thing altogether to identify a tidy solution. Many developing countries are reliant on external support and this is not always readily forthcoming. The Global Map specification is widely recognised but is difficult to implement, particularly in countries unable to access contemporary mapping systems. The specification is also becoming dated when compared with data models and formats in wide use today. The product is useful in many applications but could certainly be enhanced with new content and with new data structures, particularly structures supportive of continuous coverages and Internet applications. The online distribution of Global Map data from the ISCGM Internet site is effective but lacks the sophistication and reach of the more commercially oriented sites in existence today.

At present commitment to the Global Mapping project is high, in terms of the number of countries prepared to contribute data, in terms of the enthusiasm and active involvement of ISCGM members, and in terms of the dedication of the Geographical Survey Institute of Japan. However, participation is voluntary and as is the case with many international projects ultimate success depends greatly on continued support from stakeholders. The importance of maintaining strong relationships with kindred organizations and activities cannot be underestimated. Relationships at present are principally targeted at government agencies and non-government organizations. An obvious absence is a strong relationship with industry, which on the one hand potentially offers resources, and on the other potentially risks alignment with a particular commercial offering. The marketing of the Global Map product has not been a major requirement thus far. However as the product and its distribution channels grow in sophistication, and they most certainly will, the need for a strong strategic marketing focus will become more important. Although the Global Mapping project is presently visible and well recognised internationally, it is not the only global data gathering activity today, and is likely to be faced with many more such activities in future. The need for strong coordination with complementary global data gathering programs will become increasingly important.

3.3.4 Future Plans and Discussion

The challenges and issues identified in the previous section might appear to some to be almost insurmountable. That is not the case. Certainly they require careful management but the underlying strength of the Global Mapping project should ensure that the most important issues are properly addressed.

The Global Mapping project is clearly demonstrating the value of global cooperation. It is attracting the attention of individual countries, regional groups and international organizations like the United Nations. It is delivering a valuable global dataset and is bringing countries together in a friendly and non-political way. The project brings together national mapping agencies in an environment of cooperation and common interest, providing a forum for information exchange and institutional strengthening. Most importantly it is helping national mapping agencies adjust to the fact that they are now an important player in the global context.

ISCGM meetings are well attended and attendees gather from all parts of the globe. Meeting agendas are comprehensive, technically interesting, and outcome oriented.

The Global Mapping project is forward looking in many ways. Participating countries are being encouraged to seek continuity of availability, and low-cost and open access, to future data sources such as Landsat and ALOS satellite imagery from the USA and Japan respectively. The ISCGM is working with the ISO TC-211 to provide workshops and documentation on technical standards to Global Map participants. The ISCGM is pressing for the establishment of an industry advisory group to act as a channel with the commercial sector. A cooperative arrangement with the European group of national mapping agencies will see the availability of a milestone continuous mapping database of that important area of the globe. In addition significant progress is being made within the ISCGM to identify the market needs for global products, to define the next generation of Global Map product that meets these needs, and to identify the access regime that will need to be implemented for this sophisticated product to be widely adopted in the marketplace.

Against a background of, on the one hand, rapid globalization of economic activity and explosive growth in global information and communication technology, while on the other hand, continuing intractable pockets of regional instability and protective national policies on trade and foreign affairs, the ISCGM continues to provide an effective forum for cooperation and collective action at the global level. The Global Map continues to be an important contemporary data product that holds out the prospect of helping resolve many global problems in the future and supporting the implementation of a global SDI.

3.4 EVALUATION OF GLOBAL INITIATIVES

The GSDI initiative and the Global Map project are two of the primary SDI activities at the global level. They both aim to improve global decision-making through the better use of spatial information. They both recognise their place within the broader arena of information infrastructures. Whereas GSDI has an expansive, some might argue imprecise, vision and plan encompassing policy, standards, institutional linkages, technology, education, resources, capacity building, users, applications and by no means least, spatial information; Global Map has a relatively tightly focussed vision and plan, the creation of a fundamental global spatial information product. Notwithstanding, GSDI and Global Map are both highly ambitious in their goal for at least one common reason, the essential requirement for cooperation between stakeholders in many countries.

GSDI and Global Map emerged, not coincidentally, at around the same time in the mid-1990's. This marks the point where global challenges (particularly economic and environmental challenges) were being recognised as a mainstream political agenda (particularly in developed countries), and where information, technology and skills, used in a cooperative way, were being seriously considered (arguably for the first time) as solutions to these problems. Both activities have gathered momentum since their inception. The term GSDI is now reasonably widely known, at least in the SDI community of practitioners. GSDI conferences now regularly attract a broad base of stakeholders. GSDI-sponsored studies and outreach projects are now recognised as important value-adders, the latter particularly in developing countries. The GSDI Association, albeit in its infancy is a sign of the developing maturity of the initiative. Global Map has similarly achieved a reasonable level of name recognition, especially within national mapping organizations and most recently on the global WSSD stage. The Global Map product, despite the as yet relatively low level of country-specific updates, is beginning to be used in real-world applications, enabling the ISCGM to begin moving in new and innovative directions.

GSDI and Global Map, to a certain extent, attract the interest of similar groups of stakeholders. This is not surprising given the strategic positioning of the Global Map product as a core element of the GSDI.

In summary, the strength of the GSDI lies principally in its very credible claim to be a focal point for the growing number of SDI initiatives around the globe; and by extension, its ready ability to attract important stakeholders. The strength of Global Map lies principally in the potential relevance of its output, and the high level of commitment by its key stakeholder group, national mapping organizations. The principal weakness of the GSDI is, at present, its lack of focus in action, and its as yet immature organizational form. The principal weakness of the Global Map is the relatively low level of action by national mapping organizations (as opposed to the relatively high level of commitment), and the relatively inflexible design of the product and its distribution channel. There is tremendous opportunity for GSDI and Global Map, separately through creative partnerships with other initiatives, and by working much more closely together. Finally, there is a common threat to both the GSDI and the Global Map - failure to maintain relevance to, and commitment and cooperation amongst, stakeholders.

The strategic balance sheet for both global initiatives is currently in the black. If this continues there is a real prospect of significant benefit at global, regional and national levels. Time will tell whether or not this balance sheet position can be maintained.

3.5 REFERENCES

Borrero, S., 1999, Case Study at the National Level: ICDE, the Colombian SDI. Presented paper, *Cambridge Conference for National Mapping Organizations*, Cambridge, United Kingdom.

Borrero, S., 2002, The GSDI Association: State of the Art. *6th GSDI Conference*, Budapest, Hungary. Online.
<http://www.gsdi.org/docs/GSDI6/PPT/sb270902.htm>.
Brand, M., 1998, Global Spatial Data Infrastructure: Policy and Organizational Issues. *3rd GSDI Conference*, Canberra, Australia, Online.
<http://www.gsdi.org/docs/canberra/theme.html>.
Centre for International Economics, 2000, Scoping the business case for SDI development. *4th GSDI Conference*, Cape Town, South Africa, Online.
<http://www.gsdi.org/docs/capetown/businesscase/scoping.pdf>.
Coleman, D.J. and McLaughlin, J., 1997, Defining Global Geospatial Data Infrastructure (GGDI): Components, Stakeholders and Interfaces. *2nd GSDI Conference*, Chapel Hill, North Carolina, USA, Online.
<http://www.gsdi.org/docs/ggdiwp1.html>.
Feeney, M.E., 2002, Spatial Data Infrastructure Development to Support Spatial Decision-Making. Presentation of Working Group on DSS, *6th GSDI Conference*, Budapest, Hungary, Online.
<http://195.228.254.144/Eloadasok/Streeam3/Wednesday_16hr/Mary_Ellen_Fee ney/GSDI-DSS WG Pres.ppt>.
GSDI 5, 2001, Resolutions of the 5th GSDI Conference, Cartagena, Colombia, Online. <http://www.gsdi.org/docs/130701/gsdi5resf.html>.
GSDI 6, 2002, Resolutions of the 6th GSDI Conference, Budapest, Hungary, Online. <http://www.gsdi.org/docs/GSDI6/web/GSDI6finalres.htm>.
Holland, P.R., 1999, The Strategic Imperative of a Global Spatial Data Infrastructure. *Cambridge Conference for National Mapping Organizations*, Cambridge, United Kingdom, Online. <http://www.gsdi.org/docs/stratim.html>.
Nebert, D. D. (Ed.), 2001, Developing Spatial Data Infrastructures: The SDI Cookbook, Online. <http://www.gsdi.org/pubs/cookbook/cookbook0515.pdf>.
Petersohn, F., Primozic, K., Osborne, N. 1997, The Age of New Economics ... Based on a Global Spatial Data Infrastructure. *2nd GSDI Conference*, Chapel Hill, North Carolina, USA, Online. <http://www.gsdi.org/docs/fritz.html>.
RAND Science and Technology Policy Institute, 2001, Lessons for the Global Spatial Data Infrastructure: International Case Study Analysis, Online.
<http://www.gsdi.org/docs/Rand/GSDIfinal.pdf>.

CHAPTER FOUR

Regional SDIs

Ian Masser, Santiago Borrero and Peter Holland

4.1 INTRODUCTION

One of the distinctive features of the last decade is the emergence of regional spatial data infrastructure (SDI) organizations. This is due to the need for a Regional SDI perspective as a consequence of both the need for seamless consistent spatial data beyond national boundaries to support decision-making at this level and the lack of a coordinating body among multinational and sub-regional ongoing initiatives. This began with the creation of the European Umbrella Organization for Geographic Information in 1993 and was quickly followed in Asia and the Pacific by the establishment of a Permanent Committee for this region in 1995 under the auspices of the United Nations Regional Cartographic Conference. A similar organization for the Americas followed in 2000, after a three years process, with support from 21 nations. At the turn of the century Africa and the Middle East were the only regions of the world without such an organization. However, moves are currently under way to create a Committee on Development Information under the auspices of the UN Economic Commission for Africa.

With this in mind this chapter reviews the experiences of the European, Asian and American initiatives. The chapter is divided into four substantive sections. The first three of these contains profiles of each of these organizations that describe their history and backgrounds, their status and the issues and challenges that they face. The final section of the chapter is devoted to a comparative evaluation of the strengths and weaknesses of these organizations.

4.2 EUROPEAN UMBRELLA ORGANIZATION FOR GEOGRAPHIC INFORMATION (EUROGI)

A large part of the European land area and most of its population live in countries that are members of the European Union. Since its inception in 1958 the numbers of members of the EU has grown from 6 to 15. These countries account for nearly two thirds of the total European land area (excluding the former Soviet Union) of 4.9 million square kilometres and their combined population of 397 million represents nearly three quarters of the total European population. At the present time negotiations are in progress with a further 13 countries regarding their applications for membership. These include virtually all the former Communist states in central and Eastern Europe as well as Cyprus, Malta and Turkey. As a result the EU may have at least 25 members with a combined population of 500 million by the end of this decade.

As a result of these developments a strong impetus has been built up over the last half century towards a European wide perspective. This is evident in growing pressures for European integration, the creation of a wide range of European level institutions and the availability of funds to support a variety of regional policies. This has also been an important factor in the emergence of a European perspective in the geographic information field during the last ten years.

In 1991 the then DG XIII (now DG Information Society) of the European Commission set up an enquiry to consider the desirability of establishing a European association for geographic information to promote the use of geographic information at the European level. Four prominent members of the European geographic information community carried out the enquiry: Michael Brand (Ordnance Survey Northern Ireland), Peter Burrough (University of Utrecht), Francois Salge (Comite Europeen des Responsables de la Cartographie Officielle, now Eurogeographics), and Klaus Schueller (a leading consultant in the GIS field).

The team's initial findings were presented at a Forum in Luxembourg in Oct 1992. They argued that there was a strong European wide demand for an organization that would further the interests of the European geographic information community. They also presented a vision of EUROGI as an organization that would not "replace existing organizations but ... catalyse effective cooperation between existing national, international, and discipline oriented bodies to bring added value in the areas of Strategy, Coordination, and Services" (Burrough *et al.* 1993, 31).

With this in mind the team proposed the establishment of a European Umbrella Organization for Geographic Information at another meeting in Luxembourg in November 1993 that was attended by delegates from 14 different European countries and six pan European organizations. These delegates formally resolved to set up EUROGI at this meeting and elected Michael Brand as its Founder President together with an eight person Executive Committee to produce a work plan to address some of the most important issues that had been identified by the team (Masser and Wolfkamp, 2000).

4.2.1 Organizational Status

EUROGI is an independently funded European organization that seeks to develop a European approach towards the use of geographic information technologies. Its mission is to maximise the use of GI for the benefit of citizens, good governance and commerce. With this in mind it promotes, stimulates, encourages and supports the development and use of geographic information and technology and also acts as the voice of the wider European geographic information community (www.eurogi.org).

Its membership is drawn from two different categories of organization: national GI associations and pan European organizations. EUROGI has 22 national and three pan European members (at the time of writing). The national associations can be divided into two categories: members with voting rights and candidate members who are still in the process of building up their associations. Throughout its lifetime EUROGI has experienced a slow but steady growth in the number of its national members. Its current membership includes representatives from all 15 EU countries together with Iceland, Norway and Switzerland. Recently there has been a lot of interest from the central and east European countries that plan to join the EU. Hungary and Poland have been full members of EUROGI for some time and the Czech Republic and Slovenia have become candidate members in the last two years. In overall terms these members constitute more than 6000 organizations.

At the present time only three pan European bodies are members of EUROGI. The European association of national mapping agencies, Eurogeographics (formerly CERCO) is a full member and the European Association of Remote Sensing Companies and the Urban Data Management Society are associate members. Nevertheless EUROGI has strong links with most of the other pan European bodies in the field including the UNECE Working Party on Land Administration, the Association of GI Laboratories in Europe and the European Association of Remote Sensing Laboratories and hosts a regular meeting of these groups each year. It also has an active MOU with the Joint Research Centre for various joint activities including the organization of specialist workshops.

EUROGI is a 'stichting' under Dutch law that supports a small secretariat based at the offices of the Dutch Kadaster in Apeldoorn in the Netherlands. This consists of a full time Secretary General, an Assistant Secretary General and some contract staff. The activities of EUROGI are managed by the Secretariat in conjunction with an Executive Committee elected by its members. The President meets and chairs this four times a year. The seats on this committee are currently held by representatives from the British, Danish, French, German, Hungarian, Italian, and Swiss national associations and the European association of national mapping agencies (Eurogeographics). Under its statues and bylaws EUROGI must convene a General Board meeting once a year to approve its work plan and budget. The meeting also approves any proposed changes in the statutes and bylaws and elects its President (every two years) and its Executive Committee (on a rotational basis).

The current activities of EUROGI are closely linked to one or more of the following five strategic objectives:

- Encouraging greater use of geographic information in Europe: this is the overarching goal as it is vital to ensure that GI is used as widely as possible in both the public and private sectors as well as by individual citizens in the interests of open government.
- Raising awareness of GI and its associated technologies: there is a continuing need to raise awareness in the community as a whole regarding the importance of recent advances in both technology and their potential for an increasing range of applications.
- Promoting the development of strong national GI associations: an important element of EUROGI's strategy is to create the institutional capacity to take a lead in its formulation and implementation. This is particularly important given the need for national associations to maintain some measure of independence from government.
- Improving the European spatial data infrastructure: Although many of the main elements of a European infrastructure are already in place in different

countries there is a lack of effective mechanisms at the European level to
promote greater harmonization and interoperability between countries in this
respect.

- Representing European interests in the global spatial infrastructure debate: In
an era of increasing globalization it is essential that Europe does not evolve in
isolation.

With these strategic objectives in mind EUROGI published a consultation paper in
October 2000 entitled 'Towards a strategy for geographic information in Europe'
(EUROGI, 2000). The starting point for this initiative was the belief that positive
steps are needed to fill the current void with respect to GI strategy at the European
level (see below). This paper sets out a framework for the first stage of this strategy
and outlines a number of measures that will bring it into being. Following the
publication of this paper, EUROGI, together with the Joint Research Centre, Open
GIS Consortium (OGC) Europe and the University of Sheffield submitted a
proposal for part funding of this strategy to the European Commission as an
accompanying measure under its Fifth Framework for Research and Development
in October 2000. The research program subsequently agreed with the Commission
in connection with the GI Network in Europe (GINIE) project (Craglia, 2002).

The consultation paper also gives a useful overview of EUROGI's current
activities with reference to these objectives. General activities listed under the
heading of encouraging greater use of GI in Europe include disseminating
information about applications through publications, and presentations at
workshops and conferences. It also includes stimulating investment in GI Research
and Development. In particular there is a need for a greater emphasis on
geographic information related research topics in the European Commission's
Framework Programmes for Research and Technology Development.

Under the heading of raising awareness of GI and its associated technologies
the following activities are listed:

- Influencing key decision-makers. EUROGI is developing an integrated
lobbying strategy that involves putting into place and then maintaining a
logistical infrastructure that supports a sustainable co-ordinated lobbying
program. This program complements that of the EUROGI members at the
national level and that of other pan European representative bodies.
- Informing the population as a whole. European GI strategy must seek to
facilitate the diffusion of knowledge and experience between different
countries and between different professional groups and different application
fields. An important tool in this respect is EUROGI's strategy for developing a
GI Case Study Service.

There are two examples of activities under way in connection with the objective of
promoting strong national GI associations. The first is promoting the comparative
analysis of national GI policies. In its early years EUROGI commissioned
comparative studies of copyright and commercialization in various European
countries. In November 1999 and June 2001 it organised, together with the Joint
Research Centre and DG Information Society, workshops on Data Policy (Craglia
et al., 2000) and Cadastral issues relating to agri-environmental policies (Waters
and Dallemand, 2002). The second is pooling the experience of the national
associations themselves as an instrument for national capacity building. A lot can

be learnt from the positive (and negative) experiences of different national associations. With this in mind EUROGI commissioned a comparative study of its national member associations in 2000-1 (van Biessen, 2001).

Two specific examples illustrate the kind of activities that are under way under the heading of improving the European SDI:

- Creating a forum for the candidate nations for EU membership to discuss their accession requirements in terms of geographic information. An initial joint EUROGI/Joint Research Centre workshop on this topic was held in Brussels in November 2000 to explore these issues (Craglia and Dallemand, 2001).
- Working with other organizations to improve the quality of metadata services, given that lack of information as to what data is available in different European countries on particular topics is a major barrier to many transnational applications. A good example of this is the European Territorial Management Information Infrastructure project (ETeMII) co-ordinated by Associazione GISFORM. EUROGI, together with a number of other key stakeholders in GI such as the Joint Research Centre, OGC Europe and AGILE as well as the French, German and Portuguese national GI associations were partners in this project (ETeMII, 2002).

Finally, under the heading of representing European interests in the GSDI debate EUROGI is focussing its attention on a number of issues. Firstly, playing a full role in the organization of GSDI activities. The GSDI structure is loosely organised around the periodic rotation of responsibilities between its regional bodies. EUROGI was in charge of the GSDI web site until June 2000 and organised the GSDI 6 conference in Budapest in September 2002. Secondly, participating in GSDI working groups. EUROGI is participating in several Working Groups that have been set up under the auspices of the GSDI. These groups are essentially virtual Working Groups in that they rely on the exchange of information and opinions over the Web rather than face to face meetings.

4.2.2 Issues and Challenges

The main issues and challenges that are likely to face EUROGI in the coming years can be summarised as a SWOT analysis.

(a) Strengths

As a result of its achievements since 1994 EUROGI is widely regarded as the voice of the wider European geographic information community. It brings together representatives from both the public and private sectors and has played a major role in raising awareness and stimulating the debate on geographic information policy issues at the European level. It has also been very successful in facilitating the establishment of national geographic information associations and promoting the greater use of geographic information in most European countries.

(b) Weaknesses

EUROGI has very limited resources at its disposal. Consequently, its success as an organization is dependent on the degree to which it can mobilise the skills and experiences of its member bodies in its activities and the extent to which it can form strategic alliances with other key players in the European geographic information field such as the Commission's Joint Research Centre.

(c) Opportunities

The demise of the GI2000 initiative created an opportunity for EUROGI to take the lead in this respect with respect to the establishment of an Advisory Board for Geographic Information for the whole continent as part of the GINIE project (Craglia, 2002). EUROGI is in a good position to carry out this task. It also has a number of advantages when compared to the GI2000 initiative. These include its independent position from any government body and the fact that its membership is not restricted to the EU. The GINIE project and the European Environmental SDI initiative (INSPIRE) (van der Haegan, 2002) also provide an excellent opportunity to move towards a comprehensive European SDI.

(d) Threats

Due to the limited resources at its disposal EUROGI has suffered in the past from a lack of stability that threatened its effectiveness as an organization. The disruption caused by the periodic relocation of its Secretariat has been considerable and should be avoided as far as possible.

4.3 PERMANENT COMMITTEE ON GIS INFRASTRUCTURE FOR ASIA AND THE PACIFIC (PCGIAP)

The Asia and Pacific region is a vast geographic area of land and water encompassing a diversity of languages and cultures, political systems, social and economic priorities, and environmental pressures. Historically, matters of a spatial information nature have been considered at a regional level within the United Nations organizational framework, specifically at the United Nations Regional Cartographic Conference for Asia and the Pacific (UNRCC-AP).

The UNRCC-AP was conceived in the period immediately following the Second World War. In February 1948, the Economic and Social Council of the United Nations recommended that governments of member states stimulate surveying and mapping of their national territories. This recommendation resulted in the 1st UNRCC-AP meeting being held in Mussoorie, India, in February 1955. Conferences have been held at three-yearly intervals since the inaugural meeting. Since 1990 conferences have been held in Bangkok, Thailand (1991), Beijing, China (1994), Bangkok, Thailand (1997) and Kuala Lumpur, Malaysia (2000). The 16th UNRCC-AP will be held in Okinawa, Japan, in July 2003. The UNRCC-AP is organised by the United Nations Statistics Division.

By the early 1990's it was apparent a triennial regional conference could not properly address on its own the rapid changes taking place in the spatial information world. This issue was considered during the 13th UNRCC-AP in Beijing. The conference resolved that directorates of national survey and mapping organizations in the region form a permanent committee to discuss and agree on geographic information system (GIS) standards and infrastructure, institutional development, and linkages with related bodies around the world. The inaugural meeting of this permanent committee, the PCGIAP, was held in Kuala Lumpur, Malaysia, in July 1995.

The aims of the PCGIAP are to maximise the economic, social and environmental benefits of geographic information in accordance with Agenda 21 by providing a forum for nations from Asia and the Pacific to:

- Cooperate in the development of a regional geographic information infrastructure;
- Contribute to the development of the global geographic information infrastructure;
- Share experiences and consult on matters of common interest; and
- Participate in any other form of activity such as education, training, and technology transfer.

Members of the PCGIAP are directorates of national survey and mapping organizations, or equivalent national agencies, of the 55 members and associate members of the United Nations Economic and Social Commission for Asia and the Pacific. The committee meets annually and reports every three years to the UNRCC-AP. Committee business is performed through working groups. An executive board comprising president, vice-president, secretary, and up to 7 other members, administer the PCGIAP. The activities of PCGIAP are largely funded by member agencies. Some project activity receives specific funding from external agencies such as aid organizations and government grants.

4.3.1 SDI Components and Organizational Model

The concept of a spatial data infrastructure (SDI) of the Asia and the Pacific region (the APSDI) was developed early by the PCGIAP. It was formally recognised at the 14th UNRCC-AP in Bangkok. This conference resolved that: governments in the region should consider establishing a National SDI, and participating in the work of the PCGIAP to establish an APSDI; and, the PCGIAP should endeavour to link the APSDI to the global SDI. The PCGIAP published the details of the APSDI in 1998 (PCGIAP, 1998).

In this publication the PCGIAP outlined its vision for the APSDI - a network of databases, located throughout the region, that together provide the fundamental data needed to achieve the region's economic, social, human resources development and environmental objectives. The databases were envisioned to contain fundamental spatial datasets, and might be inter-linked electronically. Most importantly, the databases would be linked:

- By an intra-regional institutional framework that provides mechanisms for sharing experience, technology transfer and coordination of the development of the regional fundamental datasets;
- By the use of common technical standards, including a common geodetic reference frame, so that data from numerous databases can be brought together to create products and solve problems, both regionally and globally;
- By the adoption of common policies on data access, pricing, privacy, confidentiality and custodianship;
- By the implementation of inter-governmental agreements on data sharing; and

- Through a comprehensive and freely accessible directory of available datasets containing descriptions and administrative information that accords with agreed standards for metadata.

This suite of administrative and technical linkages was seen by the PCGIAP to distinguish the APSDI from a collection of uncoordinated datasets and would make it such a powerful tool for economic and social development in the region. The model of the APSDI comprises four core components – institutional framework, technical standards, fundamental datasets, and the access network. Most of the activity of the PCGIAP and its working groups at present is directed towards developing and implementing these components of the APSDI.

4.3.2 Status of Development

(a) Institutional Framework

The PCGIAP and its structural elements – the executive board, the secretariat, working groups, and ad-hoc committees – provides the leadership and organizational focus for the APSDI. The PCGIAP has met on seven occasions since its inaugural meeting: in Sydney, Australia (September-October, 1996); Bangkok, Thailand (February 1997, prior to the 14th UNRCC-AP); Teheran, Iran (February-March, 1998); Beijing, China (April, 1999); Kuala Lumpur, Malaysia (April, 2000, in conjunction with the 15th UNRCC-AP); Tsukuba, Japan (April, 2001); and, Bandar Seri Begawan, Brunei Darussalam (April, 2002). It meets next in conjunction with the 16th UNRCC-AP. The executive board has met between PCGIAP meetings. The PCGIAP maintains a comprehensive internet site (http://www.gsi.go.jp/PCGIAP/) that contains: statutes of the committee; composition of the executive board; names and contact details of members; activities of each working group; reports and resolutions of each meeting; technical papers and publications; and, resolutions of each UNRCC-AP since 1991.

Between the 1st and 4th meetings, the focus of the committee was generally on institutional and policy matters. At the 4th meeting the committee examined its future directions and agreed to shift to a more production focus. This resulted in a restructure of working groups. Working groups were restructured again at the 6th committee meeting. The present working groups are regional geodetic networks, regional fundamental data, cadastre and institutional strengthening.

The PCGIAP has developed a policy (PCGIAP, 2002) for sharing fundamental data. The policy was endorsed by resolution at the 15th UNRCC-AP. The policy applies to: specific fundamental spatial data; the collection, management and use of fundamental spatial data in the regional interest, whether its application is at national, regional or international levels; and, the use of fundamental spatial data by governments, industry and the community. The policy defines a set of principles for management of, and access to, these data. A core principle is that all sectors of the community should have easy, efficient and equitable access to fundamental spatial data. The policy recognises the importance of custodians of data, and includes guidelines on custodianship.

Given the sensitivity of some countries to sharing certain types of fundamental data, the committee has issued a special policy statement on administrative boundaries. The statement recognises that PCGIAP members are entitled to represent their boundaries data in accordance with their own stand on territorial sovereignty, and that these data are shared purely for the purposes of developing the APSDI.

The PCGIAP has entered into a co-operative arrangement with the Permanent Committee on SDI for the Americas (PC IDEA). Under this arrangement both committees have agreed to build a more effective relationship with the United Nations system. Specifically, the committees propose:

- To work with the United Nations to re-engineer UNRCC structures in the Asia Pacific and the Americas;
- Provide assistance in the development and use of the United Nations geographic database; and,
- Coordinate efforts in promoting SDI goals and objectives at the United Nations World Summit on Sustainable Development and beyond.

The cooperative arrangement also aims to promote and stimulate SDI development in each region by sharing experiences and best practices, implementing pilot projects in areas of mutual interest, and undertaking an inventory of SDI programs and projects. Finally, the cooperative arrangement encourages members of both committees to participate in GSDI and global mapping activities.

In addition to the cooperative arrangement with the PC IDEA, the committee has established liaison relationships with the International Organization for Standards Technical Committee on Geographic Information Standards (ISO/TC211), the European Umbrella Organization for Geographic Information (EUROGI), and a number of international professional and coordination bodies. These relationships aim to ensure that the APSDI is compatible with other regional and global initiatives and that the PCGIAP utilises appropriate standards from, and avoids duplicating the activities of, these bodies.

The PCGIAP has endeavoured to facilitate capacity building amongst its members. Seminars on contemporary SDI topics, presented by local and international authorities, for the benefit of members and non-members, have been conducted during committee meetings. Working groups have held workshops between meetings. The PCGIAP has established a training facility at Haikou, Hainan Islands, China, through the support of the State Bureau of Surveying and Mapping China.

The PCGIAP has made special arrangements to try to meet the unique needs of some of its members. As a result of a PCGIAP-sponsored workshop 1999, a Pacific group has been recognised by the committee as representing the interests of the island countries in this particular area. Similarly, a Commonwealth of Independent States (CIS) group has been recognised as representing those CIS countries that fall within the Asia and Pacific region.

(b) Technical Standards

Generally speaking, the PCGIAP has taken the approach of adopting technical standards and specifications developed by others. The PCGIAP draws on the work of ISO TC-211, and also the Open GIS Consortium, in this regard. In a few instances the committee has developed its own approaches. Technical standards and specifications have been adopted thus far in the areas of the reference framework, data structures, cadastral systems, and metadata and the access network.

The International Terrestrial Reference Framework (ITRF) system and the GRS80 ellipsoid have been adopted by the PCGIAP as the fundamental datum parameters for the APSDI and regional applications. A number of PCGIAP-sponsored geodetic campaigns have been conducted and these campaigns have assisted, and will continue to, assist the development of a regional geodetic network. The network will provide the framework for transforming coordinates of points on the many national geodetic datums (PCGIAP 2002) in the Asia and Pacific onto a single regional datum. Transformation parameters have been generated for countries participating in the geodetic campaigns. Work is presently proceeding in the PCGIAP on the definition of a vertical datum for the Asia and Pacific region.

The PCGIAP has adopted the global mapping specification of the International Steering Committee for Global Mapping as its initial specification for the fundamental datasets of the APSDI. This specification defines the structure of vector and raster data representing various thematic layers of geographic information. The PCGIAP has developed a draft specification for administrative boundary data.

The recently re-established PCGIAP cadastral working group is using cadastral definitions and frameworks established by the International Federation of Surveyors (FIG), and like organizations, wherever appropriate. The ISO TC-211 metadata standard, and OGC concepts of the technical architecture for online mapping and distribution of spatial data, is being used in the development of the APSDI access network.

(c) Fundamental Datasets

A fundamental dataset, in the context of the APSDI, is a dataset for which there is a demand for consistent regional coverage by a number of users. The APSDI was originally conceived to include fundamental data such as the geodetic control network, elevation, drainage systems, transportation, populated places, geographical names, vegetation, natural hazards, administrative boundaries and land use.

The PCGIAP recently committed to defining the contents of, and developing the technical specifications for, the Asia-Pacific Regional Fundamental Dataset, using the global mapping and administrative boundary pilot project specifications as a reference. The first version. of the Asia-Pacific Regional Fundamental Dataset will contain vector data no less accurate than 1:1 million in map scale, and raster data no less than one kilometre in ground resolution. PCGIAP has encouraged members to provide data at larger scales on a voluntary basis,

At present, no specific fundamental datasets of the region have been built. There is, however, global map data of the region (version 0 data) and several PCGIAP member countries have updated this data. An administrative boundary dataset has been constructed over several countries in order for the PCGIAP to assess the technical and policy issues associated with building a complete dataset of the region.

The PCGIAP is planning to investigate administrative infrastructures that might be appropriate for managing marine resources in the context of the United Nations Convention on Law of the Sea, producing an atlas of land administration and land tenure, and comparing land administration and cadastral systems across the Asia and Pacific region.

(d) Access Network

The access network for the APSDI is envisaged to be a network of clearinghouse nodes that will:

- Make data discoverable and enable access to data for regional applications;
- Provide a means to advertise data collection programs, user needs, data inventories and data quality, and will support and encourage documentation of datasets; and
- Minimise duplication of effort in data collection and processing in the region.

The node would include a web server and a metadata directory. Functionally the clearinghouse node would support identification and authorization, metadata browsing and data query, data exchange, and possibly online mapping and analysis.

Currently the PCGIAP has developed a draft guideline for APSDI clearinghouse node development, and a draft metadata profile complying with the ISO19115 core metadata profile. A prototype network of clearinghouse nodes is being actively considered.

4.3.3 Issues and Challenges

The issues and challenges facing the PCGIAP in the development of the APSDI have policy, organization and technical dimensions. In several instances they have been known for some time. For example, participation rates by member countries, designs of the work program and working groups, and availability of resources to pursue programs are issues as relevant today as they were in 1998. Notwithstanding the endorsement of a policy of data sharing at the 15th UNRCC-AP in 2000 there is little evidence yet of significant data transfers within the PCGIAP. In fact, at present, member contributions to the Global Map, under a policy of open use in non-commercial applications, would arguably be the most significant evidence of commitment to the principles of the PCGIAP policy. Fundamental data identification has generally been slower than originally planned with the notable exception of the definition of the regional geodetic datum and the specification of a regional administrative boundary dataset. One area of APSDI activity that may soon move forward at some pace is the regional data directory and distribution network, both topics of discussion at working group meetings late in 2002. None of the preceding comments should be taken to be a criticism of the PCGIAP. Rather, they point to the underlying difficulty of achieving regional outcomes when domestic influences by necessity take precedence.

4.3.4 Future Plans and Discussions

The PCGIAP will ultimately achieve its objective to create an APSDI. Over time, it will deliver a workable policy, standards, data and distribution regime, committed to by the majority of its members. This will come about not only because of actions - initiated by the PCGIAP. Rather, the actions of the PCGIAP in concert with the influence of SDI developments within countries, within other regions, and at the global level will drive the APSDI. The future plans of the PCGIAP must therefore continue to have a policy, organizational and technical dimension. A policy dimension that achieves workable, beneficial, sustainable, financially viable and globally consistent outcomes; an organizational dimension facilitates active participation of members and productive linkages to like groupings at national, regional and global level; and a technical dimension that creates useable data products and infrastructure, that improves the level of knowledge of members, and builds on the positive experiences of others. Through the work plans of its executive board and its working groups, the PCGIAP is heading in the right direction.

4.4 PERMANENT COMMITTEE ON SDI FOR THE AMERICAS (PCIDEA)

In 1997, during the Sixth United Nations Regional Cartographic Conference for the Americas, the delegates recognising the rapid global emergence of national and regional spatial data infrastructures and its contribution to maximise the benefits of geographic information for sustainable development, recommended the establishment, "within one year", of a Permanent Committee for SDI/GIS Infrastructure in the Americas, reporting to the following meetings of the UNRCC-Americas. There were as well regional projects in need of such a committee, including the Inter-American Biodiversity Information Network (IABIN), the Inter-American Geospatial Data Network (IGDN) and SIRGAS, for the establishment of a unique geodetic reference system for South America.

At its own pace, Latin-America responded to this challenge. Its promotion needed additional effort. Many understood the new drivers behind the GIS/SDI orientation of the proposed committee, while others were comfortable with the still prevailing "conventional" cartographic approach. This is not surprising. South America, for instance, is a region of significant contrasts, and at that time was notably covered by analog maps at small scales averaging 1:500.000, with significant levels of obsolescence. In a space twice as large as Europe there were ten national mapping organizations. There was also a problem of budget and one of adaptation to the ICT revolution.

In February 1998, taking advantage of a UN meeting of delegates and experts held in Aguascalientes, Mexico to define the mission and focus of the 7th UNRCC for the Americas, the delegates representing member states from the region established the Committee, in an ad-hoc manner, with Colombia elected as pro-tempore chair until full formalization of the committee was achieved, within the following year (Borrero, 1998).

However, this was not an easy goal to obtain, for several reasons. The idea of spatial data infrastructures for Latin America was difficult to digest at a time when many national mapping and surveying organizations were making "changes in old organizational structures." Specialists were in the process of redefining their own data in the digital format from a local perspective. Ideas were also not yet clear in the face of nationalistic attitudes behind previous failures concerning regional standards for mapping and geodesy. Finally, evident isolation, was expressed in low levels of participation at the global level.

A survey conducted in the region (Hyman *et al.,* 2000), along the lines of the one undertaken by Onsrud (1998), represented the following situation for Latin-America:

- GIS/SDI issues led by national mapping agencies, but challenged by other types of geographic information providers.
- NSDI initiatives in the Region were the responsibility of five national organizations, on average.
- Absence of policy concerning development of national spatial data infrastructures.
- Industry and other types of technology providers not involved in the development of SDI initiatives.
- Information layers most frequently considered as fundamental data: topographic mapping, roads, land cover and land use, administrative borders, and hydrography.
- A tendency towards cost recovery, as the main factor for pricing data.
- Legal issues, funding, lack of standards for geodata, pricing and data access considered as major constraints for the consolidation of NSDI initiatives.

Consequently, the promoters of the committee started by convincing Latin-American state members about the need for harmonised spatial data infrastructure at the local, national, regional and global levels and its contribution to economic, social and environmental sustainable development. After three years a new situation arose in the Region and there were several reasons behind this change. First, the Pan-American Institute for Geography and History (PAIGH), as well as the Directorate of Geographic organizations for South America (DIGSA) were involved in the promotion of enabling information technologies and spatial data infrastructures. Second, a wave of optimism derived from recent experiences, such as SIRGAS and the IGDN/PAIGH Electronic Atlas. Third, numerous ongoing National SDI initiatives and the impact of regional and global initiatives like the Global Spatial Data Infrastructure and Global Map.

After a three years process, with support from 21 nations, PC IDEA, was established in 2000, as the main result of the International Seminar on Spatial Data Infrastructures organised in Bogotá, Colombia. This was supported by the World Bank Information Development Program, the Centre for Inter American Agriculture Technology (CIAT), the US Federal Geographic Data Committee (FGDC) and PAIGH. As a result, with 17 SDI initiatives identified at the national level, it can be stated that Latin America is now looking seriously at the importance of having a regional spatial data infrastructure. The recognition of the importance of spatial information is growing rapidly worldwide and, though there is a limited capacity to produce and consume digital information, the region is already immersed in the Information and Communications Technology (ICT) concept.

PC IDEA was built on the experience observed in other regions, in particular, that of Asia-Pacific and Europe. The influence of PCGIAP is evident for instance in documents on statutes, procedural rules and the terms of reference for working groups (Borrero, 2001).

4.4.1 Organizational Status

In a very nationalistic region, the role of the Committee, as an Inter-American forum, is essential for the development of spatial data information capabilities at all levels, across the continent. This is why the vision of PC IDEA can be best summarised as "the end of spatial information isolation in the Americas". PC IDEA is looking to:

(a) Increase production of spatial data, impacting Research and Development and sustainable development;
(b) Migrate from local data to National SDI, leading to regional spatial datasets
(c) Locate geoinformation as one strategic sector for development, by convincing decision makers on the need to maximise benefits derived from geographic information;
(d) Increase knowledge capabilities for all in the American hemisphere community, by incrementally increasing access to data and information;
(e) Contribute to the development of GSDI and Global Mapping.

In terms of government and operations, like the PCGIAP organization, PCIDEA has an executive committee including its President, Vice President, Executive Secretariat and the chairs of the three working groups: Legal and Economics; Communication and Awareness and Technical Aspects. The last of these includes the following seven subcommittees: Information Policy, Fundamental Data, Clearinghouse, Standards, Cadastre, Geographic Names and Capacity Building.

PC IDEA reports to the UN Regional Cartographic Conferences and contributes to the development of GSDI, Global Mapping and the UN Geographic Information Working Group (UNGIWG). In 2001, the Committee signed a Memorandum of Understanding on cooperation with its sister Committee for Asia and the Pacific (PCGIAP), focusing on the promotion of SDI at the regional level and in the process of formalising relations with other relevant international organizations.

The Committee promotes the participation of all countries in the region (nowadays there are 24 national members) looking not only for the integration of a truly regional infrastructure but also for the establishment of sub-regional committees to stimulate the development of seamless data bases, especially in Central America, the Caribbean, and the Andean nations in South America. So far, there are two such sub-regional committees one for the English-speaking Caribbean Islands and the other for Central America, called PROCIG.

As for main events to date and in the future, there have been three PC IDEA Meetings, with the next three planned for Caracas, Venezuela (2002), San Jose,

Costa Rica (2003), along with a UN/FIG Regional Workshop on Capacity Building for Land Administration, and Panama City, Panama (2004).

4.4.2 Issues and Challenges

The contribution of PCIDEA to the development of SDI is evident at all levels and sectors. In particular, the Committee is having an impact on substantive elements related to: (i) Institutional capacity building, education and training; (ii) establishment of networks for the exchange of knowledge and experience dealing with geographic information; (iii) appraisal of economic aspects concerning spatial data management and infrastructure; (iv) the relation between land administration and spatial data infrastructure development as strategic policy and (v) increasing understanding of specific regional framework data needs.

PC IDEA has directly contributed to the formal establishment of NSDI initiatives in Panama, El Salvador, Ecuador, Cuba and Honduras and has encouraged activities resulting in NSDI initiatives in Nicaragua, Venezuela, Chile and Argentina. As mentioned above, at the supra national level, PC IDEA has contributed directly to the formation of the SDI for Central America, an initiative advanced with the cooperation from the International Center of Tropical Agriculture (CIAT).

As for challenges, there are many. In most Latin American countries, geographic information is produced by different public and private sector organizations, in a disparate fashion, using different technical specifications and data formats. Pricing policies are linked to cost recovery approaches and data access is limited, very frequently by simple formalities. When it comes to democratization, the fact is that a large part of the population in the Americas has limited access to spatial data and is unable to take advantage of the potential benefits offered by geo-referenced data.

The members of the Permanent Committee are the legal representatives of national organizations in charge of national mapping. In this context, there is need for a broader participation, in particular, the directors of the competent organizations on geographic information or the national entities in charge of environmental and development issues, in accordance to PC IDEA Statutes.

Looking into the future, PC IDEA will have to deal with (i) the fact that for a significant period of time there will be in the region different legal frameworks affecting spatial information policies and regulations; (ii) it will not be easy to move from an organizational structure based on key players or "champions", to a solid institutional base; (iv) support from government at highest levels and improved national coordination will vary notably, although the number of NSDI initiatives backed by sound policy concerning e-government, standards, intellectual property rights in the case of digital information, is growing in the region.

Nonetheless, the main difficulties PC IDEA is facing are directly related to the need for financial support and to attain 'sustainability' and the capacity to influence and implement capacity building strategies for SDI at all levels. As well, there is a need and expectations for PC IDEA to move the Region from a project based strategy and from a local to national and from there to regional data bases and seamless data, which are almost none-existent today. Lastly, there is need to improve the levels of coordination and integration between local, national and sub-regional SDI initiatives, in order to achieve a minimum level of harmonization from which PCIDEA can really expand.

4.5 COMPARATIVE EVALUATION

From the profiles of the three regional spatial data infrastructure organizations it can be seen that there are many similarities between them. This is particularly the case with respect to their general objectives. EUROGI's mission 'to maximise the use of GI for the benefit of citizens, good governance and commerce' has much in common with PCGIAP's mission 'to maximise the economic, social and environmental benefits of geographic information in accordance with Agenda 21'. Nevertheless there are important differences between the three regional organizations. Both PCGIAP and PC IDEA are closely linked to the UN Regional Cartographic Conferences and their membership is limited to representatives of the national mapping agencies of the countries involved. EUROGI, on the other hand, is based on a completely different model. It was set up with some assistance from the European Commission as a regional umbrella organization. For this reason its members are either national geographic information organizations such as the British Association for Geographic Information (AGI) or the Hungarian Association for Geographic Information (HUNAGI) or pan European associations such as the European association of national mapping agencies (Eurogeographics). Consequently EUROGI is able to claim that it represents the interests of over 6000 organizations in Europe through its network of national and pan European associations.

EUROGI also differs from the other Permanent Committees in that it is independent from any national or intergovernmental body. It also supports a full time secretariat from the subscriptions of its members. In contrast the two PCs are an integral part of the intergovernmental structures of the United Nations. They rely heavily on the voluntary input of their members who take it in turns to provide a secretariat to support the activities of their various working groups.

Table 4.1: The PCGIAP (Permanent Committee) Model

Strengths	Weaknesses
Status within existing UN trans-national structures	No independence
Consistency of membership	Limited to national mapping agencies
Funding in kind/secondments	Dependent on members commitment

Tables 4.1 and 4.2 summarise the strengths and weaknesses of these two models. From Table 4.1 it can be seen that the PC model has some obvious strengths. The PCs occupy a clearly defined position within existing United Nations regional structures. Their membership is relatively homogeneous and their members share common professional aspirations. They are also independent financially in that they operate through funding in kind to support their activities.

However, each of these apparent strengths has its weaknesses. Their position within existing intergovernmental structures makes it difficult for them to take a critical position in public or to lobby support for their activities. By restricting membership to national mapping agencies many of the key stakeholders from industry and research are left out of their discussions. The funding in kind model also has its weaknesses in the degree to which it is dependent on voluntary contributions from its members. This is particularly a problem in Asia and the Pacific, according to Rajabifard (2002,3) where 'out of the 55 member nations of the Asia and Pacific region, only six are active core participants, about 19 countries are occasional participants and the remaining countries have never really attended any meetings.'

Table 4.2: The EUROGI Model

Strengths	Weaknesses
Independent status	Legitimacy has to be earned
Diversity and range of membership	Diversity of membership
Permanent secretariat	Dependency on subscriptions for funding

Table 4.2 summarises the strengths and weaknesses of the EUROGI model. Its strengths lie in its independent position which enables it to criticise national and regional governments if necessary and make it possible for EUROGI to lobby for support of its ideas. Unlike the PCs, its member associations include all the main stakeholders in spatial data infrastructure activities and its permanent secretariat facilitates the flow of information between its members and also within the European Commission. This is reflected in high levels of attendance at meetings. Two thirds of its full members have also actively participated in the work of its Executive Committee which meets four times a year.

As with the PC model each of the strengths can also be regarded as weaknesses. EUROGI's independent status also means that it has to make constant efforts to ensure its legitimacy to policy makers and the EU as a whole. The diversity of its membership also means that it is often difficult to find a common position among the competing and sometimes conflicting interests of its members. Finally the overhead costs of maintaining a permanent secretariat are relatively high given its small number of member states. This means that EUROGI has also to look for support in kind to sustain its activities.

From this comparative evaluation it can be seen that both models of regional spatial data infrastructure organizations have their strengths and weaknesses. Notwithstanding this the profiles of their activities in this chapter highlight the degree to which they have been successful in raising awareness of the regional dimensions of SDI through their manifold activities and the vital role that they have played as intermediaries between the national and the global levels in the establishment of a Global Spatial Data Infrastructure and the organization of its conferences. It is also worth noting that GSDI 5 in Cartagena (May 2001) was organized by the Agustin Codazzi Geographic Institute of Colombia (IGAC) and the PC IDEA, while GSDI 6 in Budapest (September 2002) was organised by

EUROGI in conjunction with the Hungarian national geographic information association (HUNAGI). GSDI 7 which will take place in Bangalore, India in February 2004, will be organised by PCGIAP in conjunction with the Indian Government.

Although most countries in the world are covered by Regional SDI initiatives there are still one or two gaps in coverage. This is particularly the case with the Middle East and along the North African coast. With this in mind a resolution was passed at GSDI 6 tasking the GSDI to explore the possibilities of organising GSDI 8 in this region in 2005.

4.6 REFERENCES

van Biessen K. T., 2001, Models of national GI associations in Europe, EUROGI, Apeldoorn.

Borrero, S., 2001, The Role of the Permanent Committee on Spatial Data Infrastructure for the Americas -PC IDEA, *7th United Nations Cartographic Conference for the Americas*, New York, United States.

Borrero, S., 1998, Case Study of Transnational Initiatives: Latin America. *3rd GSDI Conference*, Canberra, Australia.

Burrough P. M. Brand, F. Salge and K. Schueller, 1993, The EUROGI vision, *GIS Europe* 2, (3), 30-31.

Craglia M., 2002, The GINIE project, *Proceedings 8th EC GI Workshop*, Dublin.

Craglia M., Annoni, A. and Masser, I., 2000, Geographic information policies in Europe: national and regional perspectives, Report of the EUROGI - EC Data Policy Workshop, EC Joint Research Centre, Ispra.

Craglia M. and Dallemand, J.F., 2001, Geographic information and the enlargement of the European Union, Report of the EUROGI - European Commission Workshop, EC Joint Research Centre, Ispra.

ETeMII, 2002, The road to Europe's future in spatial data infrastructure activities, www.ec-gis.org/etemii.

EUROGI, 2000, Towards a strategy for geographic information in Europe: a consultation paper, EUROGI, Apeldoorn.

van der Haegen, M., 2002. The INSPIRE project, Proc 8th EC GI workshop, Dublin.

Hyman G., Lance, K. and Ines Rey, D., 2000, Survey of National Spatial Data Infrastructures in Latin America and the Caribbean, International Centre for Tropical Agriculture (CIAT) and Agustin Codazzi Geographic Institute (IGAC), *International Seminar on National Spatial Data Infrastructures*, Bogotá, Colombia.

Masser, I. and Wolfkamp, A., 2000, EUROGI: past, present and future, *GeoEurope* 9, (8), 37-39.

Onsrud, H.J., 1998, Survey of national and regional spatial data infrastructure activities around the globe, *3rd GSDI Conference*, Canberra, Australia.

PCGIAP, 2002, Permanent Committee on GIS Infrastructure for Asia and the Pacific Home Page. <http://www.gsi.go.jp/PCGIAP/> (Accessed September 2002).

PCGIAP, 1998, A spatial data infrastructure for the Asia and the Pacific region, PCGIAP Publication No.1. Online.
<http://www.gsi.go.jp/PCGIAP/tech_paprs/apsdi_cnts.htm>.

Rajabifard, A., 2002, 'Diffusion of Regional Spatial Data Infrastructures with particular reference to Asia and the Pacific', *PhD dissertation*, Dept of Geomatics, University of Melbourne. Online.
<http://www.geom.unimelb.edu.au/research/publications/Rajabifard_thesis.pdf>.

Waters, R. and Dallemand, J.F., 2002, Cadastral data as a component of spatial data infrastructure in support of agri-environmental programmes, Report of the EUROGI - European Commission Workshop, EC Joint Research Centre, Ispra.

CHAPTER FIVE

SDI Diffusion – A Regional Case Study with Relevance to other Levels

Abbas Rajabifard

5.1 INTRODUCTION

Many countries are developing spatial data infrastructures to better manage and utilise their spatial data assets by taking a perspective that starts at a local level and proceeds through state, national and regional levels to the global level, as was highlighted in Chapter 2. Increasingly these countries are finding it necessary to cooperate with other countries to develop Regional (multi-national) SDIs to assist in decision-making that has an important impact across national boundaries.

However, the development of a Regional SDI presents different challenges than the development of the individual National SDI initiatives on which it is based, as was demonstrated in Chapter 4. This is mainly because of the voluntary participation and cooperation required between different countries and initiatives, at an international level, to form a Regional SDI. As a result, despite considerable interest and activities, the development of an effective and comprehensive Regional SDI is hampered by a lack of support from member nations which results in this initiative remaining very much an innovative concept. This is because the adoption of a new idea, even when it has obvious advantages, is often very difficult.

This chapter introduces key factors that facilitate Regional SDI development through recognising the complexity of the interaction between social and political issues. This draws on the experiences of SDI development in Asia and the Pacific region, as a selected case study, and the applicability of the theories of innovation diffusion to provide a useful framework for the study of the development of SDIs and improvement of their conceptual models.

The results and lessons learned from this research can also be used and applied in other regions, and potentially other jurisdictions such as local, state, national and global.

5.2 ASIA AND THE PACIFIC REGION AND REGIONAL SDI ACTIVITIES: A CASE STUDY

Asia and the Pacific region world with some 60 per cent of the world's population is the largest region in the and includes 55 countries as defined by the United Nations. The countries span a wide part of the globe from Iran and Armenia in the west to French Polynesia in the east, from the Russian Federation and Japan in the north to New Zealand in the south. This region has emerged as one of the most dynamic regions of the world (Fukasaku, 1995). Its rapid and sustained development has created vast trade and investment opportunities, especially for the economies of its individual nations. This region is changing fast and it is changing for the better, even recognising the economic crisis of 1997. Asia and the Pacific region has witnessed dramatic and widespread changes due to the forces of globalization, industrialization and urbanization.

One of the most significant developments for this region has been the rapid growth of regional cooperation which has directly impacted on its regional economy. Within a few years, a number of regional initiatives have been endorsed and various forms of cooperative ventures have been established. For example, the results of these developments can be seen in the formation of many cooperative organizations such as APEC (Asia-Pacific Economic Cooperation), ASEAN (Association of South East Asian Nations), ECO (Economic Cooperation Organization), SAARC (South Asian Association for Regional Cooperation), as well as various innovative types of subregional cooperative ventures, such as the activities of Australian-ASEAN Economic Cooperation Programs (AAECP) at a country level.

The Asia and the Pacific regional organizations mostly work and cooperate with each other on different areas including development assistance; human resource development; economic development; science technology transfer; political and institutional cooperation; and security issues. The aims and objectives of these organizations are mainly related to specific parts or the whole area of the region. In order to achieve their objectives, regional organizations need to access regional spatial data to identify spatial features and their characteristics to make informed decisions and to implement resulting regional initiatives. In Asia and the Pacific region, spatial data is traditionally collected and disseminated by a range of mandated national organizations according to a variety of national standards.

A major difficulty in relation to the spatial data in this region is coordination. National administrations do not systematically cooperate with their equivalents elsewhere. Due to problems in coordination, different data specifications and standards are used by member nations which make it difficult for data exchange. More generally there are a lack of common elements that could facilitate data exchange such as compatible working scales, compatible GIS software, and the completion of a regional database which could be used for standard basic information layers.

As a result, the Permanent Committee on GIS Infrastructure for Asia and the Pacific (PCGIAP) was formed through the efforts of the United Nation Regional Cartographic Conference for the Asia-Pacific region (UNRCC-AP) in May 1994 (PCGIAP, 1995). As was reviewed in chapter 4, the aims of the PCGIAP are to maximise the economic, social and environmental benefits of geographic information in accordance with Agenda 21 by providing a forum for nations across the region to cooperate in the development of the Asia-Pacific Spatial Data Infrastructure (APSDI) and contribute to the development of the global infrastructure. The Committee's vision for the APSDI is a network of databases, located throughout the region, that together provide the fundamental data needed to achieve the region's economic, social, human resource development and environmental objectives.

In this regard, the PCGIAP has developed a conceptual model for its SDI initiative that comprises four core components - institutional framework, technical standards, fundamental datasets, and access networks (PCGIAP, 1998).

5.2.1 Current Progress of PCGIAP and APSDI Development

According to the PCGIAP (2001), some important steps toward the development of the APSDI have been achieved since its establishment. For example the committee successfully implemented a regional geodetic network, defined a regional geodetic datum, developed and approved a policy on sharing fundamental data, developed guidelines on custodianship, and in particular, agreed upon the definition of APSDI.

Although these achievements are very important, and provide a valuable contribution and will form the basis for the APSDI development, there are some other issues involved in the progress of PCGIAP which need to be discussed. These issues include the low rate of participation in PCGIAP activities, the organizational structure of PCGIAP, and the APSDI conceptual model.

For example, the current rate of participation in PCGIAP activities, shows that after many years of effort the APSDI initiative still does not receive full support from all member nations and regional organizations (Holland, 1998; Mohamed, 1999; Rajabifard *et al.*, 2000).

According to the report of the PCGIAP-Taskforce group, presented at the 15th UNRCC-AP conference, April 2000 in Malaysia, the maximum number of countries participating in PCGIAP meetings is less than half of the members (this rate remained the same for the next two PCGIAP annual meetings in 2001 and 2002). This limited number of participants is an important issue which needs to be considered, discussed and resolved, before the committee moves on to developing important principle policies such as a policy on sharing fundamental data. This is important because these kinds of policies need to be accepted and supported by the majority of member nations in the region, rather than by the majority of participating nations.

The problem of participation can also be observed in many National SDI initiatives throughout the world (Masser, 1998; Onsrud, 1998). Some reasons for the limited support from certain nations, regional organizations and other relevant institutions include:

- the lack of awareness of the value of SDIs;
- the lack of understanding about the concept and nature of SDIs;
- the incompatibility of the current conceptual and organizational model with the perceived needs of the member nations; and
- the complexity of different regional issues such as diverse political, cultural and economic positions.

One of the reasons why the PCGIAP does not receive full support from all member nations is related to the PCGIAP organizational structure. This Committee is comprised of 55 nations who are represented by directorates of national survey and mapping organizations and equivalent national agencies. This structure causes problems from two different perspectives:

(a)   The PCGIAP members are mainly providers or producers of national spatial datasets and not necessarily the users of such national and regional datasets. Yet, one of the most promising advantages of SDIs is to facilitate sharing and access to spatial datasets by users. Therefore, it is essential to involve potential users of regional spatial datasets and the politicians concerned with their promotion and funding, in the development and implementation of the Regional SDI.

(b)   The organizational and political position and responsibilities of each national surveying and mapping organization (from which the members of the PCGIAP come) differ from country to country. For example, in some nations the mapping and spatial data activities are the responsibility of a civilian organization, but in other nations the mapping and spatial data activities are the responsibility of a military organization.

Another reason the PCGIAP cannot achieve full member support is due to their SDI conceptual model. As discussed in Chapter 2, an SDI comprises not only the four core components identified by PCGIAP as the institutional framework, technical standards, fundamental datasets, and access networks, but also an important additional component, namely, human resources (people). This component includes the spatial data users and suppliers and any value-adding agents in between, who interact to drive the development of the SDI.

The absence of this component results in a situation where the SDI coordinating agency, like the PCGIAP, concentrates on the four core components identified and develops their strategies to build the APSDI in such a way that ignores the interests and potential contributions of other stakeholders, such as the non-participating members and agencies. To avoid this problem, the current APSDI model and development strategy need to be modified.

5.3 SDI DIFFUSION

Getting a new idea adopted, even when it has obvious advantages, is often very difficult. It is noted that innovations may require a lengthy period, often many years, from the time they become available to the time they are widely adopted. Therefore, a common problem for many individuals, communities and organizations, as suggested by Rogers (1993), is how to speed up the rate of diffusion of an innovation.

The APSDI is an innovation for the PCGIAP member nations, who face the same challenge of fast-tracking the APSDI diffusion. Therefore, the theories of innovation diffusion provide a useful framework for the study of the development of Regional SDI and the improvement of its conceptual model in this region (Rajabifard et al., 2000).

Diffusion can be referred to as the process of communicating an innovation to and among the population of potential users who might choose to adopt or reject it (Zaltman et al., 1973, as cited by Pinto and Onsrud, 1993). Gattiker (1990) views diffusion as 'the degree to which an innovation has become integrated into an economy'. Spence (1994) describes diffusion as 'the spread of a new idea from its source to the ultimate users'. These definitions package different concepts inherent in diffusion.

Rogers has followed and documented the development of diffusion research over the years (Rogers, 1971; Rogers, 1983; Rogers, 1993; Rogers, 1995). He views diffusion as 'the process by which an innovation is communicated through certain channels over time among the members of a social system' (Rogers, 1983). In particular, he used the organization innovation process model Figure 5.1) to describe the process in which an innovation is adopted and utilised.

Rogers' definition or its variations have been used by people from different disciplines over the years (Goodman, 1993; Pinto and Osrud, 1993; Zaltman *et al.* 1973). This definition also gives rise to four elements of diffusion, namely the innovation, the communication channel, time and the social system, which constitute the foci of research activities in the past five decades. Further Rogers explains, it is a special type of communication, in that the messages are concerned with new ideas. Communication is a process in which participants create and share information with one another in order to reach a mutual understanding. So diffusion is a special type of communication, in which the messages are about a new idea. This newness of the idea in the message content gives diffusion its special character. The newness in this case as highlighted by Chan and Williamson (1999) means that some degree of uncertainty is involved in diffusion. Uncertainty is the degree to which a number of alternatives are perceived with respect to the occurrence of an event and the relative probability of these alternatives. Uncertainty implies a lack of predictability, of structure, of information. In fact, information is a means of reducing uncertainty.

Based on Rogers (1995), the organization innovation process model is generally made up of two main stages, namely, initiation and implementation and five sub-stages (Figure 5.1).

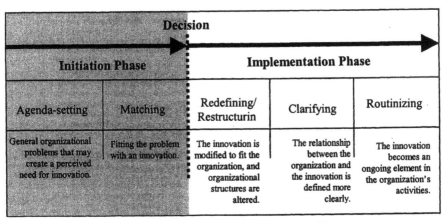

Figure 5.1: Organizational Innovation Process Model

Initiation is concerned with all activities, including information gathering, conceptualising and planning that culminate in the decision to adopt an innovation by the decision makers in an organization (Rogers, 1995). Implementation refers to the steps taken after the adoption decision that lead to utilization of an innovation prior to its ultimate institutionalisation (Goodman, 1993 as quoted by Chan, 1998). Based on Rogers' organizational innovation process model, Chan *et al.* (2001)

suggested an integrated framework for GIS diffusion research. According to this framework, any innovation like GIS and SDI is a dynamic entity that is central to the diffusion process. This entity assumes multiple identities or configurations as diffusion progresses over time, as represented by the simplified staged model of the diffusion. The characteristics of this entity may change as it passes from the initial conceptual configuration, through one or more intermediate configurations, to an actual physical configuration of GIS or Regional SDI that serves the needs of the organization or a region. Whether diffusion has failed or succeeded, there is a feedback loop to allow the process to start all over again.

Based on Chan's framework, in order for diffusion of a Regional SDI to be successful in a region, it is important to take into consideration the conceptual configuration of Regional SDI, the social system of the region as defined by the boundary, and the other external, organizational and personal factors which have an impact on the diffusion of Regional SDI. But the approach taken by PCGIAP suggests that the nature of the social system and may other factors as illustrated in Figure 5.2, are ignored.

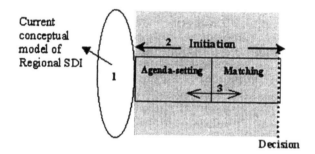

Figure 5.2: Current Approach Taken by PCGIAP for APSDI Development, Based on
Organizational Innovation Theory
1-Innovation, 2- Communication Channel, 3- Time

In this case, the number of nations not participating in the APSDI initiative, suggests that many nations are still not aware of the concept of Regional SDI or do not fully appreciate the value of the innovation as portrayed in the current conceptual model. In any case, it shows that these nations have not therefore entered into the initiation stage of the organizational innovation process model.

The social component of diffusion has been identified as an important component for the study of an innovation (Scott, 1990; Rogers, 1995). Rogers (1995) defined a social system as a set of interrelated units that are engaged in joint problem solving to accomplish a common goal. He pointed out that 'innovation diffusion is affected by different aspects of the social system'. On a similar line, Coote (1999) believes that social change is a very important issue for analysing the impact of change on an initiative in an organization. He further clarified that all organizations are about people.

Given the impact of the social component of diffusion to innovation studies, Chan (1998) pointed to the need to conduct integrated studies involving the elements of time and the social system in diffusion research. It was therefore necessary to conduct the Asia and the Pacific case study on an expanded conceptual model of the APSDI, one which occurred within a social boundary (Figure 5.3). This extended model helped to understand various features involved in the Asia and the Pacific social system as well as identifying key factors among different features of the social system, which are influencing the diffusion of APSDI initiative. This enabled the identification of a series of factors which are important for the development of a Regional SDI initiative. The next section presents and discusses these factors.

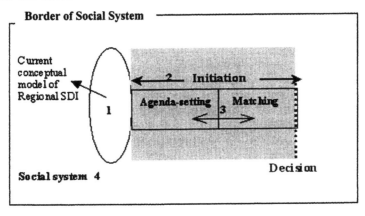

Figure 5.3: Extended Conceptual Model
1- Innovation, 2- Communication Channel, 3- Time, 4- Social System

5.4 INFLUENCING FACTORS FOR REGIONAL SDI DIFFUSION

The challenge of designing, building, implementing, and maintaining a Regional SDI lies in the structured arrangement of a substantial number of different disciplines and the examination of a large number of factors and issues. For example, arrangement and getting agreement between and from different countries, building a common communication network etc. Therefore, it is essential that SDI practitioners understand the significance of human and community issues as much as technical issues, which determine the success of SDI developments. SDI development is a complex task fraught with difficulties in sustaining a culture of sharing, a shared language, a shared sense of purpose, and reliable financing.

The Asia and the Pacific region as an example has complex social and political environments, typified by competing and often conflicting priorities and motivations. Every case in this region is unique because of its national context, language and characteristics (such as size, population, political systems, varied infrastructures and skills), the national traditional and cultural attitudes, and the people who participate, develop and use SDIs.

In order to develop a functioning Regional SDI efficiently, the Regional SDI coordinating agency must manage such diversity to gain the necessary support with which to meet their objectives. The management of such diversity can be facilitated by identifying critical social factors and processes in the acquisition, implementation, and utilization of a technology (Rajabifard, 2002).

By identifying key human and technical factors within classes of potential users, diffusion studies have the potential for directing the design strategy and efforts of SDI coordinating agencies to those jurisdictional characteristics and improvements most valued by target users. Case study research of this nature (Rajabifard and Williamson, 2000) has identified three major classes of factors, which are influencing, or contributing to the development of the Asia-Pacific Regional SDI initiative. These classes of factors are: Environmental Factors, Capacity Factors, and SDI Organization Factors as illustrated in Figure 5.4. These three classes of factors have been identified based on the discussion and investigation on the current situation in the Asia-Pacific region and also lessons learned from other researchers (Onsrud and Pinto, 1991; Onsrud and Pinto, 1993; Budic, 1993; Budic and Godschalk, 1996; Budic and Pinto, 1999; Masser, 2001). According to the following figure, the three classes of factors together effect the participation rate. Therefore, by considering these three classes of factors in the design and implementation of a Regional SDI, it can be expected that the rate of involvement and participation of member nations would increase.

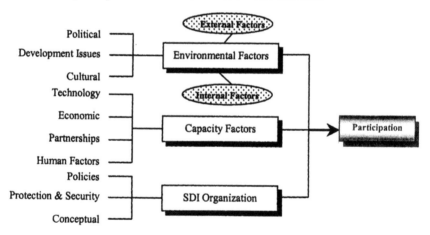

Figure 5.4: Factors Influencing the Development of a Regional SDI

5.4.1 Environmental Factors

The environment is the overall structure within which the social system operates and is characterised by internal and external factors. Therefore, the different characteristics of social systems, or communities, adopting the SDI concept can be attributed to a number of environmental factors, including the different cultures of the communities, political factors, and development issues. The external factors are those factors outside the border of the social system which affect, or could potentially affect, the performance of an organization. These factors impinge more on management levels. Some examples of external factors are Globalization (global market, global economics, other global initiatives); the GSDI initiative; and the global environment.

In terms of internal factors, they are those factors inside the border that affect both management and member levels. Therefore, determining an appropriate social border for study and analysis of a social system is very important (understanding the social system is the first step. It determines how we define implementation success and the drivers of implementation success). Some examples of internal factors are political climate; political structure and procedures; relationships with regional organizations; technological pressure; financial stability of each member nation; organizational structure of the coordinating agency (this is one of the most important factors); regional security; market pressure; and the degree of culture of data sharing.

5.4.2 Capacity Factors

Capacity building as defined by Georgiadou (2001), may refer to improvements in the ability of institutions and (government and non-government) organizations to carry out their functions and achieve desired results over time. It may also refer to the provision of foundation data, metadata standards, clearinghouse functionalities and a facilitating environment for decentralising GIS application in manageable application domains within the SDI concept. Therefore, based on this definition, capacity building for an SDI in a broad sense may refer to improvements in the ability of all involved parties to perform appropriate tasks within the broad set of principles of that particular SDI initiative.

Capacity building can be undertaken in various ways. The important issue which needs to be considered is to conduct both institutional as well as individual capacity building. In this regard, the importance of training in creating a successful environment for SDI diffusion needs to be realised. Training should be of the largest possible breadth and depth. It is not simply a matter of learning a particular concept. It goes much further than that, to a whole new way of thinking about sharing and exchanging spatial data assets, and about optimum solutions. So this is an essential issue to be considered for the success of SDI diffusion.

Capacity factors are therefore those factors that cover technology, economic factors, partnerships, and human factors. Based on the areas covered by this class of factors, one can conclude that this class of factors encompasses technological capacity, human capacity, and financial capacity. Some examples of Capacity Factors are: the level of awareness of values of SDIs; the state of infrastructure and communications; technology pressures; the economic and financial stability of each member nation (including the ability to cover participation expenses); the necessity for long-term investment plans; regional market pressures (the state of regional markets and proximity to other markets); the availability of resources (lack of funding can be a stimulus for building partnerships, however, there should be a stable source of funding); and the continued building of business processes.

5.4.3 SDI Organization Factors

These are factors related to the way that an SDI is defined, designed and implemented. This mainly includes all SDI core components, including technical and institutional issues such as access policies, access networks, technical standards, and the SDI conceptual model. Some examples of SDI Organization Factors are: the suitability and degree of complexity of the SDI conceptual model; the availability of spatial data and metadata; the integration and inter-flow of datasets from different parties (this has important implications for the ownership and control of information); access networks; and multiple trusted data sources.

5.5 SDI DEVELOPMENT MODELS

As discussed in Chapter 2, much has been done to describe and understand the components and interactions of different aspects of SDIs and their integration into the transactions of the spatial data community. However, there is still a need for descriptions to actually represent the discrepancies between the role and deliverables of an SDI and thus contribute to a simpler, but dynamic, understanding of the complexity of the SDI concept (Chan *et al.,* 2001). To this end, Rajabifard *et al.* (2002) suggested, that the roles of SDI have been pursued through two different approaches: product-based and process-based models, which contribute to the evolution, uptake and utilization of the SDI concept in different ways.

The product-based model, represents the main aim of an SDI initiative being to link existing and potential databases of the respective political and administrative levels of the community. Whilst, the process-based model, presents the main aim of an SDI initiative as defining a framework to facilitate the management of information assets. In other words, the objectives behind the design of an SDI, by any coordinating agency, are to provide better communication channels for the community for sharing and using data assets, instead of aiming toward the linkage of available databases. With this in mind, an SDI initiative can proceed by following certain steps towards the creation of an infrastructure in which to facilitate all parties of the spatial data community in the cooperation and exchange of their datasets. This facilitation can be supported by creating a clearinghouse system, metadata directory system or other forms of collecting and storing information about datasets and databases within that community. This is the prerequisite for data discovery and access. In this context, harmonised and quality metadata or clearinghouse systems are therefore essential for the facilitation tools that users are expecting. By creating such systems the coordinating agency is able to increase the knowledge infrastructure for that community by which to enable them to better identify appropriate datasets and communication links with custodial agencies. In order to take full advantage of this approach, it is important to understand the social system of the community or jurisdiction in which this approach is designed to be executed. In return, this can also facilitate the concept of partnerships.

Based on these two SDI development models, and the roles played by each SDI jurisdictional level within an SDI Hierarchy and their similarities to the organizational structure, Rajabifard *et al.* (2002) has suggested that any multi-national SDI (regional or global) can be considered similar to the strategic tier of an organizational structure, in which it is suggested that this tier and nations with federal systems are advised to adopt the process-based model of SDI development. The main reason multinational and federated nations can benefit more from using a process-based model is that SDI participation at these levels of SDI hierarchy is voluntary.

5.6 NEW STRATEGIES AND FUTURE DIRECTION

As a result of the case study, in order to facilitate a Regional SDI development, there are four major suggestions that can be derived from a consideration of the key factors important to the success of an SDI and the selection of a proper SDI model. In light of the study in Asia and the Pacific and its findings, these suggestions are proposed as central to the PCGIAP as a Regional SDI coordinating agency achieving an increased rate of participation from member nations. The adoption and implementation of these suggestions can assist Regional SDI coordinating agencies in such a way that they overcome the problem of low participation and speed up the progress in the development of their SDI initiatives. The suggestions are organizational restructure of SDI coordinating agency; redesign of the future strategy based on the study and understanding of the regional social system; modification of Regional SDI conceptual model; and adoption and utilization of a process-based model instead of product-based models. Each suggestion will be discussed and presented in more detail in the context of Asia and the Pacific Regional SDI as an example, in the next sections.

5.6.1 Organizational Restructure

As was discussed in section 5.2.1, two problems exist due to current structure of the PCGIAP as a coordinating agency for the design and development of a Regional SDI. In order to resolve their problems, the first suggestion is to revise the current structure and organization of the PCGIAP. As discussed, this Committee is currently comprised by directorates of national survey and mapping organizations and equivalent national agencies of 55 nations, which are mainly providers or producers of national spatial datasets and not necessarily the user of such national and regional datasets. However, it is essential to involve potential users of regional spatial datasets in development and implementation of the Regional SDI as one of the main promising advantages of these initiatives is to facilitate sharing and access to common and required spatial datasets by users. By involving regional users, the coordinating agency can identify and include the user needs in the design and implementation of the Regional SDI. In this regard, the involvement of politicians' concerns and regional organizations are essential in the success of a Regional SDI development. This is one of the Environmental Factors.

The other problem as discussed was that the level of political position and responsibilities of each PCGIAP member which are different from the position and responsibilities of other members. To overcome these problems the PCGIAP should restructure its organization in such a way that can also invite and involve other interested and related organizations within the region and even within each member nation as well as other regional users. As part of this restructuring, the PCGIAP should evaluate and modify the responsibilities and membership on the Executive Board and Working Groups.

5.6.2 Redesign Future Strategy Based on Social System

The second suggestion is to redesign the future strategy based on the regional social system. As was noted earlier, one of the obstacles of gaining support from certain countries and regional organizations to develop an SDI is the lack of understanding of the complexity of the interacting social, economic and political issues. Therefore, in order to develop an appropriate and functioning Regional SDI and receive support from different parties, and also to speed up the process of such development, it is important to understand the social system of the community or jurisdiction. The characteristics of the social system strongly influence the approach taken to the development of an SDI initiative. The understanding of the social system then can help with the selection of an appropriate strategy for SDI development. With this in mind, it was suggested that the PCGIAP improve its understanding about the complexity of the interacting social, economic and political issues in Asia and the Pacific region by undertaking comprehensive research in the region. This research should be carried out by a multidisciplinary group of experts including people from the academic sector, regional organizations and member delegates.

5.6.3 Modify the SDI Conceptual Model

The third suggestion is to modify the SDI conceptual model. As was mentioned above, one major obstacle in gaining support is defining the SDI and its related conceptual model. It was argued that the current SDI conceptual model adopted by the PCGIAP is incompatible with the perceived needs of the member nations. Further, it was suggested an SDI comprises not only the four core components identified by the PCGIAP as the institutional framework, technical standards, fundamental datasets and access networks, but also an important additional component, namely, the human resource-people. Therefore, the absence of this important component would cause the problem that the Regional SDI coordinating agency would just concentrate on the other four core components and develop their strategies to build the Regional SDI in such a way that ignores the interests and potential contributions of other stakeholders such as the non-participating members and agencies. To avoid this problem, therefore, it was suggested that the current APSDI model and the strategy for its development need to be modified. For this modification the fifth component – human resources, needs to be defined and presented clearly within the current APSDI conceptual model.

5.6.4 Adopting SDI Process-Based Model

The last suggestion is the adoption of an SDI process-based model instead of the current strategy for the APSDI development which is a product-based model. Selecting and using a process-based model instead of product-based one for SDI development would be a better approach to overcome some of the challenges facing SDI initiatives persisting with a product-based approach, especially in this region.

With this is mind, the Regional SDI coordinating agency should recognise the value in taking a facilitation role for SDI development rather than that of implementation of a specific data product by itself. Based on the initial aims for the APSDI development, the difficulties of coordinating many individual efforts toward SDI development, including the various stages achieved by this committee, and awareness of the value and vision of SDI development have made the objective of the APSDI development difficult to achieve.

Therefore, it is suggested that by adopting a process-based model, the PCGIAP can develop a spatial data clearinghouse system to facilitate a regional communication channel for data sharing, and at the same time contribute to the regional knowledge infrastructure instead of collation and integration of regional datasets which seems to be very difficult, if not an impossible and challenging task. Also, by improving the current conceptual model by adding one more component (people) and defining a proper border for its social system, the PCGIAP can define its future strategy by better understanding the complexity of the interacting social, economic and political issues within its border.

5.7 CONCLUSION

This chapter discussed and presented the key factors influencing the diffusion of a Regional SDI and also demonstrated the applicability of the theories of innovation diffusion as a useful framework for the development and improvement of SDI conceptual models and implementation.

The chapter first discussed the importance of identifying key influencing factors. It presented three major classes of factors namely Environmental Factors, Capacity Factors, and SDI Organization Factors, and argued that these classes of factors are influencing the development of SDIs, and together they can effect the participation rate of members.

This chapter confirmed that the adoption of a new concept or idea even when it has obvious advantages is often very difficult. It is noted that innovations may require a lengthy period, often many years, from the time they become available to the time they are widely adopted. Therefore, a common problem for many individuals, communities and organizations, is how to speed up the rate of adoption.

It is argued that the adoption and implementation of the key influencing factors for SDI and the selection of a process-based model can assist Regional SDI coordinating agencies to overcome the problem of low participation and speed up their progress in the development of their initiatives.

Although this chapter focused on the development of Regional SDI initiative in the Asia-Pacific region, the results and lessons reviewed in this chapter – especially the key factors influencing the diffusion of a Regional SDI - can also be used and applied in other regions, and potentially other jurisdictional levels.

5.8 REFERENCES

Budic, Z. D. and Godschalk, D. R., 1996, Human Factors in Adoption of Geographic Information Systems: A local Government Case Study. *Public Administrative Review*, November/December 56(6): 554-567.

Budic, Z. D. and Pinto, J.K., 1999, Interorganizational GIS: Issues and prospects. *The Annals of Regional Science* (Springer-Verlag).

Budic, Z. D., 1993, *Human and Institutional Factors in GIS Implementation by Local Governments*. The University of North Carolina at Chapel Hill.

Chan, T. O., Feeney, M., Rajabifard, A. and Williamson, I. P., 2001, The Dynamic Nature of Spatial Data Infrastructures: A Method of Descriptive Classification. *Geomatica* 55(1): 451-462.

Chan, T. O. and Williamson, I. P., 1999, Spatial data infrastructure management: lessons from corporate GIS development, *AURISA '99 Conference*, 22-26 November 1999, Blue Mountains, NSW.

Chan, T. O., 1998, 'The Dynamics of Diffusion of Corporate GIS'. *PhD Thesis*, Department of Geomatics, Melbourne, The University of Melbourne: 220. Online. <http://www.geom.unimelb.edu.au/research/publications/IPW/TOphd%20thesis%2098-12.pdf>.

Fukasaku, K., 1995, Regional Cooperation and Integration in Asia, Organization for Economic Cooperation and Development, France.

Gattiker, U. E., 1990, *Technology Management in Organizations,* (Newbury Park: Sage Publications).

Georgiadou, Y., 2001, Capacity Building Aspects for a Geospatial Data Infrastructure (GDI). *5th GSDI Conference*, 21-25 May, Cartagena de Indias Colombia.

Goodman, P. S., 1993, Implementation of new information technology. In *Diffusion and Use of Geographic Information Technologies*, Edited by I. Masser, and H. J. Onsrud. (Dordrecht/Boston/London: Kluwer Academic Publishers), 45-58.

Holland, P., 1998, Roles of the Permanent Committee on GIS Infrastructure for Asia and the Pacific (PCGIAP) and the United Nations Regional Cartographic Conference for Asia and the Pacific. Paper presented at the Special Working Group Meeting of the United Nations Regional Cartographic Conferences, Aguascalientes, 25-27 March 1998, Mexico.

Masser, I., 2001, Reflections on the Indian National Geospatial Data Infrastructure. GIS Development.net, Application, Geographic Information Infrastructure, http://www.gisdevelopment.net/application/gii/global/giigp0009.htm, (Accessed June 2001).

Masser, I., 1998, The first generation of national geographic information strategies. *3rd GSDI Conference*, 17-19 November 1998,Canberra, Australia.

Mohamed, A.M., 1999, PCGIAP and the Asia Pacific Spatial Data Infrastructure (APSDI), *Cambridge Conference for National Mapping Organizations*, 19-23 July 1999, Ordnance Survey, Cambridge.

Onsrud, H.J., 1998, Survey of national and regional spatial data infrastructure activities around the globe. *3rd GSDI Conference*, 17-19 November 1998, Canberra, ANZLIG.

Onsrud, H. J. and Pinto, J. K., 1993, Evaluating correlates of GIS adoption success and the decision process of GIS acquisition. *URISA Journal* 5(1): 18-39.

Onsrud, H. J. and Pinto, J. K., 1991, Diffusion of geographic information innovations. *International Journal Geographical Information Systems* 5(4): 447-467.

PCGIAP, 2001, Report of the 6th Permanent Committee on GIS Infrastructure for Asia and the Pacific meeting, 11-15 April, Kuala Lumpur, Malaysia, Permanent Committee home page, http://www.gsi.go.jp/PCGIAP/, (Accessed September 2001).

PCGIAP, 1998, A Spatial Data Infrastructure for the Asia and the Pacific Region, (PCGIAP Publication No. 1, Canberra).

PCGIAP, 1995, An Interim Report of formation of Permanent Committee On GIS Infrastructure for Asia and the Pacific, Kuala Lumpur, Malaysia.

Pinto, J. K. and Onsrud, H. J., 1993, Correlating adoption factors and adopter characteristics with successful use of geographic information systems. *Diffusion and Use of Geographic Information Technologies*, I. Masser and H. J. Onsrud, (Dordrecht/Boston/London: Kluwer Academic Publishers), 70: 165-194.

Rajabifard, A., 2002, 'Diffusion of Regional Spatial Data Infrastructure: with particular reference to Asia and the Pacific'. *PhD Thesis*, Department of Geomatics, The University of Melbourne, Australia. Online. <http://www.geom.unimelb.edu.au/research/publications/Rajabifard_thesis.pdf>.

Rajabifard, A., Feeney, M. and Williamson, I.P., 2002, Future Directions for SDI Development. *International Journal of Applied Earth Observation and Geoinformation*, Vol. 4, No. 1, pp. 11-22, The Netherlands.

Rajabifard, A. and Williamson, I.P., 2000, Report on Analysis on Regional Fundamental Datasets Questionnaire. *15th UNRCC-AP (UN-Regional Cartographic Conference for the Asia and the Pacific)*, 11-14 April 2000, Kuala Lumpur, Malaysia.

Rajabifard, A. Williamson, I.P., Holland, P., and Johnstone, G., 2000, From Local to Global SDI initiatives: a pyramid building blocks. *Proceedings of the 4th Global Spatial Data Infrastructures Conferences*, 13-15 March 2000, Cape Town, South Africa.

Rogers, E. M., 1995, *Diffusion of Innovations*. (New York: The Free Press).

Rogers, E. M., 1993, The diffusion of innovation model, In *Diffusion and Use of Geographic Information Technologies*. Edited by I. Masser and H. J. Onsrud. (Dordrecht/Boston/London: Kluwer Academic Publishers). 70: 9-24.

Rogers, E. M., 1983, *Diffusion of Innovations*. (New York: The Free Press).

Rogers, E. M., 1971, *Diffusion of Innovations*. (New York: The Free Press).

Scott, W. R., 1990, Technology and structure: an organizational-level perspective. In *Technology And Organization*, edited by P. S. Goodman, and L. S. a. A. Sproull. (San Francisco, Oxford: Jossey-Bass Publishers), 109-143.

Spence, W. R., 1994, *Innovation: The Communication of Change in Ideas, Practices and Products*. (London, Glasgow, Weinheim, New York, Tokyo, Melbourne, Madras, Chapman & Hall).

Zaltman, G., Duncan, R., and Holbeck, J., 1973, *Innovations and Organizations*, (New York: John Wiley & Sons).

CHAPTER SIX

National SDI Initiatives

Abbas Rajabifard, Mary-Ellen F. Feeney, Ian Williamson, Ian Masser

6.1 INTRODUCTION

Navigating the complexity of communications and relationships between sectors and agencies to achieve a common understanding of spatially-related issues across a nation is paramount for any economy, management of the environment, social issues and security. SDIs are established at this level to form a framework to share and exchange data across agencies and between disciplines to achieve these and other objectives. National SDIs create an environment where a wide variety of users are able to access and retrieve complete and consistent data in an easy and secure way. When coordinated nationally such an infrastructure provides the means to support and improve existing and potential bilateral and multilateral interactions as well as strengthen domestic institutional and commercial interactions.

The objective of this chapter is to discuss the nature of National SDIs and the particularity of their relationship to other SDI initiatives at the local, state and international levels. SDI initiatives are underway in more than half the countries of the world (Borrero, 2002; Crompvoerts and Bregt, 2003; Masser, 2002) the majority of these are formed at the national level. This has resulted in a great diversity of national initiatives and two significant waves of SDI development, from the outset of the concept, the first generation, to the more recent wave of innovative initiatives, the second generation, within a broader continuum of strategic spatial data development. These national initiatives vary in their organizational and strategic drivers, however they bear similarities in the issues and challenges faced as well as the roles they play within an SDI hierarchy. This chapter evaluates and discusses the similarities and differences between National SDI initiatives of the first and second generations. Based on this evaluation, recommendations and future directions are suggested to facilitate SDI development at a national level and its influence on other levels in the SDI hierarchy.

6.2 NATIONAL SDI – THE CONCEPT AND NATURE

The National SDI is an initiative intended to create an enabling environment for a wide variety of users to access and retrieve complete and consistent datasets with national coverage in an easy and secure way.

Due to the complexity of decision-making it is impractical for a single organization to produce and maintain the wide variety of data and information needed to inform many decisions, yet the need for a common understanding of spatially-related issues is paramount. Navigating the complexity of communications between the sectors and agencies to achieve this common understanding necessitate systems for collaboration and sharing information. SDI at the national level constitutes such a system.

The establishment of a National SDI forms a fundamental framework to exchange data across many agencies and disciplines. A National SDI can provide the institutional, political and technical basis to ensure the national consistency of content to meet user needs in the context of sustainable development. Within this framework, fundamental data can be collected and maintained through partnerships. This data will include all data necessary to understand the country, in both spatial and non-spatial forms for stakeholders.

In this context, stakeholders are the jurisdictional data providers such as national, state/province/county and local governments, commercial data integrators and end users. However a National SDI also provides support for improving existing or even establishing new bilateral and multilateral relationships and exchanges with other countries. Therefore, to maximise the benefit from investment in data collection, maintenance and accessibility from both a national perspective and that of the individual stakeholders, the national infrastructure should ensure that stakeholder efforts are focused and coordinated.

6.2.1 Motivation for National SDI Development

A National SDI ideally should provide benefits for all stakeholders at the national level. In particular the needs of the cooperating members at this level must be met but there must exist provision for aligning with other SDI levels and for the inclusion of previously non-participating groups. As the membership grows the data pool widens and there are further economies of scale and benefits realised. The benefits realised through SDI developments generally include:

- Reducing the costs of data production and eliminating duplication of effort;
- Developing applications more quickly and easily by using existing data and data development standards;
- Providing better data for decision-making;
- Saving development effort by using fundamental and standardised data, guidelines, and tools;
- Ability to perform cross-jurisdictional and cross-sectoral decision-making, analysis, and operations based on common data and understanding of issues;
- Expanding market potential through recognition and credibility as an SDI participant as well as through the formation of beneficial partnerships;
- Providing consolidated directions to vendors regarding required technical features;
- Facilitating the development of knowledge infrastructure and communication networks.

In addition to the above, some of the areas of industry and government motivating different agencies to cooperate with each other in the context of National SDI development can be summarised as:

- sustainable development,
- national economic development and cooperation,

- national mapping,
- emergency management,
- national security,
- environmental monitoring and management,
- resource management,

- urban and regional planning,
- agricultural and forestry management,
- maritime relationships.

The development of any SDI is a matter of its related jurisdictional cooperation and partnerships, awareness of the value of spatial information and SDIs as well as successful widespread data use. There must be willingness for cooperation between various stakeholders to facilitate data sharing which is crucial to the success of SDIs. The involvement of those politicians concerned with SDI development is essential. The politicians' support provides legitimacy and encourages the necessary financial investment for SDI development. Knowledge about the type of data, its location and quality is also required. This will lead the community toward a knowledge based infrastructure society. Also, spatial data need to be equitably accessible to all parties and the wider community, and widespread use of these data need to be facilitated by appropriate infrastructure such as suitable intellectual property laws and proper human resource development.

In addition, there are a number of key issues and strategies to be considered within the design process for a National SDI. These include the development and associated implementation of a national strategic vision, the recognition that SDI is not an end in itself, and the independent administration and coordination of the strategic vision. Masser (2001) highlighted four issues that are likely to need special consideration by those involved in SDI design and development. In order of priority, these are the nature of the machinery for coordination, the need to develop metadata services, the importance of capacity building initiatives and the need to promote data integration.

6.2.2 Models for National SDI Development

Two models can be identified among SDI developments based on the strategies, aims, objectives, and status of individual initiatives as discussed in Chapter 5. The product-based being to link existing and upcoming databases of the respective political/administrative levels of the community. The process-based model, presents the main aim of an SDI initiative as defining a framework to facilitate the management of information assets. In other words, the objectives behind the design of an SDI, by any coordinating agency, are to provide better communication channels for the community for sharing and using data assets.

Both these development models are relevant to National SDIs depending on the political system of the country being Federated or non-Federated (centralised). Non-Federated nations are able to take either model depending on their national spatial data strategies. From experience, however, Federated nations can capitalise on using a Process-based model because of the distributed responsibilities and cooperation towards spatial data management requiring communication channels for national sharing and use of spatial data.

Until now most national initiatives have taken a product-based development approach possibly due to a lack of alternative options and awareness of the use and advantages of alternative models characterising the development-coordinating SDI community. However, an intermediate position between the two models is also some times observed. A good example of this is the development of a particular initiative or project at the outset of an SDI development to commence momentum and support for further development activities. This sort of leverage plays an important role in facilitating the cooperation and SDI communication channels.

6.3 ROLE AND RELATIONSHIPS WITHIN SDI HIERARCHY

The relationships among different levels of SDIs are complex (as discussed in Chapter 2). This complexity is due to the dynamic, inter- and intra-jurisdictional nature of SDIs. One way to observe and map these relationships in the context of an SDI hierarchy can be to assess the impact and relationships of each component of any level of SDI on the same component of an SDI at a different level (Figure 6.1).

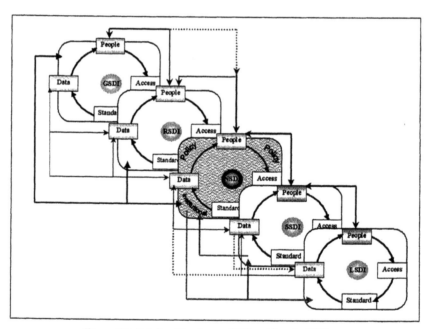

Figure 6.1: Relationships Among Different Level of SDIs

Based on this approach, Rajabifard *et al.* (2000) observed the behaviour and inter-relationships between any level of SDI on the other levels through each of the components, and demonstrated a general pattern of direct and indirect potential impacts and relationships between them (Table 6.1).

Table 6.1: Behaviour and Inter-Relationships of SDIs

SDI Components	Local SDI	State SDI	National SDI	Regional SDI	Global SDI
Policy	L→S L——N L——R L——G	S→L S→N S——R S——G	N→L N→S N→R N→G	R——L R——S R→N R→G	G——L G——S G→N G→R
Fundamental Datasets	L→S L┄┄┄→N L——R L——G	S┄┄┄→L S——N S——R S——G	N——L N┄┄┄→S N→R N→G	R——L R——S R→N R→G	G——L G——S G→N G→R
Technical Standards	L→S L——N L——R L——G	S——L S→N S——R S——G	N→L N→S N→R N→G	R——L R——S R→N R→G	G——L G——S G→N G→R
Access Network	L→S L→N L——R L——G	S→L S→N S——R S——G	N→L N→S N→R N→G	R——L R——S R→N R→G	G——L G——S G→N G→R
People	L→S L→N L——R L——G	S→L S→N S——R S——G	N┄┄┄→L N——S N→R N→G	R——L R——S R→N R→G	G——L G——S G┄┄┄→N G→R

→ Direct impact ┄┄→ Indirect impact —— No impact

L= Local SDI: S= State SDI: N= National SDI: R= Regional SDI: G= Global SDI

According to the pattern, a National SDI has a full impact and relationship on the other levels of the SDI hierarchy through its components. In terms of policy, National SDIs have an important effect on the upper and lower levels. However, policy at a global level has only a direct impact on and relationship with Regional and National SDIs. In terms of fundamental datasets, a National SDI has an important role in forming this component of the upper levels, and its datasets are created based on the datasets from the lower levels of SDIs. But the fundamental datasets at a national level can have an indirect impact on the fundamental datasets at a state level. Users at a state level might need to use national fundamental datasets for their applications before using state datasets that are in more detail. In terms of technical standards, a National SDI has a direct influence on the State and Local SDIs, and its position is important for the upper levels to decide on their strategies and standards.

An SDI at the national level therefore, has stronger relationships as well as a more important role, in building the other levels of SDI. The role of a National SDI in an SDI hierarchy displays a particularity not present in the other levels of the SDI hierarchy (Figure 6.2). This particularity is that bottom levels of an SDI hierarchy, such as local and state, have no strong links to the upper levels of the hierarchy, like to the GSDI. So, the National SDI provides a crucial link between the lower and higher levels, ensuring ongoing alignment of standards and policies for spatial data sharing.

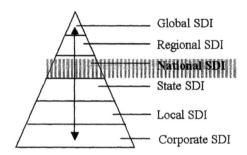

Figure 6.2: National SDI- A Particularity

Additional to the vertical relationships between different levels of SDIs (Chapter 2), there are also horizontal relationships between individual SDI initiatives within any level of an SDI hierarchy which should be taken into consideration (Figure 2.7). These relationships become more important when the respective jurisdictions are spatially adjacent and proximate. SDIs belonging to adjacent jurisdictions play more important roles and have more influence and impact on each other than on SDIs of non-adjacent jurisdictions.

6.4 GENERATIONAL DEVELOPMENT OF SDIs

Every nation undertakes to some extent the development of strategic national mapping and spatial data activities to meet their national planning and management needs. The cumulation of these activities over time has resulted in the identification of key linkages between institutional and technical aspects similar in many respects to other forms of infrastructure, and occurring in a continuum of development strategies. Based on this the concept of SDI first emerged in the mid 1980s around the need for cooperation and sharing of spatially-related information across a nation. For example, in Australia national land-related information initiatives commenced with a government conference in 1984 which led to the formation of a committee eventually responsible for SDI development (this is discussed further in Chapter 8). In the United States of America, discussion about the National SDI initiative started around 1989 primarily in the academic community (Mapping Sciences Committee, 1993) and progressed especially rapidly after the Executive Order from the President's Office was issued in 1994 (Executive Order, 1994). These, followed quickly by a number of other initiatives, characterised a first wave of SDI development. The experiences garnered from these initiatives precipitated the development of the SDI concept at other levels of the hierarchy and have facilitated a second wave of the development at national level amongst all regions of the world within the continuum of the strategic spatial data development. This section discusses the patterns of these developments.

6.4.1 First Generation of National SDI Initiatives

The first generation of National SDI development can be traced from the mid 1980s when the USA and Australia, for example, started to develop the data access relationships which became the precursor to the development of National SDI initiatives. At this time, countries developing SDI for any jurisdictional level had only very limited ideas and knowledge about different dimensions and issues of the SDI concept, rather less experience of such development.

Each country designed and developed SDI based on their own specific requirements and priorities and nationally specific characteristics. As Masser (1999) summarised, the ultimate objectives of the SDI initiatives in this first generation were to promote economic development, to stimulate better government and to foster environmental sustainability. Since then these countries have become more aware of different dimensions of SDI development and have therefore been able to identify emerging issues and challenges involved in the SDI concept. For these countries the identification of important components, the experience and resolution of different issues during this initial period was largely an iterative process of SDI development, sometimes making the adaptation to new ideas and important principles difficult given the directions to which development had already been committed.

One of the adaptations faced over the last five years, for instance, has been the changing emphasis on measuring the value of SDI development – the initial motivations were in reducing duplication, using resources more effectively, and creating a base from which to expand industry productivity and the spatial information market. As a result many countries involved in SDI development over the first generation took a product-based approach, which became the dominant model for SDI justification and development partially through a lack of awareness of other options.

A significant milestone overcome by the first generation, for whom there were few experiences and existing SDI developments from which to learn, was the documentation of researchers' and practitioners' experiences and status reports on their SDI initiatives. This achievement not only gave countries a knowledge-base from which to learn and/or develop their initiatives, providing exposure to the developmental strengths and weaknesses of different SDI initiatives, but it provided social capital to share and foster SDI developments in other countries.

Masser (1999) was an early contributor to this documentation process, evaluating and reporting on eleven first generation National SDI initiatives, based on the driving forces behind their development and the main features of the initiatives in terms of their status, scope, access, approach to implementation and resources. As part of his review, he argued that these initiatives formed the first generation of national geographic information strategies. He reviewed the experiences of these countries' strategies and considered what lessons might be learnt for the next generation of National SDI developments.

Masser's evaluation, revealed that there was a great deal of diversity in the first generation of national geographic information strategies. Insofar as status is concerned there were clear advantages associated with a formal mandate for a national geographic information strategy, provided that this was accompanied by the necessary resources to enable its implementation. Lack of dedicated resources is obviously a weak point, where initiatives are essentially outgrowths of existing coordination activities; yet this model has the advantage that it builds upon existing cooperative procedures.

Based on the number and diversity of countries falling into this first generation, Masser concluded that there would be a considerable increase in the number of national geographic information strategies being implemented throughout the world in the next ten years. For instance the first generation of National SDI initiatives included Australia, Canada, China, Denmark, Finland, France, Germany, Indonesia, Iran, Japan, Korea, Malaysia, Netherlands, Portugal, Qatar, Switzerland, UK, and USA to name a few (Masser, 1999; Onsrud, 1998; Rajabifard and Noori, 1997). It seems equally likely that the second generation will therefore be at least as diverse in character as the developments of the first generation. Masser suggested that the potential candidate for the next generation would in practice probably be three different sets of countries. The first of these being members of the first generation who will substantially restructure their current approaches as a result of the experience that has been built up over the last half-decade. Secondly, there are a number of developing countries, especially in the Asia and Pacific region including Pacific island nations, who are likely to follow the examples of Indonesia and Malaysia and develop SDIs to facilitate the planning and management of economic development and natural resources. Last, but not least, are the group of central and east European countries, such as the Czech Republic, Hungary and Poland, where there has been a massive investment recently in cadastral and digital mapping programmes as part of national restructuring activities in the post-Communist era.

6.4.2 Second Generation of National SDI Initiatives

The transition to the second generation can be marked by a change in focus on SDI development by several countries involved in developing the concept from the beginning. This was marked by a rapid increase in the number of countries becoming involved in SDI development, fostered by the definition of an SDI community where they could share and exchange experiences. This also shows the continuum of strategic spatial data development. The SDI community has been the result of increases in conferences at which SDI developments are discussed as well as increased participation at these conferences and involvement in these discussions. The Global SDI conference series (GSDI), UN Cartographic Conferences, Digital Earth, alongside the involvement of groups like the Open GIS Consortium and their different test-bed projects, and the publication of shared experiences in the SDI Cookbook (2000), are just a few examples of forums and involvements nurturing the sharing of SDI development experiences. These forums have been reinforced by the publication of new development strategies from some of the lead coordinating agencies, many of which occurred in 2000 as well as the establishment of a Centre for SDI and Land Administration research in 2001. We can consider that the second generation started from the year 2000 when some of the leading nations on SDI development changed their development strategies and updated their SDI conceptual models like USA and Australia.

Some SDI development initiatives have in recent times begun to manifest characteristics of both the product and process-based models having made an initial or partial commitment to addressing the essential balance between the product and process-based models, or being in a transitional stage - developing a more process-based approach having had product-based origins.

As a result of some of the difficulties in developing National SDI in the United States of America (Budic *et al,*. 2001; Reichardt and Moeller, 2000), at the end of 1999 the FGDC started to develop a new GeoData Organizational initiative for the geospatial data community. This new strategy appears to show that the FGDC is moving from a product-based to a process-based model of SDI development in order to neutralise difficulties arising from existing models.

Australia, have also started a transition from a product to a more process-oriented SDI development to address some of the challenges faced, particularly at a national level, under the influence of a federated political system. Whilst product-based models to dataset assembly and sharing were the focus of SDI development from 1991-1996 (ANZLIC, 1996) a transition toward process-based development has been initiated through a clearinghouse initiative - the focus of an ANZLIC workshop in March 2000 (ANZLIC, 2000).

The difficulties faced by the USA and Australia, precipitating their change in strategies in development of their National SDIs, are not unique and suggest a trend for other SDI initiatives throughout the world which face similar challenges. These challenges seem particularly prominent when the political structure of a nation is a federated system, which relies on the voluntary participation and cooperation of member agencies, a model of cooperation also seen at the multinational level.

The second generation of SDI developments characteristically fall into two groups: those countries who started to develop an SDI initiative during the period of the first generation and are gradually modifying and upgrading the initiative, as well as those countries who have decided to design and develop an SDI for their respective countries and/or have just commenced doing so (for example Wehn de Montalvo, 2001; Hyman and Lance, 2001).

The distinguishing features of the second generation include leverage of the experiences, expertise and social capital of SDI development derived from the first generation. This legacy includes:

• organizational, research and conference materials;
• forums for discussion such as workshops and conferences;
• demonstrated evolution of understanding about the concept and nature of SDIs from the first generation;
• openness of SDI community participants to sharing spatial data, experience and knowledge;
• recognition of the importance of partnerships to SDI developments; as well as
• direct learning from the first group of countries working in SDI.

These have corresponded with different researchers working on the nature and concept of SDI, identifying important components, its dynamic nature, gaps in SDI development including capacity building, socio-technical dimensions as well as technical dimensions, identifying influencing factors, as well as the adaptation of the SDI concept to respond to the developments and applications of new technologies.

Countries just beginning SDI development in the second generation now have a lot more from which to justify and gain support for their SDI development, both politically and in terms of financial and human resources. Awareness of the challenges and issues involved in developing SDI initiatives has enabled greater preparation prior to SDI development, and identification of key stakeholders involved in the exercise - providers, users, value-added-resellers, product-integrators etc. - to provide new services and produce business opportunities.

For the first generation, data was the key driver for SDI development and the focus of initiative development. However, for the second generation, the use of that data (and data applications) and the need of users are the driving force for SDI development.

In the second generation, people recognise that SDI is all about facilitation and coordination, which is fundamentally about people. One of the important outcomes of the first generation has therefore been the inclusion of people, alongside technical and institutional components, in the definition of SDI initiatives throughout the spatial data community. This has resulted in a shift to a socio-technical viewpoint in the second generation, from the first generation, which had a more techno-centric position (Figure 6.3), a trend Petch and Reeve (1999) have also identified as being more prevalent today.

Techno-Centric **Spatial data community**	**Socio-Technical** **Spatial data community**
Focus on technology	People and technology
• Technology push	• Demand pull
• Because it's possible	• Because it's needed
• Others are developing	• We need it
• Specified by technologist	• Specified by Users
• Static in nature	• Dynamic in nature

Figure 6.3: From a Techno-Centric Position to a Socio-Technical Position
(Modified from Petch and Reeve 1999)

As a socio-technical viewpoint of SDI is increasingly taken, it is recognised that even if it is assumed that SDI succeeds on a technical level, its adoption will still ultimately depend on how well implementation strategies address the respective community barriers. In the second generation, people recognise that societal issues can be critical factors in determining the success of SDIs, which has meant that the SDI coordinating agencies have had to develop a much richer and broader conception of who their communities are, how they behave, and particularly how they are likely to respond to the introduction of a new SDI

SDIs was measured in terms of their
and from

now measured in many respects, including in terms of its support for spatial decision-making (Feeney *et al.*, 2002), its criticality to national security and emergency management, and in terms of its intrinsic value – who can afford not to have it?

The first generation of users are active participants in the second generation, as they are improving and modifying their existing initiatives in accordance with the transition factors, such as the balance and definition of components and the SDI conceptual model, development strategies, as well as implementation timeline/ priorities, funding models which incorporate partnerships and varying models for participation, and benefit sharing. These are lessons that new countries entering SDI development are able to take into consideration for SDI design and implementation. This generation has recognised the value of independent and permanent committees to guide the development of SDI initiatives. Interoperability with other SDI initiatives has been an additional motivator as working together has demonstrated the value of having different levels of SDIs (operating in an SDI hierarchy). This generation has also progressed work on cooperation towards international standards and advanced understanding of the need to look beyond individual countries' needs when adopting standards and technical specifications, because decision-making needs increasingly to transcend national borders, for social, environmental, economic and political issues. In the second generation, these types of initiatives become easier to integrate with each other to form the building blocks for broader SDI development between the jurisdictions.

In summary, in the second generation SDI development has been relatively quick due to the concept gaining momentum and the existence of early prototypes, clarification on many initial design issues, increased sharing and documentation of experiences to facilitate implementation and face the complexity of decision-support challenges.

6.4.3 Comparative Analysis

There are many similarities and differences between the first and second generations of SDI development as described in Table 6.2. Understanding the evolution of SDI development through these facilitates awareness of future directions to improve participation and design of SDI.

Developing a successful SDI initiative depends at least as much upon issues such as political support within the community, clarifying the business objectives which the SDI is expected to achieve, securing sufficient project funding and enlisting the cooperation of all members of the community, as upon technical issues relating to spatial data quality, standards, software, hardware and networking. Therefore, developing a successful National SDI within a political and/or administrative level must be seen as a socio-technical, rather than a purely technical exercise.

Table 6.2: Similarities and Differences Between First and Second
Generations of SDI Development

Similarities & Differences	1st Generation	2nd Generation
Nature	Explicitly National	Explicitly National within the hierarchical context and therefore more flexible for cross jurisdictional collaboration.
Development Motivation	Integration of Existing Data	Establishing the Linkage between People and Data
Expected Outcomes	Linkage into a Seamless Database	Knowledge Infrastructures, Interoperable Data and resources
Development Participants	Mainly Data Providers (eg. NMOs)	Cross-Sectoral (providers, integrators, users)
Funding/Resources	Mainly no specific or separate budget	Mostly include in National Mapping program, or having separate budget
Driving/coordinating Agency	Mainly National Mapping Organizations	More independent organizational committees/ Partnership groups
Awareness	Low awareness at the beginning, gradually learning more	More aware, knowing more about SDI and its requirements
Capacity Building	Very low	Communities are more prepared to engage in on-going activities
No of SDI Initiatives	Very limited	Many more
SDI Development Model	Predominantly Product-based	Increasingly Process-based, or hybrid Product-Process approach depending on the jurisdiction
Relationship with the other SDI levels and International Initiatives	Low	Much more
Measuring the Value of SDIs	Productivity, savings...	Holistic socio-cultural value as well, measuring the expense of not having

The development of any SDI is a matter of its related jurisdictional
cooperation and partnerships. Based on the comparisons above (Table 6.2) both
generations of SDI development are explicitly national, however, the second
generation of National SDIs embrace models of cooperation that facilitate far

greater inter-jurisdictional exchange. There must be willingness for cooperation between various stakeholders to facilitate data sharing which is crucial to the success of SDIs.

One feature of SDI design that has been adapting to this requirement has been the coordination model. In the first generation, the coordinators of SDI developments were predominantly National Mapping Agencies, which have been one of the key fundamental data providers. However, the role of coordination generally has a much broader scope than that of a single agency/stakeholder and therefore benefits much more from second generation models which are generally independent and designed to be representative of different stakeholder groups, functioning in a more consultative capacity. In terms of the motivations and expected outcomes for SDI development, the first generation was more data-focussed, concentrating on data integration, due in part to the strengths and skills of the lead coordinating agencies. However, the issues of the second generation are access to the data and its applicability to the decision and problem environment. With this in mind the expected outcomes of the second generation are knowledge infrastructure, interoperability of data and resources, other than linkages of seamless digital data. In this regard second generation initiatives are designed to take greater advantage of linkages between people and data. This has resulted in an increasing number of initiatives moving from a product-based to a process-based model of SDI development.

Funding and resources to secure the implementation of SDI is always an important issue. In the first generation the principle support for SDI development came from the National Mapping Agencies as part of their national mapping programs. However, lack of specific and independent budgets for SDI development activities places significant limitations on the longevity of development efforts and the long-term nature of the infrastructure. Learning from these experiences the second generation have taken a more strategic development approach and made individual business cases for SDI funding and sought political involvement to secure the importance of their mandate. The politicians' support provides legitimacy and encourages the necessary financial investment for SDI development.

Whilst first generation developments of SDI grappled with the newness of the innovative concept, its definition and understanding, the second generation have been in a much better position to take advantage of the understanding and design experience gained throughout the first decades of development. The first generation had support of a limited number of countries able to see the benefits of potential SDI developments. The second generation have been able to take advantage of the challenges faced throughout the first generation to gain support for the concept, understanding and awareness, to attract a much greater participation from countries in the development of National SDIs, as well as SDIs in other jurisdictions. The awareness of SDI as a concept and its advantages have lead to nations building technological capacity, human capacity, and financial capacity.

6.5 RECOMMENDATIONS AND FUTURE DIRECTIONS

Despite the international interest and activities toward SDI development, SDI remains very much an innovation even among practitioners. There are still doubts regarding the nature and identities of SDI, particularly in connection with how they evolve over time to meet user needs. Therefore, an SDI at a national level has a crucial role in the development and implementation of the other levels of SDI in the hierarchy, due to the particularity of its influence on these other initiatives. Those countries that are able to develop an efficient National SDI will be well placed to contribute to the development of the Regional and the Global SDI initiatives.

This chapter concludes that SDIs are a much-needed tool to better facilitate data sharing as well as jurisdictional cooperation and partnerships. However, an understanding of key SDI principles, such as the jurisdictional hierarchy of SDIs and their dynamic nature, are also important but not fully understood.

The socio-technical issues including those which are political and organizational, are the major impediments to the widespread and successful use of National SDIs rather than technical issues. An example of the difficulty arising from political and organizational issues is the diversity of the policies amongst member jurisdictions, especially in countries with a Federal state system.

There are also some other factors which influence the initiative of a National SDI and make it difficult to prepare an environment for implementation by a large number of potential stakeholders. These factors include, but are not limited to, the lack of awareness of the potential usefulness of spatial data and SDIs, social and cultural diversities, the differences in administrative systems and responsibilities, as well as the total land area of the member jurisdictions.

Participating member jurisdictions may be at different stages of economic development which can also create difficulties. Another challenge arises from differing legal and administrative structures between jurisdictions. The lack of a mandate or policy on spatial information is another example of a difficulty that retards development of SDI. This in turn causes unnecessary expenditure. What is required is an institutional framework to set up and maintain stable standards and procedures for creating and maintaining spatial data within a jurisdiction. Such an institutional framework must create a favourable business environment for the collection of national spatial data that is easily identifiable and accessible.

6.6 REFERENCES

ANZLIC, 1996, National spatial data infrastructure for Australia and New Zealand. ANZLIC Discussion Paper, Commonwealth of Australia, Online.
<http://www.anzlic.org.au/anzdiscu.htm> (Accessed November 1998).
ANZLIC, 2000, Outcomes of the ANZLIC Clearinghouse Workshop, 3-4 May 2000, Adelaide, Online.
<http://www.anzlic.org.au/news/workshop/outcomes.htm> (Accessed November 1998).
Borrero, S., 2002, The GSDI Association: State of the Art, *6th GSDI Conference*, Budapest, Hungary.
Budic, Z. D., Feeney, M.E., Rajabifard, A. and Williamson, I.P., 2001, Are SDIs Serving the Needs of Local Planning?, Case Studies of Victoria, Australia and Illinois, USA, *Computers in Urban Planning and Urban Management Conference*, 18-21 July 2001, Hawaii.
Crompvoerts, J. and Bregt, A., 2003. World status of National Spatial Data Clearinghouses, *URISA Journal*, Forthcoming.

Executive Order, 1994, Coordinating geographic data acquisition and access, the National Spatial Data Infrastructure. *Executive Order 12906*, Federal Register 59, 1767117674, Executive Office of the President, USA.

Feeney, M., Williamson, I.P. and Bishop, I.A., 2002, The role of institutional mechanisms in SDI Development that supports decision-making. *Cartography Journal* 31(2): In Press

Hyman, G. and Lance, K.T, 2001, Adoption and implementation of national spatial data infrastructure in Latin America and the Caribbean, *5th GSDI Conference*, Cartagena de Indias, Colombia.

Mapping Sciences Committee, 1993, *Toward a Coordinated Spatial Data Infrastructure for the Nation*, (Washington D.C: National Academy Press).

Masser, I., 1999, All shapes and sizes: the first generation of National Spatial Data Infrastructures, *International Journal of Geographical Information Science* 13, 67-84.

Masser, I., 2001, Reflections on the Indian National Geospatial Data Infrastructure, GIS Development.net, Application, Geographic Information Infrastructure, Online. <http://www.gisdevelopment.net/application/gii/global/giigp0009.htm> (Accessed June 2001).

Masser, I., 2002, A comparative analysis of NSDI's in Australia, Canada and the United States. Report for the GINIE project, Online. <www.ec-gi.org/ginie> (Accessed October 2002).

Onsrud, H.J., 1998, Survey of national and regional spatial data infrastructure activities around the globe, *3rd GSDI Conference*, 17-19 November 1998, Canberra, Australia.

Petch, J. and Reeve, D., 1999, *GIS Organizations and People, a socio-technical approach*, (UK: Taylor & Francis).

Rajabifard, A., Feeney, M. and Williamson, I.P., 2002, Future Directions for SDI Development. *International Journal of Applied Earth Observation and Geoinformation*, Vol. 4, No. 1, pp. 11-22, The Netherlands.

Rajabifard, A. Williamson, I.P., Holland, P., and Johnstone, G., 2000, From Local to Global SDI initiatives: a pyramid building blocks. *4th GSDI Conferences*, 13-15 March 2000, Cape Town, South Africa.

Rajabifard, A. and Noori Bushehri, S., 1997, Development of National Spatial Data Infrastructure in Islamic Republic of Iran, *Global Mapping Forum*, Gifu, Japan.

Reichardt, M. E. and Moeller, J., 2000, SDI Challenges for a New Millennium-NSDI at a Crossroads: Lessons Learned and Next Steps, *4th GSDI Conferences*, 13-15 March 2000, Cape Town, South Africa.

SDI Cookbook, 2000, Developing Spatial Data Infrastructures: The SDI Cookbook, Version 1.0, Prepared and released by the GSDI-Technical Working Group, Online. <http://www.gsdi.org/pubs/cookbook/cookbook0515.pdf> (Accessed July 2001).

Wehn de Montalvo, U., 2001, Strategies for SDI implementation: A survey of national experiences, *5th GSDI Conference*, Cartagena de Indias, Colombia.

CHAPTER SEVEN

State SDI Initiatives

Don Grant and Ian Williamson

7.1 INTRODUCTION

What constitutes a State SDI is complex. State SDIs are closely tied to land administration activities (cadastre, land registration, land use planning, land valuation and supporting land markets) and associated natural resource management objectives within a jurisdiction and require medium (1:10,000 to 1:25,000) to large scale (1:5,000 to 1:1,000 or larger) spatial data, which includes cadastral or land parcel data as a key component. These forms of SDI are significantly different from National SDI which tend to be the major focus of SDI conferences and workshops and which are based on small scale datasets (1:25,000 or smaller) as a central component.

The SDI characteristics described above are most typically exhibited in State SDI initiatives in a country which is a federation of States where land administration is a State responsibility. A detailed example of this level of SDI is shown in Chapter 9 where a case study of the SDI in the State of Victoria, Australia is described. However this level of SDI could include SDIs in non federated countries or countries which are federations of States where land administration is a federal function.

While this chapter will focus on State SDIs in countries such as the USA, Canada, Russia, Australia and Germany where land administration is a State or provincial responsibility, as mentioned it could include the medium to large scale component of National SDIs, such as in most of the Central and Eastern European countries, England, Scandinavia, New Zealand, The Netherlands and Thailand where land administration (as distinct from land management) is primarily a national responsibility. There are also many countries which are also federations of States where aspects of land administration (such as cadastral surveying and cadastral mapping) are a national responsibility, with aspects of land administration also being a State responsibility (such as land transfer and land registries), such as in Malaysia.

So for simplicity this chapter uses the term *State SDI* to encompass all those State SDIs and components of National SDIs which focus on land administration or cadastral data and support medium to large scale data. These SDIs have a particular role in providing a link between Local Government SDI (large scale data) and National SDI (small scale data) in the SDI hierarchy. State SDIs in this context exhibit specific characteristics and it is these characteristics and experiences which this chapter will discuss.

There is currently a diversity of SDI initiatives within State jurisdictions, however, there is similarity between the strategies, organizational models, issues and relationships between these initiatives. The initial similarity that is identified is the heavy reliance for comprehensive and relevant SDIs where the cadastre or land parcel data is a key component and on the inclusion of datasets from Local Government (such as land use and street addresses). While the powerhouse for decision-making and the creation of the context for SDI policy may well flow down from Federal initiatives, it is the State governments which provide the management and coordination for the many jurisdictional datasets. This is particularly the case as many countries complete medium to large scale datasets as distinct from the digital small scale datasets of one or two decades ago. The result is that users want to access these medium to large scale datasets, and increasingly want them on a national scale which means aggregating State datasets to produce national street address datasets or national cadastral datasets for example. This presents a whole range of issues and experiences concerned with interoperability, jurisdictional responsibility, privacy and cost recovery.

The reality is that the relevance of spatial information to the community and the decisions which impact on the lives of everyday citizens are significantly sourced from large scale cadastral, topographic or Local Government datasets. It is the richness of this large scale information and people-relevant data which breathe life into SDIs.

This chapter will evaluate the similarities and differences between State SDIs and will in particular discuss the importance and role of partnerships, organizational and institutional models, and funding mechanisms.

7.2 THE NATURE OF SPATIAL INFORMATION AT A STATE LEVEL

Before considering the issues concerned with and components of a State SDI, some discussion of the nature of spatial information at a State level is useful. The importance and character of spatial information has been discussed at length in many forums including throughout this book. This importance of spatial information has been emphasised since estimates that between 60 and 80 percent of all data held by government agencies can be classified as geographical in nature. Masser (1998) referred to the many features which information has in common with other resources such as land, labour and capital and describes the claims that information possesses a number of characteristics that make it inherently very different from these traditional resources. Further, he considered geographical information as a resource, a commodity, an asset and as an infrastructure.

When spatial information is considered to be an asset, with all the characteristics of information, the role of government and the public interest come into focus. The aggregation of spatial information from a range of agencies also becomes an issue (the broad issue concerned with the ease of aggregating and exchanging spatial datasets between different organizations or jurisdictions is termed "interoperability") and the concept of custodianship (the process of identifying the organization responsible for maintaining the integrity of a specific dataset within the SDI) arises. Much of this information is not collected to perform agency tasks at a State level. It is often the result arising from the performance of agency tasks eg road networks, natural resources management, land registration, land valuation or cadastral databases. This information is not primarily gathered for resale as a commodity and the costs for much of this information are met through taxes imposed on the public in the interests of good government which much of this spatial information being considered a public good.

There has been considerable discussion and debate over the years about the nature of spatial information as a public good. Some of the most extensive investigations in this area have been undertaken by the Ordnance Survey of Great Britain (Ordnance Survey, 1996 and 1999). As stated by Ordnance Survey (1996) information is a form of "public good" and is often associated with "external" benefits. These "external" benefits include:

- support to the defence of the realm;
- support to the emergency services to assist with civil disaster and accident mitigation;
- input to better business and policy-making decisions; and
- ease of coordination between users, thus saving time and resource.

In a similar manner OXERA which was commissioned to prepare a report on the economic contribution of the Ordnance Survey of Great Britain (Ordnance Survey, 1999) stated the economic gains from the use of geographic information are not limited to gains made in the commercial sphere. There are wider societal benefits arising from the role that Ordnance Survey plays in contributing to:

- national security and defence;
- education;
- the rights of ordinary citizens to access basic location information; and
- leisure pursuits

In the Australian context, where most public interest mapping is carried out by other than a national mapping agency (either by the States or for defence purposes), there is a similar justification. Achieving this objective is enhanced by a consistency of data to meet natural disasters and emergencies, coverage of an entire jurisdiction, a long term and planned maintenance regime, consistent standards, quality evaluation and a transparent process of expenditure open to public scrutiny.

These then are some of the advantages of orderly and jurisdictionally managed spatial data acquisition and assembly which are just as relevant at a State level as at a National level. Efficient use of resources and a planned balance between social needs and priority of areas mapped is also more likely to occur when effective jurisdictional coordination is in place.

Another issue in State SDIs is the definition of core datasets. While there is no clear consensus with respect to the definition of these core datasets which are the fuel for State SDIs, there is also not much argument as to what they are. The pattern is similar but the range and content reflects local priority and emphasis. What is clear is that spatial data acquisition at both State and National government levels, and usually Local Government levels, is long term and is the result of many decades of investment by government for public good purposes using general or consolidated revenue from the community tax base. Also significant is that, although there is often no statement or claim as to the ownership of these datasets, it is clear that there is no debate as to the funding source, the liability and the responsibility for the creation and maintenance of the core datasets. It rests with government.

7.3 STATE SDI - ORGANIZATIONAL ISSUES

The organizational shape of State SDIs in particular is the manifestation of history (in many cases State Departments of Lands date back over 100 years), institutional frameworks, management innovations and economic reforms. There are other drivers, both of a social and economic nature. Some are transparent and are the reasoned result of sound accountable considerations and others are the result of political decisions based upon nothing more than the need to augment an agency portfolio or balance ministerial responsibility.

In particular, the current institutional and organizational models are, in many cases, the result of actions driven by management ideology of the late 1980's and early 1990's. This approach favoured "downsizing" of both the private and public sector. What was actually needed then was not so much cutting, but a consolidation of years of unmanaged incremental growth in services provided by the public sector. These services were generally paper based and with the advent of electronic processes, little consideration was given to serious re-engineering of these processes. Technological innovation had impacted the public sector, societal demands had grown and performance indicators and efficiency and effectiveness had become the norm of management in public sector agencies. In many cases, these agencies unwittingly competed with each other at the State level to provide information and build datasets to meet new and unprecedented community expectations.

A simple example of this incremental growth and duplication could have been, for instance, commercial and domestic address sets. The service delivery of, say billing destinations for utility accounts, of the need to compile electoral rolls and jury lists and the every day Local Government business of property rating in hundreds of municipalities created large repositories of common but inconsistent, incomplete or overlapping datasets with little ability or apparent need to match each to the other. Accordingly, there were silos of information created and maintained and probably operational functions that were performed which were duplicated or replicated in the name of service to the community by government agencies; agencies ignorant, in many cases, of the like performance by other agencies in the jurisdiction.

The solution to this over servicing by the public sector, as a result of incremental growth over a half century, should have been a calm look at the content rather than the configuration of those agencies and an equally calm and reasoned re-allocation of roles and activities. Instead, there was the same "downsizing" which the private sector had suffered earlier in the name of efficiency and effectiveness. Whilst the authors are unaware of the measured results of the public sector cuts, it is reasonable to assume a like result to the private sector which showed little increase in productivity and profits, and a severe drop in staff morale (Drucker, 1993).

In fairness, however, some dramatic change was necessary in the land related information functions within the public sector at the State level to break down the discrete fiefdoms which were often a hang over from colonial legislation (particularly the land registration and cadastral surveying and mapping functions). These changes were essential to prepare the way for an all-purpose, expandable array of integrated land related information. New organizational structures were necessary. In some jurisdictions this re-engineering of form, content and responsibility is well advanced. But even in those cases there is much refinement to be done to maximise the promise of the SDI vision at a State level. See for example the experience of the State of Victoria in Chapter 9 and in Williamson *et al.* (1998).

More recent initiatives in South East Asia and some parts of Europe have seen similar attempts to rationalise land related functions. There is usually a level of acceptance for this rationalization of activities until the reality of the reform starts impacting and the financial implications and loss of influence become evident.

Even when an autonomous land registration agency, with combined land registry and cadastral surveying and mapping functions is created, as in the case of Greece where the organization was created to oversee the development of the Hellenic Cadastre, the lack of transparency and an absence of effective management resulted in gross overspending, charges of corruption, considerable European Union and press criticism and the replacement of the Board of Management (Kathimerini, October 2, 2001, Athens News October 5 and 12, 2001). There are other examples of ineffective structures in Central and Eastern Europe (land administration structures in this context are discussed in Bogaerts *et al,* 2002) and elsewhere as the process of restitution and allocation of ownership unfolds and opportunities for corrupt practices emerge.

What is becoming clear is that the institutional arrangements which support State SDIs must clearly identify the roles of the three sectors in the spatial information industry, namely the government, private and academic sectors. Quite often the role of the academic sector is forgotten with the result that appropriate research and development is not carried out, as well as a poor commitment to education and training in the broad SDI area, which results in a lack of capacity in establishing and maintaining the State SDI. The roles of the government, private and academic sectors should be clearly described in the State SDI vision and the associated strategic plan for the SDI, which is usually a three to five year plan.

At the same time as getting the operational levels for the State SDI organised it is important to establish a high level policy advisory body involving the three sectors with ministerial reporting. Recent experience shows that while government plays a critical leadership role in establishing and maintaining the State SDI, without the support of the private and academic sectors of the industry recognised in the institutional structure, the achievement of the SDI vision will be extremely difficult, if not impossible.

It would seem that institutional configuration, as mentioned earlier, is the result of many influences and when these are compounded by an immaturity of public administration and a disregard for good governance, such institutions are not likely to be effective (also see Chapter 11 for more detail). As with all major organizational modifications, institutions are the reflections of society and there are conditions which must be in place before success is likely.

7.4 THE OPERATION OF STATE SDI

Key issues to be considered when reviewing the operation of a State SDI include:

- the range of information sets,
- the product,
- the access policy, and
- the value-adding regime.

Initially it is asserted that core spatial datasets should be seen as a government infrastructural asset or public good and their collection and maintenance funded to an agreed level to prevent system collapse and to meet the reasonable expectations of a developed society. This cost can be readily ascertained and expressed in budgetary terms. It is this cost which should be recognised by any policy, which deals with the supply of spatial information.   In the absence of widespread agreement or government endorsement that the cost of provision should be treated as an infrastructural government asset, increased on-going Departmental budgetary assistance will be required and duplication is likely.

7.4.1 Range of Datasets to Support State SDI

As already mentioned, the range of core information sets in a State SDI may differ due to local circumstances but the ownership of them, as a government asset, is clear. A major policy issue which influences all builders and users of spatial information is whether the organization fits in the *supply* family or the *user* family and if in the latter, to what extent of use is the information put. But the primary distinction to be made is between *Infrastructure* and *Business System* datasets (Chan and Williamson, 1999). Broadly, the infrastructure set encompasses both Core and Thematic government datasets. The core datasets are those to which most users of spatial information relate whilst the thematic sets have specific attributes. For example the themes of geomorphology and soil types may be considered for a land use application but they both rely to some extent on the core datasets such as topography and the cadastre. The scope of these core datasets is fairly universally accepted with variations at the margins. The broad forms of spatial data at a State level are shown in Figure 7.1 and are discussed in detail below.

The thematic datasets are generally held within government or at least relate to traditional government activities and can run into the hundreds of themes. They, like all spatial information sets, require custodianship to be defined and are more readily able to conform with sharing protocols. Generally they are linked strongly to the objectives of the agency and are technologically enhanced systems, which allow the parent agency improved performance and service delivery to government and the community. Most thematic datasets within government and many within Local Government are in the supply chain and deserve special consideration in any information access policy.

The Business Systems can also be divided into two streams. The normal government stream of business systems may include for instance, the conduct of elections and any districting process that accompanies or precedes an election. The call for spatial information in this instance is great. It seems reasonable that this information should be provided at a minimum cost since it is provided to encourage or assist the process of good governance. Therefore within the business system model it is maintained that any policy should discriminate between those government bodies that can maximise the use of government generated spatial information for an increased financial position and those government agencies, which simply use the spatial information to better perform the charter which they are required to meet.

The non-government business systems also attract two categories. There are those which enhance their operational capability, perhaps reverse engineer their own or the government datasets to create systems of market advantage or simply introduce efficiencies which bring financial gain - all this without seemingly revealing any spatial information products to the market place. Under the remaining sub-set of non-government business systems there is the array of value added resellers (VARs) that makes no pretence about its business.

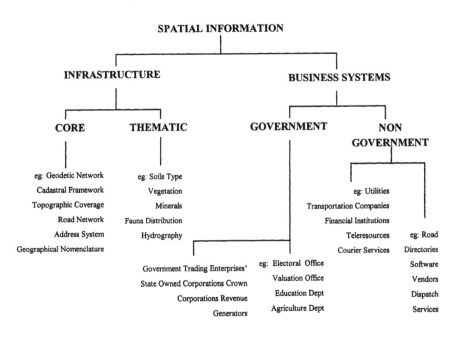

Figure 7.1: Spatial Datasets at a State Level

7.4.2 The Product

Secondly, it is essential to have a clear identification of the "product" and a reasonable appreciation of the user group, not limited by the comfort zone of the producer. While this applies to all SDIs it is particularly important at the State level with all the related business systems. The product is almost certainly the result of a marriage of two or more datasets. Further, it seems reasonable to claim that an increase of datasets, over the basic two sets, provides an exponential number of opportunities for application to expanding decision-making processes.

An example of a "product" was that developed through cooperation between the custodians of the Australian State SDIs and the National Bureau of Census and Statistics. The Public Sector Mapping Agencies-Australia Ltd (PSMA) had a clear definition of its original product - a national census map. As awareness grew within the national data user community of the utility of such a dataset and opportunities were identified through the PSMA management, the charter of the PSMA was changed and changed again. This evolution reflected, not only the growing social and community expectations, but the capability of the maturing PSMA to more adequately create new products from the aggregation of State and federal datasets not previously available. Importantly it showed the growing importance of State SDIs in providing National SDI products. It also highlights the problems of interoperability between different State datasets. So while each State has total jurisdictional authority for its spatial datasets, State SDIs are increasingly being pressured to meet national objectives (see Chapter 8 for more detail).

7.4.3 Access Policy

Thirdly, a State wide data access policy should be adopted to provide a rational and consistent basis for data access that will facilitate the efficient production, supply and use of spatial information held by the State. Increasingly State SDI strategies are incorporating a "point and click" vision based on the cadastral parcel base, the use of the Internet and increasingly wireless applications (also see Chapter 16) and associated technologies.

The policy must differentiate between infrastructure and business systems described above and foster the industry opportunities whilst not creating monopolies at the expense of the public interest. Whilst accepting the reality of capitalist market forces operating in a socialist or democratic society, the aim of a sound and responsible access policy should be a spatially enabled community and not an enlargement of the information gap.

All government agencies and private sector agencies must be given the opportunity to hold and maintain geospatial information related to their individual responsibilities. However, this data must use, and be compatible with, the nominated basic fundamental datasets for which there is a clear responsibility for acquisition and maintenance. Having created the stimulus, government has a responsibility to not delay the commercial opportunities of the private sector but has the right to recoup, by appropriate means, adequate return for the value adding which the private sector has the right to exploit. It may be necessary to introduce an accelerated program to acquire the fundamental or infrastructure data necessary for effective management of the State and to generate these commercial opportunities. This fundamental information is as basic as topographic mapping, a State address file, property or cadastral mapping, census enumerator definition, an intelligent road network and suburb definition.

7.4.4 Value Adding Regime

Fourthly, as argued in this chapter, the future of spatial information use at a State level will find itself increasingly in the hands of the private sector. Gradually the information products will move from government to the private sector on the continuum shown in Figure 7.2. More outsourcing of both traditional government collection and maintenance of spatial information activities will occur. Thematic systems will continue to be developed and maintained within or by government and shared within the supply family. Profitable Government Trading Enterprises will be privatised as to function with the inherent information bases either lost to government or retained as a State asset. Government information sets supporting a business system of government will be provided at minimum cost. In this environment, value added resellers (VARs) should be encouraged to develop new products and services through a range of financial initiatives.

VARs fill a most valuable role in converting, aggregating, enhancing and marketing any and all of the spatial and aspatial information sets to the world. It is they who have most experience in the identification of market need. In the provision of data from government to the VARs, experience to date would suggest that a combination of financial arrangements is more appropriate than any set formula. The essential conditions to apply to this flexibility are non-exclusivity and consistency of approach in relation to each product being supplied to the VAR.

It is when considering access policy and the positioning of the VARs in the spatial information market place that a clear understanding of funding mechanisms is essential. The mechanisms must have all the qualities of probity and transparency expected of good governance but, in simple terms, they are all positioned somewhere on the continuum of total government or total private sector funding. Funding options are discussed in more detail in Chapter 13.

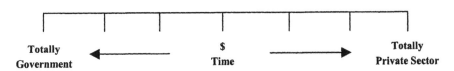

Figure 7.2: Spatial Data Infrastructure – Funding Models

7.4.5 Integration of State SDI Operations

This section summarises the discussion about the integration of State SDI operations and attempts to suggest a development pathway for State SDIs as they become more developed and sophisticated and move more into the private sector.

The positioning of State SDI activities on the government-private sector continuum in Figure 7.2, at any time, is based on the prevailing assessment of value at a State level. Just as the potential energy in an object has no value until it is acted upon to create usable energy, so too is data, by itself, of little value unless it is applied to creating an outcome (product) or solving a problem.

It is widely recognised that the future wealth of society is dependent on knowledge capital rather than physical capital. Many organizations now have almost no physical assets on their balance sheets, and their net worth, as indicated by share price, is predicated mostly on intellectual capital or knowledge - including systems, copyright, patents and intellect. SDIs contain fundamental building blocks as explained in Chapter 2, which will assist in building knowledge capital through systems and policies, which are underpinned by rigorous data.

The fundamental information sets at a State level have been traditionally government created and maintained. The information sets were usually the result of administrative action - for instance, a subdivision of land or a land transfer - or to provide a basis for government action - say, urban planning initiatives or property taxation assessment. The pricing imposed by government for these transactions can be administratively, economically or politically motivated. The success of the outcome depends on the balance between these motivations. Government, in addition to the economic drivers, must focus on the social and environmental outcomes whereas the private sector is not so limited in its choice of business strategies.

Government is of course restrained or encouraged in its decision-making by the political reality of staying in power. A shift on the continuum between government and private sector positioning (Figure 7.2) has come about because of at least three factors. These are:

- the reassessment by government of its core business,
- the growing emphasis on social and community issues, and
- the awareness by the private sector of new financial opportunities.

These opportunities were previously government functions. These range from activities outsourced to the private sector like construction and maintenance of a State digital cadastral data base so generating contract income; retail opportunities like value adding through the use of, say, road networks for despatch operations and build, own, operate and transfer (BOOT) activities which offer medium to long term investment opportunities and combinations of these initiatives.

The earlier claim that the positioning of funding models, at any time, is based on the prevailing assessment of value, can of course be now revisited to consider value to whom? The value of information sets to government has already been considered and the opportunities that these sets offer to the private sector have been briefly explored. However, it is noted that soft or conceptual infrastructures such as SDIs are less tangible to decision-makers and investors, therefore funding models in such areas tend to focus on cost minimization to Government and revenue maximization to the Private Sector.

There are also risks as SDI activities move along the government-private continuum. While activities rest in the government sphere they can be altered as required to reflect changes in circumstances and political trends. However when SDI activities move into the private sector the profit motive is obviously the driver and as such it is difficult to change contracts or to initiate public good activities without the injection of government funds. So while there is an inevitability about moving along the continuum, care must be taken.

There are other factors which must be considered in the move along the continuum and these factors start to create the concept of sharing and partnerships.

Some of these factors are current political philosophy, bureaucratic will, private sector maturity and commercial entrepreneurship. Government policy must be able to demonstrate adequate access to data, a clear understanding of intellectual property rights, effective relationships between agencies and non-restrictive legislation. While there are others, for effective creation and delivery, partnerships currently present the best opportunities.

7.5 PARTNERSHIPS IN STATE SDI

Partnerships are critical to the development of SDIs and can be both inter jurisdictional and intra-jurisdictional (see Jacoby *et al.*, 2002 for a good example of intra-jurisdictional partnerships). Inter-jurisdictional partnerships are principally jurisdiction based but interact between all levels of government in the SDI hierarchy, while intra-jurisdictional partnerships is that array of partnerships or derivative relationships which involve all and any parties that have an interest in the creation, maintenance, manipulation, distribution and analysis of spatial information within the jurisdiction.

While much of the focus on partnerships has been between the different levels of government in the public sector (between Local and State Government or State and Federal Government), the complete vision of the SDI, particularly at the State level, will only materialise through the cooperation and collaboration of the private and public sectors. Both sectors may be focussing on slightly different "profits", however both can achieve their outcomes by clearly defining roles and relationships in their partnership. In the broadest context, the most appropriate model is that of the public sector wholesaling data, captured in the course of performing government activities, and the private sector value-adding these data into products which meet the market's needs. This does not mean that government relinquishes all retail activities, but it does require an assessment of which services and products should become the province of the private sector. If the separation of responsibilities is not embraced, the creation of the SDI will be delayed significantly because of the capital costs of creation of datasets.

7.5.1 Types of Partnerships

Global, regional and National SDIs have been considered in earlier chapters. This section deals with jurisdictional partnerships between the private and the public sector within a State SDI. The jurisdictional type of partnership is that union of effort caused by at least two catalysts:

- The inability of individuals or individual agencies to achieve discrete goals without the support of others, or
- The inability of achieving a pan-agency objective without the sharing of individual agencies' resources.

As mentioned earlier, the continuum on the funding model (Figure 7.2) is a shift in responsibility of financial contribution. This changes as confidence grows, as the market matures and business risk-taking becomes more acceptable. As government chooses to vacate a previously held position it can also offer incentives to the private sector. It can, for instance, demonstrate and guarantee for a period, a reasonable return on investment. This is more likely when the private sector is encouraged to consider the value adding products rather than any shift of the ownership of fundamental datasets. Any shift of ownership at a State level can jeopardise the aggregation of national datasets as with the transfer of the road network dataset in one jurisdiction in Australia. Seen at the time, perhaps as an opportunity to fill a void in the local datasets, any arrangements, which transfer ownership of fundamental datasets, should be avoided at a State level, or if transferred, only after the risks have been identified.

The key to these partnerships is the need to ensure sustainability and, if for no other reason, the fundamental datasets should remain in the ownership of the State. It is essential to ensure there are adequate funds generated by partnership arrangements for the creation of new data, the improvement of the existing data and the maintenance of the datasets in an atmosphere of fairness to all parties based on contribution and risk. There are of course different and changing emphases on priorities of datasets and different approaches to stimulate the private sector in various jurisdictions. But whatever these emphases are, the managers of the State SDIs should not lose sight of the national needs nor put obstacles in the way of the national interest.

7.5.2 Derivative Relationships – Licensing, Royalties and Value Adding Resellers

Enabling the private sector to perform retail functions, and to create additional fundamental datasets, if they are economically viable, does not infer a loss of "control" by State government. Some States and jurisdictions have very successfully entered into licensing agreements which protect the intellectual property (knowledge) of government, whilst deriving benefits from returns on investment which may be of a monetary nature, or by negotiation with the private sector, of a social or policy nature. The fact that the private sector has identified some activities, previously undertaken by government, as markets where they can economically participate, including mapping, data capture and digitising, has enabled State governments to focus their diminishing resources away from these products, into more strategic activities, whilst still ensuring that the needs of citizens are satisfied.

The information components of the SDI can be licensed to private sector organizations for distribution digitally to end-users such as surveyors, lawyers, mappers and developers, and to create value-added products. Such products include derived road centrelines, component datasets, vehicle navigation suites, GPS maps, Internet-based location services, marketing programs, hardcopy maps, vehicle routing programs and demographic mapping. The range of potential applications is extensive.

One of the main advantages of developing licensing arrangements with the private sector is that products are created and enter the market under market-pricing arrangements, not artificial prices levied by government based on non-economic input data. The licensing agreements are structured to ensure a defined return to government, normally based on a share of revenues. There is a need to calculate the "value" of government intellectual property in any partnership or value adding arrangement. Normally a reasonable approach is to estimate the likely cost of capturing data if the private sector had to start anew *less* the value to Government of the applications it uses the data for - regulatory, taxation, registration *plus* social and environmental benefits.

Therefore, the more that citizens and organizations buy these products the more revenues will flow back to government as a return on investment. Government must carefully consider numerous factors when licensing datasets, including privacy, intellectual property, other legal issues, maintenance and distribution mechanisms. However suitable contractual clauses in the license agreements can obviate any potential transgression of policy.

The royalties, which return to government from the licensed use of data to create value by the private sector should continue for the life of the agreement, thus creating a significant revenue stream. Importantly, the data can be licensed many times, with value-added resellers providing competing and often complementary products to satisfy market demands.

7.5.3 An Evolving SDI Concept Based on Partnerships

There is a role in the distribution and value chain for both government and private sectors at a State level. Accessibility to data is a fundamental pillar of any SDI and the most appropriate model to be adopted to ensure optimum accessibility to data by citizens is to shift all but the most sensitive activities to the private sector. Government should retain those functions, which require a high degree of risk management in the maintenance of an SDI and its critical components. This model can only be achieved by gradual implementation and evolution over time. This is not to imply the privatization of functions necessarily, but the separation of some functions under licensed out-sourced conditions. Business functions globally are in continuous transition to this model, with a host of examples to demonstrate the model.

Whatever the models are that have and will emerge, the shape and form of State SDIs is like the closely related land administration environment. It is dynamic and subject to continual change. Certainly, conceptual thinking is required for the development of ideas leading to the use of data for spatial planning and effective decision-making. But this process will be confronted with concepts like liability, pricing, equity share, maintenance and privacy. As fast as they are considered, society is transformed, legislation is changed and public demands take on a different character. Institutional re-engineering, economic reform measures like outsourcing and private sector delivery of traditional public sector functions, changed the perception of the ownership of assets and, with that, data and intellectual property.

Accordingly, the earlier concepts of information, in some cases, become unfashionable. What was accepted philosophy in the early 1990's - cost recovery, public domain, and free data exchange - is no longer relevant. The rate of change of responsibility for service provision, the liability of government for performance and behaviour and the exploding expectations of society due to increased environmental, social and community awareness are increasing exponentially. They will continue to increase as the early vanguard of new and readily available information foreshadows the enriched spatial information of the rapidly emerging and all consuming data banks of the society wide SDIs. Accordingly new public/private sector relationships will emerge.

7.6 EVALUATION OF STATE SDI INITIATIVES

The organizational and funding models and the concepts of partnerships and alliances which have been considered in this overview of State SDI initiatives, are functions of maturity and the evolution of agencies and attitudes. There are attitudinal differences born of ideologies, a contrasting sense of values and different views which can be found at each level of government. There are specific views held by those who see SDIs as process or product driven, by members of those jurisdictions which are a federation of States as distinct from non federated jurisdictions, by economic rationalists versus the cultural traditionalists and by those driven by policy development rather than operational achievement.

Issues of policy and finance are important. However, as the emphasis for detailed information is sought, there is the nagging issue of trust between all levels of government. Local Government is suspicious of an overemphasis on policy and even, in some cases, the development of metadata. Of course there is a lack of awareness, as to National SDIs, in some areas of Local Government. But perhaps Local Government is more aware of relationship management than, say, national agencies, which in the main, are removed from the reality of everyday decisions which impact on the person in the street. And in federated States, Local Government has pressures imposed by both upper levels of government and guidelines and standards that do not necessarily reflect local needs. To enrich the National SDI with relevant data sourced from Local Government, where the State SDI plays the key role, it is not only a matter of building a pyramid of data, but also of building a pyramid of trust (Harvey, 2002).

Joffe (2002) claims that Local Governments' data sales policies are actually impeding the distribution of digital geographic data for three principal reasons:

- Local Governments are not experienced in marketing hence public awareness of the availability of the data does not reach potential users or buyers.
- Local Governments are not experienced in fulfilling data request orders hence many users' needs for special purpose data products - in a timely manner - go unfulfilled.
- In some cases, the cost of local digital geographic data is prohibitive to potential users.

However, Craig (2001) describes the successes in the State of Minnesota, USA, through Local Government initiatives. He praises the State for the foresight in funding GIS datasets through the Legislative Commission on Minnesota Resources (LCMR), a pool of money mainly filled from cigarette tax revenues and proceeds on the State lottery; LCMR has invested nearly $17 million in land use and natural resource information since 1991. He discusses the large range of data which exists in Minnesota and the sense of collegiality that supports substantial sharing of that data. In describing six case studies he lists two aspects as the key to the creation of a successful spatial data infrastructure; institutions with proper missions and individuals with high levels of motivation.

Further, he claims that institutions that create data and institutions that coordinate and disseminate data are key to success and that it is critical to have individuals who are called beyond narrow institutional mandates and are motivated to share their data with others; factors behind this motivation appear to be idealism, enlightened self-interest, and trust in those receiving the data. Perhaps the very existence of land related datasets came about as a result of such people as these; the sharing of information, the communion of ideas in the spatial information family and the growth and maintenance of datasets being generated by public spirited champions with a mutual trust and a will to share resources.

To evaluate State SDI initiatives is to grapple with the political environment in which they are nurtured, the relationship between all levels of government (national above and local below) and the relevance of the products. But most of all it is about people.

7.7 OTHER ISSUES AND FUTURE DIRECTIONS OF STATE SDI

Initially it is asserted that core spatial datasets at a State level (large to medium scale datasets where cadastral data is a key) should be seen as a State infrastructure asset and their collection and maintenance funded to an agreed level to prevent system collapse and to meet the reasonable expectations of a developed society. This cost can be readily ascertained and expressed in budgetary terms. It is this cost which should be recognised by any policy, which deals with the supply of spatial information.

Jurisdictional wide data access policy should be adopted to provide a rational and consistent basis for data access that will facilitate the efficient production, supply and use of geospatial information held by the State. The policy must differentiate between infrastructure and business systems and foster the industry opportunities whilst not creating monopolies at the expense of the public interest. Whilst accepting the reality of capitalist market forces operating in a socialist or democratic society, the aim of a sound and responsible access policy should be a spatially enabled community and not an enlargement of the information gap.

Whereas there should be an emphasis on partnerships it is clear that the future of spatial information use will find itself increasingly in the hands of the private sector. Gradually the creation and delivery of information products will move from government to the private sector. More outsourcing of both traditional government collection and maintenance of spatial information activities will occur at the State level. Thematic systems will continue to be developed and maintained within or by government and shared within the supply family. Profitable State Owned Corporations (SOCs) and Government Trading Enterprises (GTEs) will be privatised as to function with the inherent information bases either lost to government - often inadvertently - or retained as a government asset. Government information sets supporting a business system of government will be provided at minimum cost. In this environment, value added resellers should be encouraged to develop new products and services through a range of financial initiatives.

As stated in the earlier chapters in this book, SDI is still an evolving concept. This is very much the case at the State level. While this chapter discusses some of the key issues to be considered in establishing a State SDI, there are a wealth of additional emerging issues which are impacting on the design and operation of State SDIs. These include:

- the growing awareness of the critical role that SDIs play in emergency response. To some extent the requirements of emergency response have not often been factored into the design, funding and institutional arrangements for State SDI, but it is inevitable.
- An over-riding issue in State SDIs concerned with the integration of spatial data from State SDIs into national datasets, is interoperability. This is expected to be one of the major issues confronting SDI policy makers and operational personnel in the years ahead.
- As has been mentioned earlier in this chapter, issues of intellectual property, confidentiality and privacy are almost latent issues which will become more and more important in the years ahead.
- Today most of the wider community does not understand "spatial data", and even maps are a challenge for much of the general public. However society world-wide and particularly in the developed world is becoming spatially enabled with spatial data (or location) becoming a key business driver. The message is clear that the spatial information industry has to make a much greater effort to raise the awareness of the role of spatial information in society. However this is not an easy task.
- While the role of the academic sector in developing and operating SDIs was raised earlier in the chapter, there is still relatively little attention to research and development, and capacity building in support of SDIs world-wide. Again this is a major challenge.

Earlier in this chapter we discussed the inroads of micro economic reform on the spatial information sector, discussed the operation and funding of State SDIs and reviewed the economic rationalist's approach to the intricacies of the provision of spatial data. Too much fruitless time has been devoted to the debate about data pricing. What is important is the issue of access to, and the relevance of, spatial data to the community. Protocols for sharing and the secure and continuous provision of this data should be the focus of our attention. Put these in place and, at the same time, question the rationale for the debate.

7.8 REFERENCES

Bogaerts, T., Williamson, I.P. and Fendel, E.M., 2002, The Role of Land Administration in the Accession of Central European Countries to the European Union. *Journal of Land Use Policy*, Vol. 19, 29-46.

Chan, T.O. and Williamson, I.P., 1999, The Different Identities of GIS and GIS Diffusion. *International Journal of Geographical Information Science*, Vol. 13, No. 3, 267-281.

Craig, W.J., 2001, Spatial Data Infrastructure in Minnesota: Institutional Mission and Individual Motivation, *International Symposium on Spatial Data Infrastructure*, University of Melbourne, Australia, 19-20 November, 2001.

Drucker, P. F., 1993, Really Reinventing Government, *The Atlantic Monthly* February, 1993.

Harvey, F., 2002, Potentials and Problems for the Involvement of Local Government in the NSDI, *6th GSDI Conference*, Budapest, Hungary, September 16-19, 2002.

Jacoby, S., Smith, J., Ting, L. and Williamson, I.P., 2002, Developing a Common Spatial Data Infrastructure between State and Local Government-An Australian Case Study, *International Journal of Geographical Information Science*, Vol 16, No 4:305-322.

Joffe, B., 2002, ODC-The Open Data Consortium: A project to standardize data distribution, 14 August, 2002. On-line.
<http://www.directionsmag.com/article.php?article_id=235> (Accessed February, 2003).

Masser, I., 1998, What is Geographic Information and why is it so important? Chapter 2, *Governments and Geographic Information*. (London: Taylor & Francis).

Ordnance Survey, 1996, Economic aspects of the collection, dissemination and integration of government's geospatial information. A report arising from work carried out for Ordnance Survey by Coopers and Lybrand May 1996, On-line.
<http://www.ordnancesurvey.co.uk/literatu/external/geospat/> (Accessed February 2003).

Ordnance Survey, 1999, The Economic Contribution of Ordnance Survey GB. OXERA (Oxford Economic Research Associates). Final Report, On-line.
<http://www.ordnancesurvey.co.uk/literatu/external/oxera99/index.htm> (Accessed February 2003).

Williamson, I.P., Chan, T.O. and Effenberg, W.W., 1998, Development of Spatial Data Infrastructures – Lessons Learned from the Australian Digital Cadastral Databases. *GEOMATICA*, Vol. 52, No. 2, 177-187.

PART THREE

Australian Case Study
from National to Local

CHAPTER EIGHT

Development of the Australian Spatial Data Infrastructure

Drew Clarke, Olaf Hedberg and Warwick Watkins

8.1 INTRODUCTION

The Australian SDI (ASDI) provides an interesting case study of a National SDI initiative, operating in the context of both international and local initiatives. Australia has three levels of government: national (termed Commonwealth), state/territory (eight) and local (727), and a vigorous private sector spatial industry. Each level of government has distinct public spatial responsibilities, while the private sector provides services to both the government and commercial sectors. Australia is also an active participant in Regional and Global SDIs through the Asia-Pacific and GSDI initiatives.

The development of the Australian (national) SDI is described in this chapter. The Australian state/territory and local governments collectively produce and use considerably more spatial data than the Commonwealth. In many ways therefore, the ASDI comprises numerous Local SDIs, plus the private sector, all operating within an agreed national framework. The story of the ASDI is therefore about the establishment, content and management of this national framework. The framework operates through consensus – the Commonwealth has not sought to legislate or in any way force the other players to conform to a national policy or standard. This approach can be seen as a hindrance to ASDI development, but it is also a strength as it requires consultation and cooperation which help build commitment.

The key body that is driving development of the ASDI is ANZLIC – the Spatial Information Council. ANZLIC's background and structure is outlined in Section 8.2, and its conceptual model for the ASDI is described in Section 8.3. The ASDI organizational model, outlined in Section 8.4, comprises numerous public and private sector stakeholders. Several new bodies have emerged in the last few years, greatly strengthening the public, private, professional and research SDI organizational infrastructure. The role of the Spatial Information Industry Action Agenda is also outlined in Section 8.4. Some current ASDI implementation projects are described in Section 8.5, and the chapter concludes with some comments on current ASDI issues and future directions in Section 8.6.

8.2 ANZLIC – THE SPATIAL INFORMATION COUNCIL

ANZLIC was originally established in January 1986 as the Australian Land Information Council (ALIC) by agreement between the Australian Prime Minister and the heads of the state and territory governments, in response to a clear and growing need to:

• coordinate the collection and transfer of land-related information between the different levels of government; and
• promote the use of that information in decision-making.

In the late 1970s, all jurisdictions in Australia and New Zealand had been confronted by similar administrative and technical issues in relation to their computerised land information databases. Cost-efficient access to compatible land information was required to enable effective decision-making by governments. There had been minimal coordination on a national scale although there was some informal communication between land information managers.

ANZLIC arose from the need to focus national coordination of land information management. A national conference, 'Better Land Related-Information for Policy Decisions', held in 1984 and attended by representatives from the three levels of government, recommended that a peak national coordinating council be formed. This council would comprise the heads of the various land information councils in each of the jurisdictions and would be given the role of promoting and developing a national strategy to facilitate the exchange of land information.

With the support of the Prime Minister, State Premiers and the Chief Minister of the Northern Territory, ALIC was formed and held its first meeting in March 1986. The Commonwealth Government, all Australian states (except Queensland) and the Northern Territory were represented. Queensland and the Australian Capital Territory were represented for the first time as observers in 1989 and were subsequently accepted as full members. New Zealand was represented on ALIC from 1987, with the same participating rights as the Australian members. In November 1991, New Zealand formally became a full member and the Council was renamed ANZLIC. While New Zealand and Australia share technical and policy experiences through ANZLIC, New Zealand is developing its own National SDI and is not part of the Australian SDI.

Although the language used to describe its activities has changed considerably, it is clear from its objectives and activities that ANZLIC has been developing the Australian Spatial Data Infrastructure since 1986. The constitutional structure of Australia, with a Commonwealth Government and eight state/territory governments (with major spatially related responsibilities) has always meant that ASDI development required inter-government cooperation. ANZLIC has been the vehicle for this cooperative approach.

While its origins and initial focus were largely in land administration (cadastral) systems, ANZLIC has long recognised that the SDI concept was much broader, encompassing all forms of social, economic and environmental information. The term 'spatial information' was adopted by ANZLIC to replace the awkward (and misleading) 'land-related information', and the tag 'the Spatial Information Council' is now used to emphasise its broader SDI role.

8.3 ASDI CONCEPTUAL MODEL

The conceptual model of the ASDI was first defined by ANZLIC in a 1996 discussion paper (ANZLIC, 1996). The paper outlined a vision of fundamental datasets built by governments and accessed by the community, all in the national interest. The rationale was conventional, based largely on the 'better information enables better decisions' model. An earlier study commissioned by ANZLIC (Price Waterhouse, 1995) provided the economic justification, demonstrating a benefit-to-cost ratio of 4:1 for the $1 billion investment in spatial data over the previous five years. This study used a costs-avoided methodology, and did not attempt to value the broader social or environmental benefits.

The core concept is that of a distributed model, comprising eight state/territory and one Commonwealth SDI, integrated through common technical standards. Dataset custodians have specific rights and responsibilities within this model. The distributed nature of the ASDI is further replicated in the jurisdictions, with multiple agencies providing SDI elements. There is only one physical information system that could be described as 'pure' ASDI, that being the Australian Spatial Data Directory (ASDD) which supports spatial data discovery across all jurisdictions and agencies.

The 1996 model, outlined in Table 8.1, provided an effective framework for the development of the ASDI over its early years. ANLIC reviewed the model in 2002, and commenced a new stage in ASDI development. The 2002 ASDI vision and definition are:

- Vision: Australia's spatially referenced data, products and services are available and accessible to all users.
- Definition: The ASDI comprises the people, policies and technologies necessary to enable the use of spatially referenced data through all levels of government, the private and non-profit sectors and academia.

The core concepts of the 1996 model remain, but the new vision and definition provide a better balance between the producer (supply side) and user (demand side) sectors of the spatial data community.

8.4 ASDI ORGANIZATIONAL MODEL

At the national level the key organizational elements of the ASDI are ANZLIC, the Public Sector Mapping Agencies, the Australian Spatial Information Business Association, and the Spatial Sciences Coalition. The Spatial Information Industry Action Agenda (*Positioning for Growth*) is also an important part of the organizational infrastructure. Key ASDI stakeholders have submitted a bid for Commonwealth funding to establish a Cooperative Research Centre for Spatial Information which (if funded) will undertake research contributing to the ASDI. These organizational elements are described in the following sections.

Table 8.1: Extract from ANZLIC's 1996 ASDI Discussion Paper
(Initial) Conceptual Model for the Australian Spatial Data Infrastructure

---

The primary objective of a national spatial data infrastructure is to ensure that users of land and geographic data who require a national coverage, will be able to acquire complete and consistent datasets meeting their requirements, even though the data is collected and maintained by different jurisdictions. The issue, therefore, is to determine what is required of jurisdictions and their datasets, to enable them to meet national needs.

ANZLIC envisages a distributed network of databases, linked by common standards and protocols to ensure compatibility, each managed by custodians with the expertise and incentive to maintain the database to the standards required by the community and committed to the principles of custodianship.

ANZLIC believes that a national data infrastructure will provide the institutional and technical framework to ensure the required national consistency, content and coverage to meet national needs. The infrastructure also ensures that all jurisdiction efforts are focussed in the national interest, thereby maximising investment in data collection and maintenance from a national perspective. Finally, such an infrastructure will help achieve better outcomes for the nation through better economic, social and environmental decision-making.

ANZLIC has developed a national spatial data infrastructure model that comprises four core components - institutional framework, technical standards, fundamental datasets, and clearing house networks. These core components are linked as follows:

- INSTITUTIONAL FRAMEWORK - defines the policy and administrative arrangements for building, maintaining, accessing and applying the standards and datasets;
- TECHNICAL STANDARDS - define the technical characteristics of the fundamental datasets;
- FUNDAMENTAL DATASETS - are produced within the institutional framework and fully comply with the technical standards;
- CLEARING HOUSE NETWORK - is the means by which the fundamental datasets are made accessible to the community, in accordance with policy determined within the institutional framework, and to the agreed technical standards.

The four components of the ASDI model are described in the paper in some detail, and a draft list of around 40 fundamental datasets, organised into five themes (primary reference, administration, natural environment, socio-economic and built environment), is proposed.

8.4.1 ANZLIC Organizational Model

ANZLIC comprises one representative from each of the eight Australian state and territory governments, one from the Australian Commonwealth Government, and one from the New Zealand Government. Each of these representatives is the head of the spatial information coordinating body in their respective jurisdictions, ensuring that ANZLIC represents all the public sector spatial data agencies. ANZLIC activities are defined through a strategic plan and a series of annual work plans, with around four Council meetings per year and regular e-mail correspondence within a senior contact group that supports the members.

ANZLIC has created two Standing Committees, ASDI and Industry Development, each chaired by a Council member and comprising broad public and private sector membership. The ASDI Standing Committee, chaired by the Commonwealth Member, has primary responsibility for consulting ASDI stakeholders and advising ANZLIC on ASDI policies and implementation projects. The Intergovernmental Committee on Surveying and Mapping (ICSM) reports to ANZLIC through the ASDI Standing Committee, and is responsible for the development of national geodetic, topographic and cadastral standards.

Until 2001, ANZLIC was supported by a part-time Executive Officer provided by the Commonwealth. Members made a modest annual contribution to support administrative activities, and jointly funded specific projects such as consultancies, publications and workshops. During 2000 and 2001 the ASDI Standing Committee operated with a dedicated Project Officer, seconded from a State and funded jointly by the members, to undertake some specific ASDI development projects. This arrangement was very successful, and in 2001 ANZLIC decided to permanently expand its resources through establishment of a National Office, comprising an Executive Director and two project officers. The new National Office, funded through increased member contributions, will provide the resources to pursue both the ASDI and industry development goals of ANZLIC.

8.4.2 Public Sector Mapping Agencies

The Public Sector Mapping Agencies Australia (PSMA) consortium was originally created in 1993 as an unincorporated joint venture between the nine mapping agencies of the Commonwealth, states and territories, to respond to an Australian Bureau of Statistics tender for the provision of mapping services for the 1996 Census of Population and Housing. Although the mapping agencies were all represented through ANZLIC, a more commercial structure was required to bid for the Census contract. PSMA Australia was awarded the contract and broke new ground with the delivery of an integrated national topographic dataset augmented with a representation of the nation's cadastral framework.

As word of this initiative spread, PSMA received strong demand to create or provide national datasets for other organizations with national interests, both in the private and public sectors. As demand continued to grow, it was recognised that the PSMA structure was inappropriate if the organization was to explore other opportunities for use of its national datasets. After consideration of options by ANZLIC and PSMA, the parties agreed to form PSMA Australia Limited – an unlisted public company, limited by shares and owned by the governments of Australia. Incorporation was achieved in June 2001.

As a government owned company, PSMA Australia Limited is in a unique position, able to unlock and integrate data held in individual jurisdictions and supply it to value added resellers (VARs) as national datasets. VARs then add ideas and innovation to develop products and services to meet user demands.

A fundamental principle underpinning the operations of PSMA Australia Limited is that it adopts operational business rules that minimise any impact on existing revenue streams of individual jurisdictions. The focus for PSMA is the development of national datasets and value adding and distribution by licensed VARs. Requests for less than national datasets are directed to the relevant jurisdiction. There are four datasets currently licensed by PSMA Australia Limited:

- Detailed Topography: underpinned by a road centreline dataset, with over 30 feature types in hydrography, transport and points of interest themes.
- Basic Topography: includes highways, main roads, coastline, major drainage, railway lines and key locality points.
- National Cadastre: a database of Australia's 10.4 million registered land parcels, each with five key attributes.
- Points of Interest: a database with over 130,000 cultural points of interest with feature code and name attribution.

The PSMA fulfils an important role in the ASDI, by building national datasets from Commonwealth and state/territory data, and facilitating their use through a network of commercial value-adders. The PSMA Chair is a member of ANZLIC, and PSMA is represented on ANZLIC's ASDI Standing Committee.

8.4.3 Spatial Information Industry Action Agenda

Action Agendas are a key element of Australian industry policy. They position specific industry sectors to realise the opportunities of international markets and new technologies, overcome impediments and barriers, and encourage sustainable economic development and national growth. They achieve this by providing a framework where industry and government work together to identify and capture opportunities and address industry-specific impediments to growth so that the benefits of broad-based reform flow to individual companies. Around 30 industry Action Agendas have been supported by the Commonwealth since 1996, with the spatial industry selected in 2000 because of its strategic role in the economy.

The spatial information industry action agenda, *Positioning for Growth* (Commonwealth of Australia, 2001a), was launched in September 2001 by the Minister for Science, Industry and Technology. It was developed by a steering group of nine industry leaders, two from the public sector and seven from industry. The Chair of the ANZLIC Industry Development Standing Committee was a member of the Action Agenda steering group. Positioning for Growth established an industry vision - Australia will be a global leader in the innovative provision and use of spatial information.

Five goals were identified:

- develop a joint [business, academic and government] policy framework;

- improve data access and pricing;
- increase effective research and development;
- evaluate and reform education and skills formation; and
- develop domestic and global markets.

Implementation of the actions underpinning these goals, many of which refer to the ASDI, is continuing. The Action Agenda process has brought the public and private sectors of the spatial information industry closer together, and has highlighted the importance of the ASDI to industry development.

8.4.4 Australian Spatial Information Business Association

ANZLIC does not purport to directly represent private sector spatial interests, although one jurisdiction (Queensland) has included the private sector in their coordination structure. The formation in 2001, through the Action Agenda process, of the Australian Spatial Information Business Association (ASIBA) has provided a representative body for the private sector.

The primary purpose of ASIBA is to represent industry views to governments and to promote adoption of spatial information products and services. Members are businesses engaged in commercial spatial information activities, and must have at least some portion of private ownership. ASIBA operates at both the Commonwealth and state/territory political levels, and has produced a number of Industry Position Statements and responses to government reviews. ASIBA is a member of ANZLIC's ASDI Standing Committee, ensuring a direct voice for industry in ASDI development.

8.4.5 Spatial Sciences Coalition

There are five key professional associations with an interest in the ASDI: the Institution of Surveyors, Australia, the Mapping Sciences Institute, Australia, the Institution of Engineering and Mining Surveyors, Australia, the Australasian Urban and Regional Information Systems Association, and the Remote Sensing and Photogrammetry Association of Australia. In 2001 the associations began developing a unification proposal that would address systemic problems in each group. The resulting proposal describes a model for a new body, to be named the Spatial Sciences Institute, that is constructed around Commissions representing the areas of interest of the five associations. Members will vote on the unification proposal in late 2002, and if approved the new Institute will be established in 2003.

The role of the Coalition and proposed Spatial Sciences Institute in representing the spatial professions is well recognised. The Coalition is contributing to implementation of the Action Agenda, and the proposed new Institute should provide a strong voice for the professions in ASDI development.

8.4.6 Australian Spatial Information Education and Research Association

Australian Spatial Information Education and Research Association (ASIERA) was established at the joint Australian Urban and Regional Information Systems Association (AURISA) and Institution of Surveyors Conference 2002. The key objectives of ASIERA in the short to medium term are firstly to expand its membership from the nine traditional surveying and mapping schools in Australia and New Zealand to between twenty to thirty schools and departments which have teaching programs in the spatial information area or undertake research in this area; secondly to establish how ASIERA can best contribute to the wider industry and specifically how it can interact with the Australian Spatial Information Business Association and ANZLIC. Lastly it is hoped to be able to quantify the size of the education and research sector including the personnel (staff, postgraduates and undergraduates), the annual turnover of the sector and its export value.

8.4.7 Spatial Information Cooperative Research Centre

Cooperative Research Centres (CRCs) are long-term collaborative ventures between researchers from universities and other government research agencies, and private industry or public sector research users, which support research, development and education activities of national economic or social significance. CRCs are selected through a competitive process, with successful bids receiving substantial funding from the Commonwealth Government for seven years.

The need for a CRC was specially identified in the Spatial Information Action Agenda as critical to the implementation of the Agenda. As a result, key elements of the Australian spatial information community provided a model of how the academic, government and commercial sectors are collaborating within the ASDI context through the establishment of a 'CRC for Spatial Information' (CRC-SI) in 2002. The core partners comprised spatial information agencies from the Commonwealth and two state governments (Victoria and New South Wales), three universities (Melbourne, New South Wales and Curtin), and a consortium of private sector companies. The CRC-SI operates from five centres around Australia.

The CRC-SI supports national research priorities and enhances the outcomes of many others CRCs in Australia which rely on spatial data and related science and technology. The importance of spatial information was identified in December 2002 when the Prime Minister of Australia announced the National Research Priorities. As part of the National Research Priorities he identified frontier technologies for building and transforming Australian industries. He specifically stated "frontier technologies for building and transforming Australian industries is about fostering creativity and innovation by supporting leading edge research in areas such as information and communication technology (ICT), bio- and geo-informatics, nanotechnology and biotechnology ... Support for these areas of research will help stimulate vibrant new industries and ensure our future competitiveness".

The identification of geo-informatics (for geomatics, geomatic engineering, GIS and spatial information science) alongside ICT, nanotechnology and biotechnology is an important recognition of the role that spatial information will play in the development of Australia.

The objectives of the CRC-SI include a close involvement of spatial information users in research activities, stronger collaboration between industry academic and government researchers and research users, development of a long-term research agenda for spatial information, more efficient research training, and enhanced commercialization of spatial information technologies. The research program includes spatial information technologies, decision support systems and spatial data infrastructures. The central theme is to develop a 'virtual Australia' by uniting research and commercial innovation in spatial information technologies. The CRC-SI will greatly strengthen the research base for ASDI development.

8.5 ASDI IMPLEMENTATION

Much of the early ASDI development was through the ongoing spatial information programs and operations of the various stakeholders, with an ANZLIC ASDI work plan that addressed a number of technical and policy projects that went beyond any single jurisdiction. Some of these national projects are described below.

8.5.1 Fundamental Datasets

One important early ANZLIC initiative was the fundamental datasets scoreboard project, which was designed to focus attention on the availability of key reference ASDI datasets. Ten themes were selected for audit – administrative boundaries, cadastre, elevation, land use, place names, roads, soils, street addresses, vegetation and water (refer to Table 8.2). These datasets are used by multiple GIS applications, and underpin many spatial products and services.

The scoreboard methodology involved four steps for each theme. First, user requirements were identified, and tests specified to determine compliance. A general criteria model was developed, covering data access, quality, maintenance and supply. Second, the 'best' national dataset currently available was identified, and agreement obtained from the custodian to participate in the scoreboard project. Many of the selected datasets had been built by integration of state, territory and Commonwealth data. Third, the characteristics of the selected dataset were audited against the user criteria. This involved considerable dialogue between the custodian and the auditor, each party seeking to understand the others perspective (a valuable process in itself). Finally, the results were published and reviewed by ANZLIC, and the report provided to the dataset custodian. Custodians were encouraged to address any low-scoring areas.

The first dataset to go through the entire process was the PSMA roads dataset, which served as the pilot for the methodology. The PSMA Board reported that they found the evaluation report to be very useful, and have commenced addressing issues arising. Assessment of five other datasets commenced in 2001, but further scoreboard work was suspended in 2002 while the new ASDI work plan was being developed. However the data quality improvement work initiated by the scoreboard project is expected to continue for many years.

As with many projects of this nature, the detailed results are less important than the process itself. The process required ANZLIC consultation with users, dialogue between the custodian and ANZLIC, and internal review by the custodian. It is significant that existing national datasets could be readily identified for each of the ten selected themes – this would not have been possible in 1996 when the ASDI conceptual model was first developed.

Table 8.2: Fundamental Datasets Included in the Scoreboard Project

Theme	Dataset	Custodian
Administrative Boundaries	Australian Standard Geographical Classification	Australian Bureau of Statistics
Cadastre	Cadastral Lite, 2001	PSMA Australia Ltd
Land Use	Land Use of Australia 1996/1997 Version 2	Department of Agriculture
Place Names	Gazetteer of Australia, 2000	Geoscience Australia
Soil	Australian Soil Resources Information System	CSIRO
Roads	PSMA Roads Dataset - ROSD (Detailed)	PSMA Australia Ltd
Elevation	GEODATA 9 Second Digital Elevation Model	Geoscience Australia
Street Address	PSMA Australia Geocoded National Address File	PSMA Australia Ltd
Vegetation	Information products of the National Vegetation Information System Data in the National Vegetation Information System	Environment Aust. State/Territory agencies
Water	Australian Water Resources Assessment 2000	Department of Agriculture

8.5.2 Australian Spatial Data Directory (ASDD)

In 1995 ANZLIC established a Metadata Working Group (AMWG) to design and develop a national spatial data directory system. The AMWG established standards, protocols and the technical architecture for the directory. The ASDD was formally launched in 1998 and now contains approximately 30,000 metadata records on 22 distributed nodes (21 public sector and one private). The directory is accessed hundreds of times per month, although reliable usage statistics are not currently available. The focus for the first three years of ASDD operations was on development of the distributed directory architecture, and on metadata collection. Given the key role of a directory in any SDI model, ANZLIC decided to audit the ASDD in 2001. The key questions posed for the audit were:

• What records are currently in the ASDD, and what is their quality?

- What records are NOT in the directory that should be?
- Who is using the ASDD and is it meeting their needs?
- Who is NOT using the directory, and why not?

The ASDD audit methodology involved statistical analysis of 25,000 records from 15 nodes, to determine completeness of the various metadata fields; qualitative analysis of 600 representative metadata records (200 environmental, undertaken by the National Land and Water Resources Audit, and 200 each with economic and social themes) to assess compliance with the ANZLIC Metadata Guidelines (Version1, July 1996); and a user survey, run through the website (155 responses) and distributed to the spatial information community (50 responses). The results were presented in a comprehensive report to ANZLIC in October 2001. Some of the key findings are outlined in Table 8.3.

The ASDD audit concluded that ANZLIC's objective of maximising community data access through the directory has had only limited success to date. On the negative side, the ASDD is largely unknown by the target audience, the content is of highly variable quality, insufficient resources have been allocated for maintenance, and there has been minimal private sector involvement. On the positive side, those who contribute to the directory are also the major users, most users find what they are looking for, and the technical architecture has proven to be robust. There were however significant differences between the audit results for individual jurisdictions, indicating some systemic metadata management issues (both positive and negative).

The audit recommendations were all accepted, resulting in two parallel actions:

- Jurisdictions will review the coverage and content of records on their nodes, and allocate resources to upgrade both (outcomes to be reported to ANZLIC).
- ANZLIC National Office will promote the ASDD to the broader spatial information community, and formulate a plan for its future development.

Table 8.3: Directory Audit Findings

Coverage: despite containing 30,000 records, there are still many gaps in the ASDD coverage. Particular weaknesses include natural resources, utilities, social and health data, and data from local government, the private sector and academia.

Content: quality is not well documented in up to 40% of records, 50% contain poor or no descriptions of the dataset content, purpose and history, while only 33% have a currency of 2001 (13% are more than three years old).

Users: the ASDD is used by the private sector (34%), state and territory government agencies (27%), Commonwealth Government agencies (12%), academia (16%), local government (6%) and overseas (5%). The industry sectors are very diverse. Most users accessed the system less than once per month (71%), but 11% used the ASDD more than 5 times per month. The major reason for non-use was lack of awareness of the system.

Satisfaction: most users were satisfied (27% described it as comprehensive and current, 62% as reasonably descriptive and quite up-to-date), with only 11% dissatisfied (too brief and out-of-date). When they last used the ASDD, 59% of users found the dataset they were looking for.

8.5.3 ASDI Clearinghouse Definition

The clearinghouse component of the ASDI is not well defined, beyond the central role of the ASDD. ANZLIC organised a workshop in May 2000, with participants from across the spatial information community, to further develop the concept.
The workshop developed an 'ASDI Distribution Network' model (ANZLIC, 2000) incorporating three components sitting between the data and the users:

- Technical: data discovery, access and transfer facilities (including the ASDD).
- Institutional: pricing, copyright and licensing arrangements.
- Products, Services and Solutions: including service providers, value-added resellers, information brokers and product integrators.

This more holistic model is now guiding development of the data discovery, access and usage components of the ASDI. An ASDI Technical Working Group has been established under the ASDI Standing Committee to further develop the technical architecture for the distribution network. This is likely to include a prototype distributed data access system.

8.5.4 Pricing Policy

The desirability of a single national spatial data access pricing policy has long been recognised by ANZLIC, and several attempts have been made to develop and implement such a policy. These have largely failed, due to the wide variation in cost recovery and information policy between the jurisdictions at any point in time, and the inherent difficulty in getting nine governments to simultaneously agree. However, ANZLIC has been able to develop a set of spatial data access principles which emphasise easy, efficient and equitable community access and the maximising of net benefits, and there is much less variation between jurisdiction pricing policies today than at any time over the last ten years.

These principles have been published on the ANZLIC web site. ANZLIC also agreed in October 2001 to commence work on a model national spatial data access and pricing policy, to provide a reference point for further policy work in the jurisdictions and for consultation with ASIBA. The Spatial Information Industry Action Agenda makes several recommendations regarding data access and pricing, aimed at improving accessibility. Key recommendations include development of a common [national] approach to data access, pricing and copyright policy, support for the new Commonwealth policy (outlined below), development of a privacy code of practice, and improved on-line data availability.

Two related developments in 2001 underlined the trend towards a more open pro-user spatial data pricing policy within Australian governments.

- Natural Resources Data - ANZLIC recently concluded a 'Data Access and Management Agreement' with the National Land and Water Resources Audit (NLWRA), which defines the management arrangements for 38 existing natural resources datasets, and a framework for addressing 12 other national datasets currently under development. The NLWRA, a program of the Commonwealth Government's Natural Heritage Trust, has developed key natural resource spatial datasets by integration of state, territory and

Commonwealth data and contracted data collection. The ANZLIC-NLWRA Agreement provides for community access to most of these data for non-commercial purposes, at no charge, over the internet.

- Commonwealth Policy - a benchmark for spatial data access and pricing has been set by the Commonwealth Government, with a new policy that makes designated fundamental spatial datasets available at no cost over the internet for all users and all purposes, without any copyright licence restrictions on commercial value-adding. Other forms of data supply such as packaged CDs will be charged at the marginal cost of transfer (Commonwealth of Australia, 2001b). The Commonwealth will be seeking a reciprocal data access agreement with the states and territories (for government use only), and will be promoting its new policy as a model for ANZLIC. The new Commonwealth policy has, not surprisingly, been welcomed by ASIBA.

These two national developments, and the many changes made by state and territory governments over the last two years to improve community access to spatial data, herald a new era in ASDI pricing policy development.

8.6 CURRENT ISSUES AND FUTURE DIRECTIONS

ANZLIC has made considerable progress in development of the ASDI since publication of the initial conceptual model in 1996. Key national datasets have been developed, the ASDD has been implemented, and pricing policies are converging. However, much remains to be done, and it could be argued that development has barely kept up with changes in the SDI environment, let alone closed the gap to real implementation.

There are two key limitations on ANZLIC's ability to develop the ASDI, both of which arise from the intrinsic requirement for cooperation among nine governments:

- Funding: ANZLIC cannot fund the ASDI as such. Its individual members (the nine Australian governments) can only fund their own components, and will (quite properly) have regard to their jurisdiction priorities in committing resources.
- Policies: ANZLIC cannot create binding national policies. It can develop models, standards and protocols, but cannot mandate their acceptance or use.

Given these two limitations, it is perhaps remarkable that ANZLIC has achieved as much as it has. As a result of the high level of cooperation and goodwill engendered through ANZLIC, the nine Australian jurisdiction SDIs are more similar than different, and elements of the ASDI exist throughout the spatial information community.

While progress has been made in implementing most elements of the ASDI, three parallel developments have emerged which impact on the model.

1. *Technologies* – the rapid development of the key SDI-enabling technologies of the internet and distributed database systems, and the emergence of powerful application technologies such as hand-held GPS, desk-top GIS and high-

resolution satellite imagery, have all challenged conventional technical models of spatial data infrastructures.

2. *Spatial Market* – the spatial information industry is poised for rapid growth. The Action Agenda reported global expenditure in the sector as $34 billion per year and growing at 20%, with Australia comprising over $1 billion and expected to double by 2005. This will significantly affect demand for ASDI functionality and services.

3. *Business Models* – the public sector has rapidly moved from an inhouse mode of program delivery, to a policy where all non-core functions are outsourced to the private sector. This raises the question of how much ASDI-related activity is really core business, and whether public-private partnership models for program delivery may be more appropriate.

These limitations and external developments, along with the results of the dataset scoreboard, ASDD audit, clearinghouse definition and pricing policy projects described above, the development of the Action Agenda, formation of ASIBA and incorporation of PSMA, have all underpinned the new ASDI vision and definition developed in 2002. Five ASDI development priorities for ANZLIC have now been identified:

- ASDI Governance: improving institutional arrangements to remove barriers to the access and use of spatial information.
- Data Access: development of nationally consistent data access arrangements.
- Data Quality: implementing the recommendations from the ASDD audit.
- Interoperability: implementing an Open Systems approach across the ASDI.
- Integration: development of standards that support spatial dataset integration.

Consideration of these priorities raises the issue of performance indicators. How do you measure success in the development of a National SDI? The dataset scoreboard, directory audit and pricing policy projects outlined above provide an indication of progress, but also reveal that these fundamental elements are still incomplete. However the dynamic nature of SDIs may mean that the concept of 'complete' is inappropriate. One high-level definition of a successful ASDI would be an environment in which the community (government agencies, businesses and citizens) has efficient access to the fundamental spatial data needed for economic, social and environmental development. This definition, similar to the new ANZLIC ASDI vision, focuses on the end-user and uses (outcomes of the existence of a functioning SDI), rather than the inputs. A key issue for National SDI builders such as ANZLIC is to identify which inputs that they can influence (data, access systems, pricing policies etc) that will yield the most beneficial outcomes in the shortest time.

Whatever choices are made in developing the ASDI, a strong national institutional framework is essential. Recent developments, particularly the establishment of an ANZLIC National Office, incorporation of PSMA and formation of ASIBA, have all greatly strengthened the institutional framework. The spirit of cooperation that exists within ANZLIC, and the willingness to question orthodox views and openly review progress, bode well for the future development of the Australian Spatial Data Infrastructure.

8.7 REFERENCES

ANZLIC, 1996, Spatial Data Infrastructure for Australia and New Zealand, Online. <www.anzlic.org.au/asdi/anzdiscu.htm> (Accessed August 2000).

ANZLIC, 2000, Proceedings of Clearinghouse Workshop, Adelaide, Australia, Online. <www.anzlic.org.au/archive/> (Accessed August 2000).

Commonwealth of Australia, 2001a, *Positioning for Growth – Spatial Information Industry Action Agenda*, www.isr.gov.au/agendas/sectors/siiaa/index.html

Commonwealth of Australia, 2001b, Spatial Data Access and Pricing – Report of the Commonwealth Interdepartmental Committee. Online. <http://www.csdc.gov.au/cwlth_access_pricing_policy.htm> (Accessed August 2001).

Price Waterhouse, 1995, Australian Land and Geographic Data Infrastructure – Benefits Stud. Online. < www.anzlic.org.au/archive/benefits.htm> (Accessed April 2000).

CHAPTER NINE

State SDI Development: A Victorian Perspective

Bruce Thompson, Mathew Warnest and Cathy Chipchase

9.1 INTRODUCTION

Victoria, in partnership with local government and other Australian states and territories, is driving spatial data infrastructure (SDI) development through unique intra- and inter- jurisdictional partnerships at the local, state and federal levels of government. Although the benefits of partnerships between levels of government are well recognized within the SDI concept, the number of successful partnerships that have been forged has been limited (Jacoby *et al.*, 2002). To reinforce SDI concepts, this chapter will introduce Victoria's SDI initiatives and successful partnerships led by the principle agency responsible for land administration in the State, Land Victoria. The chapter demonstrates how Land Victoria is meeting the spatial information requirements of local and state government, private sector, and the community. It also sets out the Victorian view that a comprehensive SDI has both transactional and 'whole dataset' perspectives, and that these approaches need to be tightly integrated.

9.2 STATE SDI IN VICTORIA

In Australia, the eight individual states and territories (with some minor variations) assume responsibility for land administration functions including land title registration, administration of freehold and Crown lands, valuation, survey and cadastral infrastructure and mapping. The land administration functions within Australia's states and territories are dependent on sound SDI and complete, high quality information resources. Victoria recognizes that development of the State's spatial information resources will enhance the land administration systems and increase confidence in the land market that underpins the State's economy.

9.2.1 Victorian Land Administration

Land Victoria is Victoria's land administration agency. Land Victoria is a division of the Department of Sustainability and Environment (DSE) (formerly the Department of Natural Resources and Environment). DSE is responsible for balancing the development and protection of Victoria's natural and cultural resource base. Its role includes resource and industry development, land identification and administration, and the protection, conservation and management of Victoria's natural and cultural environment.

The formation of Land Victoria coincided with state and territory and national initiatives towards modernization and automation of land administration in Australia (Jacoby *et al.*, 2002). It provided the opportunity for the State to re-engineer its land administration orientated business processes and to streamline relevant legislation (Newnham *et al.*, 2001). Rationalization would consequently reduce costs to government, reduce transaction costs for business and consumers, increase the competitiveness of the State and increase industry growth due to more efficient use and availability of land information. Government review at that time concluded that Land Victoria should 'maximize the value created from land' by leading, facilitating, negotiating or doing, depending on the ability of the land market to deliver a function and the nature of the value it creates (Newnham *et al.*, 2001). The concept of supporting and encouraging the spatial information industry (particularly the private sector), rather than assuming all roles, has remained central to spatial information policy in Victoria.

The re-engineering of business processes through the creation of Land Victoria coincided with the Victorian Government's commitment to the National Competition Policy (NCP), in accordance with the Competition Principles Agreement signed by the Council of Australian Governments in April 1995 (Southbridge, 1997). NCP was the basis for many reforms undertaken throughout Australia with jurisdictions taking advantage of Commonwealth support to review dated and obsolete legislation particularly those related to land administration (Williamson *et al.*, 1998). The business process review and the State's support of NCP based review programs, led to recognition that government's vision for electronic service delivery was a key driver through which land administration reforms could be achieved. The reform-based projects, some of which are discussed later in conjunction with the Land Exchange (Section 9.5.1), have not only improved land administration business processes and procedures but have assisted in establishing comprehensive SDI for the State (Newnham *et al.*, 2001).

9.2.2 Roles and Responsibilities for Developing Victorian SDI

The information and services flowing from spatial information benefit all Victorians. The Victorian spatial information industry includes all levels of government, the private sector, academia and the community. Accordingly, all have a role in developing and implementing the Victorian spatial information infrastructure (Victorian SDI). Land Victoria maintains the overall responsibility for development and implementation of the strategy, with key roles being played by the Victorian Geospatial Information Reference Group (GIRG), by the Australian Spatial Industry Business Association (ASIBA, see also Section 8.4.4) and the recently formed Australian Spatial Industry Education and Research Association (ASIERA, see also Section 8.4.6).

GIRG members, who are representative of the range of participants in the spatial information industry, contribute their personal and industry experience to the development of Victorian SDI and to the implementation of the Victorian spatial information strategy. GIRG's primary role is to provide independent advice to government and to act as a dissemination and feedback mechanism for the spatial information industry with regard to the development and implementation of government policy. In conjunction with Land Victoria personnel, GIRG, ASIBA and ASIERA fulfill the major roles in the collaborative development of Victoria's SDI.

9.2.3 State Spatial Information Policy

A comprehensive Geographic Information System (GIS) planning study was undertaken by Tomlinson Associates for Victoria in 1992 (Chan and Williamson, 1995; Jacoby *et al.*, 2002). The report (OGDC, 1993a) encompassed approximately 40 state agencies and reviewed 270 datasets, recommended an investment of AUD 56 million over a six year period in return for a fully discounted benefit to the State of up to AUD 312 million.

The Tomlinson review was a seminal study for the spatial information industry in Victoria. It demonstrated the importance and value of land information. It was perhaps the first time the State fully recognized the potential value of investment in the establishment of SDI and that in order to harvest that opportunity, reprioritization was necessary.

Another important outcome of the Tomlinson report (OGDC, 1993a) was that the digital cadastral map base was the State's highest priority spatial dataset. Identified in the review were several requirements for content and quality for the digital cadastral map base (OGDC, 1993b):

- That the map base should be an authoritative statewide coverage of the distribution of land units in the State, with standards for accuracy, quality and data transfer;
- Clear, current and correct identification of both parcel and property units;
- Supporting land unit attributes, eg. area and dimensions, with other attributes accessible via linkages to various agency databases;
- Centralized dataset maintenance and the communication of updates to land unit boundaries, unit identifiers and related attributes;
- Support for the communication of land information between agencies; and the ability to perform topologic (spatial) analysis.

Perhaps most importantly, the Tomlinson review led to the formulation of the 'point and click' vision, which has since formed the core of Victoria's spatial information strategy. The 'point and click' vision refers to the use of spatial objects as an index, or means of locating associated information. For land, this means being able to use the parcel/property base to access a range of relevant information, including address, sales, titles, valuations, asset and planning attributes.

These early outcomes and the recognition of the need for statewide SDI led to the development of a Victorian spatial information strategy (Land Victoria, 1999a). The Victorian spatial information strategy is the mechanism by which the Victorian Government continues to look ahead and improve awareness and access to spatial data and, in particular, its framework datasets. Tomlinson's initial review has been followed by two spatial information strategies, each with a three-year lifespan, with a third for the period 2003-2006.

The first strategy was the Victorian Geospatial Information Strategy 1997-2000 (VGIS 1997-2000) (Land Victoria, 1997), which aimed for the creation and maintenance of high-quality framework datasets for Victoria, and for growth in private sector businesses working with spatial information. In consultation with the State's central budget agency (the Department of Treasury and Finance), the strategy set out a comprehensive pricing and licensing policy for spatial information. The policy was geared for cost recovery through the application of licence fees for spatial data and was carefully balanced to achieve two conflicting requirements: that access to and use of spatial data should be maximized, and for cost not be an impediment to use; and that the quality, reliability and currency of the State's fundamental spatial information infrastructure be assured, through application of revenue raised from user license fees.

The second strategy is the Victorian Geospatial Information Strategy 2000-2003 (VGIS 2000-2003) (Land Victoria, 1999a), which further consolidated the creation and maintenance of high-quality framework datasets for Victoria, and established best practice information management principles for spatial information, including custodianship, metadata, access, pricing, licensing and spatial accuracy.

The third strategy, the Victorian Spatial Information Strategy 2003-2006 (VSIS 2003-2006) (Land Victoria, 2002a) will be focused on online service delivery, and maximising whole-of-government and whole-of-sector approaches to spatial information.

9.2.4 Victorian Spatial Information Strategy

The objectives of the Victorian spatial information strategy are to contribute to Victoria's development, wealth creation and protection of natural resources through: maintenance and enhancement of a comprehensive spatial information resource; maximising value and capability of this spatial information resource into Victoria's information systems and processes (Land Victoria, 1999b). These objectives are based on the strategy vision that "all Victorians can access and use the spatial information they require" (Land Victoria, 1999a). As previously noted, the strategy is reviewed and recast every three years. The strategy for 2000-2003 (VGIS 2000-2003) comprises eight key components, outlined in Figure 9.1: Framework Information, Key Business Information, Custody, Metadata, Access Infrastructure, Pricing, Spatial Accuracy, and Awareness (Land Victoria, 1999a).

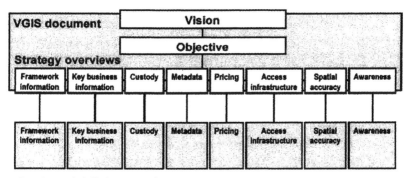

Strategy detail documents

Figure 9.1: Key Strategy Detail Documents of VGIS 2000-2003 (Land Victoria, 1999a)

(a) Framework Data

A framework approach towards developing SDI at the national level was put forward by the coordinating body for geographic information across the United States (FGDC, 2001). The framework consists of seven themes of digital geographic data that are commonly used and the procedures and guidelines that provide for data integration and sharing, institutional relationships, business practices that encourage the use and maintenance of that data. "The framework represents 'data you can trust' the best available data for an area, certified, standardised, and described according to a common standard. It provides a foundation on which organizations can build by adding their own detail and compiling other datasets" (FGDC, 2001).

Land Victoria has adopted a similar approach with the Victorian spatial information strategy. The framework data component of the strategy identifies eight fundamental datasets for Victoria with the addition of address information to the framework employed by the United States (FGDC, 2001). Framework data is that considered fundamental to the development and operation of Victoria's SDI, in that other (business) information cannot be created or maintained without it. The eight framework datasets are presented and discussed later as the components of Vicmap.

(b) Key Business Information

Key business information is spatial information (other than framework datasets) considered valuable to the development and operation of Victoria's SDI. It refers to value-added products created from or maintained with framework data, and to other products that make use of framework data for context. Key business information is developed and maintained primarily for commercial gain, or which has a revenue component. Pricing of key business information is determined by market forces and is generally specific to a single or few application areas. Whereas framework data should be priced so that its cost is no impediment to use, business information will be priced according to normal commercial principles.

(c) Custody

The Victorian strategy recognizes custodianship is the heart of spatial information management because it provides accountability for information products and identifies authoritative sources, giving users a measure of consistency and certainty. Custodianship is regarded as a means of eliminating unnecessary duplication, managing information on behalf of others and can be attributed to the provision of a sound spatial information infrastructure, assisting the creation and management of spatial information products, and facilitating the acquisition of information products.

The State's custodian agencies manage information products as trustees in a partnership with national, regional and local providers and users to enable the integration of spatial information for the benefit of the entire community. Custodianship reinforces the concept of one agency being ultimately responsible for an information product that others might use, giving users confidence in the level of integrity, timeliness, precision and completeness of that information product.

(d) Metadata

Metadata is the key management mechanism for Victoria's spatial information environment. Metadata is simply defined as 'information about information'. Metadata facilitates fundamental information management at three levels; Discovery–enabling users to locate and evaluate information; Management–enabling custodians to better manage their spatial information; and, Utilization–enabling users to access and manipulate information by means of automated and distributed systems. Additional metadata recorded can consist of detailed information about data collection methods, information about the accuracy of sources or the processing history, and archival procedures. Metadata can be in the form of descriptions of the content, quality and geographic extent, information about projection and scale, and summary descriptions of content and quality.

(e) Access Infrastructure

Access infrastructure should provide a simple, effective means of locating and obtaining spatial information. Non-digital (paper-based) information and non-Internet forms of access and distribution are accepted as traditional forms of access. However, the primary method of locating and obtaining spatial information is proposed to be via the Internet. Simple and effective access maximizes use. The starting point for access is metadata, proving a simple and effective means of locating information. The Victorian and Australian Spatial Data Directories (VSDD, 2002; ANZLIC, 2002) provide a comprehensive index of Victorian spatial information. This allows users to determine the best available information for their purpose, and then, either directly via the Internet, or via some other delivery mechanism (digital or non-digital medium) access their chosen datasets. Progress towards the objectives of an automated identification and retrieval system will be iterative, with development of comprehensive metadata capability as a priority, followed by progressive development of online delivery capability.

(f) Pricing and Licensing

Spatial information is used in both the public and private sectors to support business operations or to inform business decisions, and for applications such as emergency services and infrastructure planning. As such it has a clear value, which in most cases is the primary determinant of price. Within the Victorian spatial information strategy three categories of spatial information are considered and the pricing policies are based on several principles:

- Spatial information provided on a commercial basis, by either the public or private sectors - pricing is a matter to be determined by the provider;
- State government spatial data for which a price has been set down in a statutory or regulatory process - the pricing will be as specified in the statute or regulation; and
- Any other state government spatial information - is to be determined in accordance to government policies and guidelines on the setting of fees and charges. There is the opportunity for agencies to generate revenue to provide for development and maintenance of spatial information to the standards required by users.

Pricing and licensing provides the commercial and legal platform required to safeguard the interests of both providers and users alike. Other relevant considerations are community service obligations, offsets, promotion of industry development; and the need to ensure that price is not an impediment to use.

Licensing spatial information is undertaken to ensure effective management of risks associated with the use of spatial information; and to control the use of licences and other agreements that clearly set out the terms and conditions of for use of the spatial information. This translates to the pricing of framework information being the cost of ongoing maintenance and continuing development to meet user needs. The initial cost of construction is deemed a sunken cost.

(g) Spatial Accuracy

Spatial Accuracy refers to real-world features represented as digital data and the ability to accurately reflect their real-world spatial relativities both within and across underlying framework datasets and key business information. The aim of the Victorian spatial information strategy is to achieve certainty in specifying the level (standard) of spatial accuracy and certainty in evaluating the confidence level for spatial information derived from merging two or more datasets (Land Victoria 1999a).

Guidelines have been established to minimize the effects of any existing (historical) inaccuracies and to progressively develop a network of datasets that will seamlessly interrelate both horizontally and vertically. The anticipated outcome is a level of client confidence, whereby any user will know that they may choose any of the framework or key business information datasets, knowing them to be fully integrated, seamless in structure and meeting a level of accuracy suited to their required application.

(h) Awareness

The Victorian spatial information strategy aims to educate the community of the benefits of spatial information, to ensure that spatial data is easily obtainable and easy to use. Promotion of the strategy as the government standard for the management of spatial information is an ongoing task. The strategy is promoted through GIRG forums, joint government-industry NewTech conferences held throughout the State, meetings with managers of spatial information datasets, various spatial information industry conferences and forums held in Melbourne.

To augment the strategy promotion and broaden public awareness about spatial information, Land Victoria launched the Land Channel website in April 1998, as an online 'one-stop-shop' for land information in Victoria. Land Channel sought to deliver government information and services in terms of subjects, themes and life-events, rather than government structures. A life-event, such as buying or selling a house, would bring together, in a single location and logical sequence, all the elements of government information available to assist a member of the public in the task. Land Channel provides access to spatial information both for viewing and purchase. Its content is land information centric, with valued contributions from other government departments and agencies, including: Department of Infrastructure, Department of Justice, State Revenue Office, Sustainable Energy Authority, Parks Victoria, local councils and utilities.

Other strategy initiatives promoting the awareness and value of spatial information include:

- The development of a Victorian Geospatial Information User Group;
- Statewide primary/secondary spatial education programs;
- Continuation of Land Victoria's research and development partnership with tertiary institutions in Victoria;
- Training for business, with particular emphasis on the marketing and market research sectors; and
- Targeting application of spatial information for key high profile applications/projects/events.

9.3 LINKAGES TO THE NATIONAL SPATIAL INFORMATION SECTOR

The Victorian spatial information sector includes all levels of government, private industry and academia. The information and services flowing from spatial information benefit all Victorians. Accordingly, all have a role in the implementation of the Victorian spatial information strategy and development of State SDI. Implementation of the strategy requires the commitment and active involvement of all stakeholders and involves participation in three key strategies discussed in this section. It is expected that all levels of government and other sectoral entities will participate by mutual agreement, for mutual benefit.

9.3.1 Growing the Private Sector

The Federal Government through Senator Nick Minchin, the Federal Minister for Industry, Science and Resources, released the Spatial Information Industry Action Agenda to Parliament in September 2001 (ISR, 2001a). Land Victoria was actively involved in the scoping and preparation of the Action Agenda (ISR, 2001b). Land Victoria's industry development goals, particularly the data service provider and the market access initiatives, and pricing and access strategies were a significant influence on the direction of the Action Agenda.

The Victorian spatial information strategy is fundamentally concerned with the promotion and development of the spatial information industry—it will not achieve its objectives without the development of a robust and considerably expanded spatial information industry. It has a particular focus on industry development within its framework information, access infrastructure and pricing strategies. These strategies are, in turn, focussed on enabling online delivery, and in providing simple and efficient access to government spatial information and services.

At the state level, industry development and growth is actively pursued. The strategy promotes private sector growth by providing simple and effective access to affordable, quality spatial information. The Victorian Government maintains a policy to utilize the private sector where possible for the provision of services to government and the public. Vicmap Property, Address, Roads, and Hydrography are all datasets maintained by the private sector. This allows Land Victoria to focus on the development and promotion of these while allowing the private sector the opportunity to expand and grow their businesses through active participation in the maintenance of these fundamental datasets. Land Victoria acts as a data wholesaler, encouraging the private sector to act as data service providers and value added resellers. Only where resellers are unable or unwilling to service the market does Land Victoria step in.

9.3.2 Participation in Standards Development

To foster wider usage of the framework datasets, Land Victoria has undertaken a number of initiatives that will further improve their content and detail and provide direct benefits back to the State. Active involvement in standards development leads to consistent data structures, ease of exchange, higher quality content and greater acceptance. Programs such as the Property Information Project and Rural Addressing have improved the awareness, functionality and quality of information within the framework datasets, providing significant additional benefits in day-to-day service delivery, public safety and further development of state and local government infrastructure.

Participation in standards development related to spatial information is an important aspect of the Victorian spatial information strategy. Land Victoria personnel actively participate on a number of state and national bodies and are involved in the development of statewide, Australian and New Zealand and International standards. Land Victoria personnel also participate in Standards Australia's IT/4 Geographical Information Sub-Committee. IT/4 acts as a reference group providing input on any proposed Australian/New Zealand standard related to geographical information. IT/4 members through Standards Australia also provide input on proposed International (ISO) Standards related to geographical information. There are presently more than 27 proposed ISO standards relating to geographic information in various stages of progression and a number of Australian/New Zealand standards under development or review.

9.3.3 Jurisdictional Linkages

Land Victoria staff participate in the Intergovernmental Committee for Surveying and Mapping (ICSM) working groups, developing spatial information standards that will be adopted across Australia and New Zealand, several of which will become formal Australian/New Zealand standards. Examples of such work include Geographic Names, Street Addressing, Topography and the Cadastre. Land Victoria maintains representatives on ANZLIC and its steering committees and technical working groups. Land Victoria is a member of PSMA Australia Inc. (PSMA), a collaborative organization of all Australian public sector mapping agencies of the states and territories to create national spatial information products based on the best available data from each of the jurisdictions. PSMA is a registered business, where each jurisdiction is an equal shareholder and each provides a representative on the PSMA board of directors. Other associations include active membership on the Metadata and GDA (Geocentric Datum of Australia) geocentric datum working groups.

9.4 KEY STATE SDI INITIATIVES

The Victorian spatial information strategy, in conjunction with the Spatial Information Action Agenda and standards development, form key linkages between State and National SDI developments in Australia as well as towards local and state implementations. In Victoria, they have found practical expression in a number of key initiatives, including the Victorian Online Title Systems (VOTS), Vicmap, the Property Information Project (PIP), online service delivery (Land Channel), and GPSnet. Although in some respects unrelated, these initiatives have been fundamental to the development and advancement of Victoria's SDI.

9.4.1 Victorian Online Title System (VOTS)

Almost five million paper titles and other land-related records, some dating back to 1863, were converted to electronic format and are now accessible remotely by all Victorians, enabling instant access to information on any piece of freehold land throughout the State. This includes ownership, dimensions, location, any restrictions or caveats on its use, and any mortgage details. This initiative (completed in 2001) has provided significant time and cost savings to Victoria's land administration system, to the property industry and to Victorian consumers.

However, its real potential lies in achieving a fundamental step in advancing Victoria's SDI toward e-government as well as e-commerce.

9.4.2 Vicmap

Victoria has comprehensive spatial information resources. The core of this resource is Vicmap, a suite of eight integrated spatial information products developed from the State's eight framework information products. Tomlinson's nomination of the property cadastre as the key spatial information resource of the State (OGDC, 1993a) presumed inclusion of the truly fundamental geodetic control network for the State, and the value of the cadastre has been substantially enhanced by the inclusion of additional framework information products. For example, water quality and salinity, and water rights are now key issues facing the Victorian Government and community. Tackling these issues will require the concerted consideration and application the cadastral property base, hydrographic and imagery data. For this and similar reasons, Vicmap, comprising geodetic, cadastral, property, address, roads, hydrographic, administrative, elevation and imagery, is more valuable as an integrated product suite than the sum of its individual components might indicate. The Vicmap products are:

- *Vicmap Control* and *Position* – control and positioning services based on Victoria's extensive survey ground mark network, and on GPSnet, Victoria's GPS base station network;
- *Vicmap Property* – spatial and aspatial information for Victoria's 2.7 million properties and 2.5 million land parcels;
- *Vicmap Address* – access points or property addresses for Victoria's 2.7million properties;
- *Vicmap Administrative* – state, local government, electoral, locality, postcode, and parish/county boundaries;
- *Vicmap Transport* – transportation networks in Victoria, including all roads trafficable by four-wheeled vehicles, and rail and tram networks;
- *Vicmap Elevation* – and bathymetry) – those elements required for the description of height including spot heights, contours, and digital elevation models;
- *Vicmap Hydrography* – all surface water features;
- *Vicmap Imagery* – satellite imagery and a range of aerial photography.

Although not yet fully and completely aligned, the maintenance of these datasets is now conducted in a fully coordinated manner, meaning that corrections or enhancements in, for example, Vicmap Hydrography will be fully aligned to the relevant elements in Vicmap Elevation and Transport.

9.4.3 Property Information Project (PIP)

The Property Information Project (PIP) is the foremost success of Land Victoria's relationship based enterprises, and represents best practice example of cooperation across levels of government. Victoria's 78 local governments are working in partnership with Land Victoria to create a uniform view of all property information within the State. The objectives of PIP were to establish a common SDI between all local governments and the State (Jacoby *et al.*, 2002). Prior to PIP there was little to no commonality between local government data and that of the State.

Land Victoria funded the matching and reconciliation of each local government rates database with the cadastral map base, creating a property layer. Each local government is provided with the fully maintained and periodically updated cadastral/property base at no cost. In return local governments agree to adopt Vicmap (the cadastral and property base), allow key property information to be fed into the map base; and advise Land Victoria of all proposed plans of subdivision and changes to property information (e.g. new street addresses) (Jacoby *et al.*, 2002). Unlike other largely statute driven state initiatives, local government involvement in the PIP is voluntary. This is the most important characteristic of the project and, from Land Victoria's perspective, also PIP's greatest risk (Jacoby *et al.*, 2002). Without legislative backing, the uniform local government participation necessary for the success of the project relied solely upon local governments being able to identify and achieve benefits sufficient to justify their involvement.

The clear benefit to the State is the means to populate Vicmap, the State's re-engineered cadastral map base, and deliver a key component of SDI for Victoria (Jacoby *et al.*, 2002). There are over 500 agencies and organizations that utilize the State's cadastral/property base including planners, valuers, estate agents and other land professionals. These users require the most up-to-date information. The most critical of these users are in the emergency services sector, particularly the State's computer aided dispatch system for police, ambulance, fire services and the State Emergency Service. The benefit to local governments has been the consistency and reliability of exchange of information with state government, utilities and others and, to some extent, savings in data acquisition and maintenance costs.

Overall, Land Victoria has been able to dramatically decrease the amount of duplicative maintenance activity occurring throughout the State with the adoption of the PIP centralized data maintenance model. PIP's success ultimately represents a fundamental institutional arrangement upon which to build the vision of an SDI to support the spatial information requirements of local and state government, private sector, and the Victorian community.

In mapping every property in Victoria, PIP encourages the adoption of a standardized system of street addressing. The project meets a growing demand for simplified access to land records using common keys such as council property numbers and street addresses. This lays the framework for municipalities to market their local services and property information.

9.4.4 Online Service Delivery – Land Channel

Online service delivery has become fundamental to contemporary business practice. An Internet site has the capacity to act as a virtual nucleus, effectively gathering from otherwise unconnected participants or systems all the information elements necessary to support a process or complete a transaction. The process of coordinating and integrating unconnected participants and systems is essentially an information management exercise, with most of the complexity revolving around non-technical issues. The emerging business-to-business and business-to-consumer models are the prime examples of the first generation of online service delivery.

Land Victoria has implemented online delivery of spatial information products and services through Land Channel website (http://www.land.vic.gov.au) and the Land Channel Services initiative. Land Channel is now an integral part of the Victorian government's online service delivery, providing the community with a whole-of-government portal for land related information across the State. Land Channel Services is specifically devoted to spatial information products and services. Land Channel now generates in excess of 400,000 page impressions per month, and more than 55,000 property reports (the most common spatial information service provided) are generated each month. This level of use and application demonstrates the value of online service delivery for spatial information, and also provides a powerful marketing and awareness device for the spatial information industry.

9.4.5 GPSnet

GPSnet is rapidly developing as a key component of Victoria's SDI. It provides a fundamental spatial reference system that is open and accessible but at the same time meets consistent standards of accuracy, service and quality. GPSnet in its current arrangement is for a network of continuously operating GPS base stations located across the State to support reliable and accurate Global Positioning System (GPS) position determination. The move to augment the traditional physical ground marked geodetic network with GPS infrastructure began in 1995.

Under the umbrella of the Victorian spatial information strategy (VGIS 2000-2003), a geodetic strategy was implemented to develop a geodetic framework that provides 'easy access to accurate positioning information to a consistent standard across the entire state for all users'. This strategy is being implemented in concert with national infrastructure coordination led by the Australian GNSS Coordination committee (AGCC) in order to maximize use and application for all users of satellite positioning.

The development of GPSnet has been coordinated and facilitated by Land Victoria through a private-public partnership approach ensuring that duplication of infrastructure has been kept to a minimum and at the same time delivering an infrastructure that is capable of multi-modal operation. The modes currently being exploited include surveying, mapping and weather research, and can be extended to such applications as timing and synchronization, automation of construction and precision agriculture. It is expected that many more users will continue to come from the land, water, environmental, natural resource management, transport, fleet management, emergency response and navigation industries as well as the recreational sector. The preferred positioning technology for these sectors is largely GPS, with many prepared to erect their own GPS infrastructure in the absence of a coordinated and cooperative network such as GPSnet.

GPSnet records, distributes and archives GPS satellite correction data for accurate position determination, and allows users to determine positions that are compliant with the Geocentric Datum of Australia (GDA). The GDA has been developed and implemented across Australia since 2000 with GPS compatibility as a specific objective. GDA compliance has been achieved by linking GPSnet to a national network of base stations called the Australian Fiducial Network (AFN) which in turn is a part of a larger regional network of base stations called the Australian Regional GPS Network (ARGN) operated by Geoscience Australia (a federal government body). Land Victoria aims to cover the entire State with GPS infrastructure at an average spacing of approximately 160km in rural Victoria and 50km in the greater Melbourne metropolitan region. This level of station density is expected to require approximately twenty GPSnet base stations.

The trend to wireless applications and location based services will drive the need for GPS infrastructure, and the critical role of GPSnet in enabling Victoria's SDI will continue to grow in importance. GPSnet is managed online and delivered to users and via the Land Channel portal on the Internet. Research is being conducted on advanced base station networking solutions to deliver very high accuracy solutions wirelessly to users over large areas of operation.

9.5 VICTORIA'S SDI: NEXT STEPS

The development of Victoria's SDI has in large part progressed from an infrastructure building stage to an application development and integration stage. The key infrastructure elements include VOTS, Vicmap, PIP, and GPSnet, discussed in 9.4. Some key infrastructure building components remain: completion of rural addressing, the full implementation of standard parcel identifiers, and the complete integration of Crown and freehold parcel bases.

Land Channel represents the first provisional steps towards broad-scale application and use of this comprehensive infrastructure. The future for Victoria's SDI lies in the development of the Victorian spatial information strategy for 2003-2006 (VSIS 2003-2006), and with the development and implementation of Land Exchange, the foremost infrastructure and application integration initiative. The Victorian spatial information strategy (VSIS 2003-2006) will focus on comprehensive whole-of-government and whole-of-sector initiatives aimed at providing comprehensive SDI for Victoria, and for cementing SDI as part of Victoria's economy and economic resources.

9.5.1 Land Exchange

The Victorian Government has approved the establishment of a Land Exchange to operate as a new gateway to Victoria's AUD 28 billion-a-year property industry (Land Victoria, 2002b). The Land Exchange will be a first for Australia, providing business and the community with a single point of online access to integrated state and local government land information and transactions. Designed to operate as a comprehensive online market place, where parties can exchange land-related information and perform transactions via the Internet in a safe and regulated environment (refer to Figure 9.2).

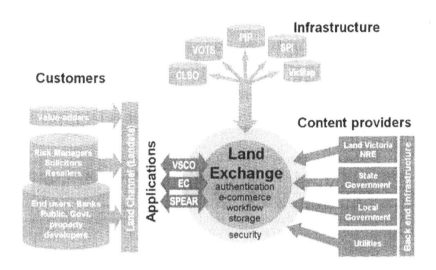

Figure 9.2: Land Exchange Inter-Relationships Between Partners and Projects
(Land Victoria, 2001)

The benefits of Land Exchange to Victoria will be considerable through savings of more than AUD 200 million per annum to users of land related information, including vendors and purchasers and the property, legal, conveyancing, finance and development industries; faster and easier property dealings throughout Victoria (Land Victoria, 2002b). The Land Exchange will provide a coordinating avenue for customers to access and use the content without the responsibility, custodianship and revenue generated by content providers being transferred to the Land Exchange. Development of the Land Exchange will not involve stopping 'over-the-counter' services currently provided by government agencies nor compromise existing statutory obligations with regard to land-related information and transactions.

Several key projects remain to be completed and integrated to enable the Land Exchange to work providing the technical framework and environment for the 'point and click' vision. The key projects include:

- *Electronic Conveyancing (EC)* – online conveyancing system for transferring ownership in land, including online settlement and lodgement and registration of interests.
- *Vendor Statement Certificates Online (VSCO)* – Online delivery of vendor statement certificates on planning, conservation, water, power, roads, and heritage trust matters that are necessary for most property transactions.
- *Streamlined Planning via Electronic Applications and Referrals (SPEAR)* – An online system to manage the planning, building and subdivision applications and referral process. SPEAR will automate the application/planning permit process so that applicants can electronically lodge

their applications and information with the council and councils can electronically refer the applications to referral authorities.

- *Standard Parcel Identifiers (SPI)* – Standard parcel identifiers that give every property parcel in the State a unique identifier, enabling connectivity between the Titles Register and Victoria's digital map base, Vicmap.
- *Crown Land Status Online (CLSO)* – Full and final integration of the State's crown land holdings with the freehold land to form a single integrated cadastral map base for the State.

The first three are strategic applications to form the core of Land Exchange; the last two constitute the bulk of the remaining infrastructure necessary to enable Land Exchange.

A fundamental characteristic of Land Exchange is that it is transaction orientated. Most SDIs retain a strong focus on datasets and on the entities and agencies that maintain, develop and distribute datasets, with a commensurate expectation that transactions will occur externally to the SDI. Land Exchange seeks to internalise SDI functionality into 'mainstream' business applications and processes. The reasons for this are twofold: it provides savings and benefits, and it promotes spatial information and assists in ensuring its value is recognized and supported.

9.6 CONCLUSION

Land Victoria's spatial data initiatives, which incorporate the Victorian spatial information strategy and the Land Exchange framework projects are driving SDI development within the State. Through active participation in national forums, including PSMA, ICSM and ANZLIC, as discussed in Chapter 8, Land Victoria is contributing significantly to SDI development nationally.

The immediate and potential benefits flowing to all Victorians from spatial information have been recognized. The Victorian spatial information strategy for 2000-2003 has proven successful in setting a vision for the future, in providing best practice information management principles, and in outlining the strategies required to realize the vision. The eight framework datasets are complete or nearing completion. Custodianship arrangements have been established and their ongoing maintenance and development are progressing. The private sector in Victoria is growing and is repositioning to take advantage of the new opportunities presented by development of the State's SDI initiatives. These achievements can in part be attributed to the success of Land Victoria initiatives such as Vicmap, PIP, VOTS and GPSnet, and to its active participation in standards development. In particular, PIP highlights how the formation of successful partnerships facilitates and ensures successful achievement of common objectives in the spatial information industry. Online service delivery through the Land Channel web site will continue to build community awareness and provide access to Vicmap and other spatial information products and services, and act as the first point of entry to locate further land and property information.

The Victorian Government supports the spatial information private sector and recognizes that private sector growth returns benefits to the wider community. To this end, the Victorian Government through Land Victoria will continue to consult with, encourage contributions from and partner with the private sector. Similarly, the growing importance of the academic sector in research and development, in education, skills development and in promoting awareness mean that Land Victoria will continue to develop its partnership with the academic sector.

9.7 REFERENCES

ANZLIC, 2002, ASDD Website, Geoscience Australia. 2002, Online. < http://www.auslig.gov.au/asdd/about.htm> (Accessed June 2002).

Chan, T. O. and Williamson, I.P., 1995, A Review of the GIS Planning Methodology for Victoria - Its Relevance to Real Life Situation, *Proceedings of the 5th South East Asian and 36th Australian Surveyors Congress*, 16-20 July, Singapore : pp. 185-212.

FGDC, 2001, Framework - Overview: what the Framework Approach Involves, Federal Geographic Data Committee, Online. < http://fgdc.er.usgs.gov/framework/framework.html> (Accessed November 2001).

ISR, 2001a, Media Release: Senator Nick Minchin, Federal Boost for Spatial Information Industry. Office of Senator Nick Minchin, Department of Industry Science and Resources, Canberra.

ISR, 2001b, Spatial Information Action Agenda "Positioning for Growth". Canberra, ACT, Department of Industry Science and Resources, Canberra : p. 123.

Jacoby, S., Smith, J., Ting, L. and Williamson I.P., 2002, Developing a Common Spatial Data Infrastructure between State and Local Government - An Australian Case Study. *International Journal of Geographical Information Science.* Vol. 16, No. 4 : pp. 305-322.

Land Victoria, 2002a, Draft Victorian Spatial Information Strategy 2003-2006, Land Victoria, Department of Natural Resources and Environment, Online. <http://www.land.vic.gov.au/spatial/> (Accessed November 2002).

Land Victoria, 2002b, Go Ahead for Land Exchange, Department of Natural Resources and Environment, Online. <http://www.land.vic.gov.au/> (Accessed September 2002).

Land Victoria, 2001, Land Exchange Business Model, Department of Natural Resources and Environment.

Land Victoria, 1999a, Victorian Geospatial Information Strategy 2000-2003, Land Victoria, Department of Natural Resources and Environment, Online. <http://www.land.vic.gov.au/spatial/> (Accessed September 2002).

Land Victoria, 1999b, Geospatial Information Custodial Guidelines for Victoria, Victorian Geospatial Information Strategy 2000-2003, Land Victoria, Department of Natural Resources and Environment, Online. <http://www.land.vic.gov.au/spatial/> (Accessed September 2002).

Land Victoria, 1997, Victorian Geospatial Information Strategy 1997-2000, Land Victoria, Department of Natural Resources and Environment.

Newnham, L., Spall, A. and O'Keeffe, E., 2001, New Forms for Government Land Administration - Land Victoria, A Case Study of the Trend Towards Combining

Land Administration Functions and the Resulting Benefits to the Community. *FIG Working Week 2001*, Seoul, Korea.

OGDC, 1993a, GIS Strategy Report 1993, State Government of Victoria - Strategic Framework for GIS Development. Tomlinson Associates Ltd for Office of Geographic Data Coordination, Department of Treasury and Finance.

OGDC, 1993b, Report No.2, GIS Planning - Land Status and Asset Management, State Government of Victoria - Strategic Framework for GIS Development. Tomlinson Associates Ltd for Office of Geographic Data Coordination, Department of Treasury and Finance.

Southbridge, 1997, National Competition Policy Review of the Surveyors Act 1978. Melbourne, Victoria, Department of Natural Resources and Environment : 93.

VSDD, 2002, Victorian Spatial Data Directory, State Government of Victoria. Online. <http://www.land.vic.gov.au/web/root/domino/cm_da/lcnlc2.nsf/frameset/spatial > (Accessed December 2002).

Williamson, I.P., Chan, T.O. and Effenberg, W.W., 1998, Development of Spatial Data Infrastructures - Lessons Learned from the Australian Digital Cadastral Databases, *GEOMATICA*, Vol. 52, No. 2 : pp. 177-187.

CHAPTER TEN

SDI Development: Roles of Local and Corporate SDIs

Tai On Chan and Rick Whitworth

10.1 INTRODUCTION

Researchers and various agencies have attempted to capture the nature of SDI in definitions and models produced in various contexts. In particular the umbrella and building block views of SDI provide both a top-down and bottom-up approach to model SDI development in a hierarchical manner (Rajabifard *et. al.*, 1999). The productional perspective of SDI also provides a conceptual tool that draws on the GIS literature to model SDI development (Chan *et al.*, 2001). In this perspective SDI is described as a composite entity that is made up of modules of corporate GIS that assume the role of either a business process or an infrastructure in a spatial information industry.

This chapter applies these concepts to the cases of SDI development at a local level involving the Council of Greater Geelong and the Department of Natural Resources and Environment (DNRE), both in the State of Victoria in Australia. The chapter concludes by highlighting how experience of SDI development involving the two organizations can be applied to managing SDI development in general.

10.1.1 The Concept of SDI

Chan and Williamson (1999a) argue that SDIs exist as a hierarchy. Rajabifard *et. al* (1999) published two views, an umbrella view and a building block view that describe the nature of the hierarchy that inter-connects SDIs at corporate, local, state, national, regional and global levels. As it is more efficient for other levels of SDIs to draw on spatial datasets from the corporate SDIs, the corporate SDIs form the base level in the hierarchy of SDIs as illustrated in Figure 10.1.

These simplified views do not model the many complex relationships that operate between jurisdictions, but they have significant relevance in contrasting some key factors involved in the relationships between corporate, local and state levels.

The umbrella view represents the necessary organizational framework, technical standards and access network required to support data sharing at the lower levels. This top-down view addresses the need to have mechanisms at each level to coordinate all of the components required and to recruit the participation of those below.

The building block view supports the assumption that more detailed data is required at the lower level with a less detailed generalised view operating at higher levels. In moving down the hierarchy, jurisdictions contribute more to collecting and maintaining the extra detail and quality that they need to operate at their particular level.

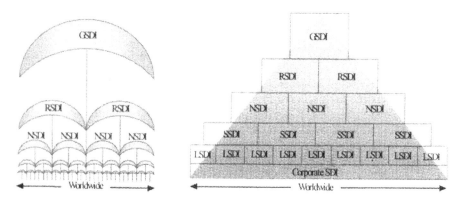

Figure 10.1: Hierarchy of SDI: Umbrella View (left) and
Building Block View (right) (Adapted from Rajabifard *et al.* 1999)

Within the local level, Chan *et al.* (2001) identify similarities between corporate GIS and SDIs within organizations, including the relationships of organizational or 'corporate' SDI between different organizations. Figure 10.2 illustrates such a model in which the building blocks are individual corporate SDIs that are both users and/or suppliers of spatial data and technology. These corporate SDIs interact progressively with one another as members of the spatial information industry, in various production processes of a jurisdiction to fulfil its range of business objectives—social, economic and environmental.

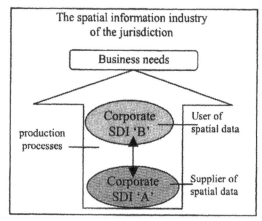

Figure 10.2: Productional Perspective of an SDI

The interaction of the corporate SDIs are said to have a 'productional perspective' (Chan *et al.*, 2001). The production processes are represented by the large arrow in Figure 10.2 and are also referred to as functional areas in this chapter. The interacting corporate SDIs can be visualised as a network of value-adding chains of suppliers and users of spatial data/technology.

While the model has been applied to SDI development in Australia, Chan and Williamson (1999a) argue that this means of SDI development is underpinned by special partnership arrangements. The value and need for partnership arrangements is clearly evident in both of the following cases of the City of Greater Geelong and Department of Natural Resources and Environment in Victoria, Australia.

10.2 LOCAL SDI DEVELOPMENT AT GEELONG

The development and use of spatial data across local authorities in the State of Victoria, Australia is varied. Size, density, local expertise and urban versus rural location are all factors that relate to the ability of authorities to embrace the technology. The following is a case of development of SDI at Geelong in Victoria.

10.2.1 History of Geelong SDI Development

In most urban areas throughout Australia, the cadastre was originally captured by the regional utilities responsible for water supply and sewerage in order to meet the need to automate accurate spatial recording of their underground assets. In the decades before the nineteen nineties the Melbourne Metropolitan Board of Works for example, undertook a massive program to digitise its large scale property-based reticulation plans.

The solution undertaken by a regional authority, the Geelong and District Water Board and its impact on the development of a Local SDI within the region is of particular interest. This authority took a specialised approach to solving the problem by capturing an integrated and highly accurate form of coordinated cadastre in reticulated areas.

By 1995, economic rationalism resulted in some major administrative changes that assisted development of the SDI in Victoria. Responsibility for the delivery of water and sewerage services in metropolitan Melbourne was split up into retail agencies and the management of the State Digital Cadastral Database eventually became the centralised responsibility of Land Victoria, a Division of the Department of Natural Resources and Environment of the State Government of Victoria, with data maintenance services provided by a private contractor.

At the same time, the areas of responsibility of the rural water authorities were also redefined with more consideration given to catchment management boundaries and serviceable regions. Barwon Water was created in the Geelong region and it now provides water and sewerage services over an area of 4,600 km^2, with 130,000 properties and 250,000 customers in five Councils including the City of Greater Geelong.

By that time, Barwon Water was well advanced with the development of its Property and Facilities Information System (ProFIS). Up until this time Barwon

Water departments had been responsible for completing all sewerage and water reticulation design and construction work within its area of jurisdiction and the ProFIS data specification was based on very high standards of accuracy and data integrity capable of use for engineering design and construction. Data capture in reticulated areas was based on a form of highly accurate coordinated cadastre with a full audit trail on all spatial data entities and their attributes and capable of providing transaction updates.

There was also a major move toward rationalising Local Government by creating new regions of responsibility. The State's total number of Councils was eventually to be cut by more than half to 78. The first Council to be created in 1995 by a turbulent amalgamation of six former municipalities was the City of Greater Geelong. It remains Victoria's largest Council and services the State's largest provincial city. It has a major port, both heavy and light industry, low to medium density residential development, rural general farming, coastal holiday resorts, sensitive environmental wetlands, state reserves, national parks and sanctuaries. The Council serves over 100,000 rateable properties covering an area of 1,300 km^2.

Common casualties caused by downsizing during any major rationalization are loss of record keeping knowledge and information. From the outset this possibility, combined with Geelong's large number of widely distributed administrative, operational and customer service centres highlighted the need to integrate corporate information with a high-speed communication network.

Through a cooperative arrangement, a highly significant feature of this network was a 10-megabit communication link between the newly created City of Greater Geelong and Barwon Water, and workstations installed in a number of Council offices to access Barwon Water's Property and Facilities Information System, ProFIS. This link enabled a high level of collaboration between the departments of the City of Greater Geelong and Barwon Water for the development of a Local Spatial Data Infrastructure (LSDI). Over the following four years both organizations received a range of benefits from this arrangement including:

- Council access to a wide range of Barwon Water spatial datasets and services
- Development and use of an interface for Council's Planning Unit to capture and maintain the statutory planning scheme on ProFIS
- Interfaces to allow batch plotting high quality Planning Maps, create Planning Certificates, create and track Planning Amendments
- Barwon Water access to current Statutory Planning data in the Geelong area.
- Regional cooperation in processing new development, creating street, suburb and property address information
- Use of the parcel information by Council consultants to extract key parameters required for property valuation throughout the region

In 1997, the Property Information Project (PIP) initiated by Land Victoria (see also Section 9.4.3 in Chapter 9), established a common geospatial infrastructure between Local and State Government based around the digital cadastral mapbase (Jacoby *et al.*, 2002). Land Victoria offered funding for Councils to reconcile their rating databases with the cadastral base. The project was based on a number of stages, the final one being an agreement for supply of maintained cadastral data to

Councils in return for updates of their property address details. This initiative, offered to Councils on a voluntary basis was accepted with enthusiasm.

10.2.2 Current Geelong SDI (Components and Organization Model)

In 1999, Council re-evaluated its position by undertaking the development of a Spatial Information System (SIS) Strategy based on an extensive user needs analysis, industry evaluation and business case. This resulted in the acceptance of the business case to develop its own in house Spatial Information System with Barwon Water as the key data service provider. The primary objective of the strategy is the delivery of all available spatial datasets to users in the region through an interface that is fully integrated with other Council data and applications as shown in the model in Figure 10.3.

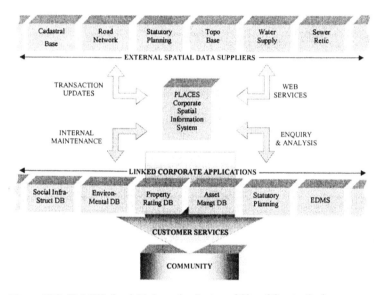

Figure 10.3: PLACES, Spatial Information System of City of Greater Geelong

Council and Barwon Water have entered into a renewable five year Data Service Agreement with Barwon Water as the supplier of Council's key fundamental datasets. The high level of local collaboration between Council and Barwon Water is recognised and supported by Land Victoria. Council continues to support Land Victoria's initiatives with regard to the PIP project and the development of the State SDI.

The SIS Development Project is located in the IT Unit within Corporate Services and is managed by the SIS Steering Committee, comprised of Council's four General Managers and six key Unit Manager / stakeholders. The Committee is chaired by the SIS Project Sponsor who is the General Manager City Development.

SIS development is supported by a Technical Working Group of twelve staff who represent a cross section of Power Users (involved in data maintenance and analysis) and General Users. This group is chaired by Council's Spatial Systems Administrator and supported by the Council Cartographer and the Barwon Water GIS Coordinator as required.

Council's Spatial Information System PLACES, is named after its fundamental data themes and primary user group sub-systems:

- Planning
- Land & Property
- Asset Infrastructure
- Community & Social Infrastructure
- Environment
- Services

Key components of the system are shown in Figure 10.3, and include:

- External Spatial Data Suppliers
- Corporate Spatial Data Base
- System Interfaces
- Linked Corporate Applications
- Community

The spatial data provided by external suppliers includes environmental data, census data, contours, aerial photography and street directory images. The cadastral base, road network, water and sewer data, survey mark data and statutory planning data are supplied by Barwon Water and are maintained by a weekly transaction update process by Council. Statutory planning data is also supplied by the Department of Infrastructure following any amendments to the statutory planning scheme.

The Corporate Spatial Data Base is stored and managed as an Oracle 8I ArcSDE Geodatabase. All feature classes are stored as members of feature datasets to control user access, data maintenance and security. Extensive feature datasets have been captured and are being–maintained by two System Administrators and fifteen Power Users based in a number of departments. Council datasets include those displayed in Table 10.1.

Table 10.1 Corporate Spatial Database Council Datasets

Theme	Feature Data
Statutory Planning	Amendments to all Statutory Planning Zones and Overlays
Environmental and Recreational	Significant Biodiversity Sites and Areas Areas of Roadside Significance Areas for Agricultural Development Public Open Space Leisure: Golf Courses, Arts/Sports Centres, Libraries…
Community and Social Infrastructure	Education: Universities, Colleges, Schools, Kindergartens… Community: Aged Accommodation, Childcare/Healthcare/Senior citizen Centres, Community Halls/Houses, Meals on Wheels, … Emergency: Ambulance, Hospitals, Fire Stations, Police, SES, … Transport: Rail and Bus Routes, Stations/Stops, Bike Paths…
Asset Infrastructure	Road Segment Hierarchy and Pavement Management Footpaths Kerb & Channel Road Furniture & Appurtenances Council Properties Drainage Infrastructure
Services	Customer Service Centres and Depots Waste Collection Zones and Days Waste Disposal Areas
Topography	Features and Contours
Georeferenced Plans	Engineering Drawings Plans of Subdivision
Property Valuation	Valuation Sub Market Groups

System Interfaces for transaction updates, data maintenance, web services and inquiry and analysis are provided by a three-tier system, listed in Table 10.2.

Table 10.2 System Interfaces for Transaction Updates

Tier	Interface	Functions	No Users
1.	Administrator	Data Mgt, Transactions, Security & QA	2
2.	Power User	Data Maintenance & Analysis	15
3.	General User	General Enquiries	380

A key purpose of the Local SDI is the vertical integration of datasets, based on their spatial elements, that is enabled across all linked Corporate applications – such as cadastral, road, environmental, community, administrative, assets and social infrastructure. More than 50 of the spatial data layers are actively linked to Council's Corporate databases. Procedures for maintaining data linkages between datasets are operated by the two system administrators and the fifteen power users.

The community served includes the local community, businesses, other agencies and SDI participants and private contractors. All available spatial information within the region is now on line on the desktop to all internal departments. Datasets linked to Council's corporate applications are also providing high levels of data inquiry and analysis.

Of Council's 650 personal computer users, over 380 have been trained to operate the general user interface and all predicted benefits for the SIS Project have been realised. In particular, there has been a significant improvement in the quality of reports involving map-related information and in the speed of retrieval and quality of information available to customers at service centres.

10.2.3 Linkages of Greater Geelong SDI Through Corporate SDI

The SIS described in Figure 10.3 represents the integrated corporate SDI of the Council. In the process, the linkage of the Council's corporate SDI with a number of contractors, Barwon Water and the State Government Departments (including Departments of Infrastructure and Natural Resources and Environment) contributed to the creation of Local and State SDIs respectively.

At this stage most of the benefits of the SIS have been the result of providing and integrating data available from Local and State SDI's on line internally to staff. The most obvious effect of integrating data from various sources through commercial arrangement with contractors and partnership arrangement with Barwon Water and State Government departments has been increased user awareness and expectations. This has reached the point where the Council now has customers and contractors who not only want instant high quality large scale maps, they also expect us to copy and hand over massive digital datasets that are not the Council's to give!

Outside commercial considerations of data use and distribution, the introduction of the Privacy Act 2000 in Victoria, in line with national directives toward protecting the individual's right to privacy have implications on how Local Governments use and exchange any data that may relate to an individual's personal information.

Geelong's most immediate SIS challenges are at the local level rather than other SDI Levels. These include training and increasing skills at the power user level to promote the capture and maintenance of more Council spatial datasets and improve the quality of existing ones. This is important for Council to fully realise the return on its investment in the technology.

Other areas important for capitalising on the investment in SDI are the development of complementary data projects, including mobile data capture and remote access. These projects include the use of mobile units in conjunction with GPS and the delivery of spatial services directly on-line to customers via Council's Website as part of E-Business.

10.3 CORPORATE SDI IN MULTI-LEVEL SDI DEVELOPMENT- THE CASE OF DNRE

The Department of Natural Resources and Environment (DNRE, now referred to as Sustainability and Environment) has significant responsibilities for the development, conservation and protection of Victoria's natural and cultural resources. Over 4000 staff in more than 200 locations across Victoria deliver services to a diverse community and a wide range of stakeholders throughout the State. It undertakes activities in 10 programs - fisheries, minerals and petroleum, agriculture, catchment and water, conservation and recreation, forests management, fire management, land management and information, energy, aboriginal reconciliation and respect (DNRE, 2001). DNRE has been the product of a series of restructuring and downsizing of State Government departments over the past two and a half decades.

When DNRE was first formed in 1996, the diverse background of its departments, particularly the different rates of uptake of GIS and other technology gave DNRE a culture of significant program autonomy and decentralization of the management of GIS. This was made more complex as each DNRE program had a head office for policy development, policy implementation was done through different groups of functional staff in six regions, with each having significant autonomy. This resulted in the development of pockets of GIS expertise in different programs, institutes and regions, which are managed, largely independent of one another. Despite the fragmented nature of GIS development, some of these programs, institutes and regions still have advanced GIS capabilities that make them key players in SDI initiatives all over Victoria and Australia. This is illustrated by the activities of two Programs described below.

10.3.1 Catchment and Water

Catchment and Water is a Program responsible for developing the water and catchment management policies for Victoria. It also administers state and federal grants to different statutory and voluntary organizations such as catchment management authorities and local land care groups to implement the policies through effort such as salinity management. Often as a result, significant spatial data is collected and kept in business units across Catchment and Water. Good examples are the Water Quality Data Warehouse and the multi-million dollar flood plain management database created for Victoria.

With Federal funding from the National Heritage Trust, Catchment and Water assigned one of its business managers to undertake two related projects, namely Regional Data Net and Catchment Activities Management System, to manage the grants allocated to and the outputs of different catchment and land care organizations in Victoria. Through the Regional Data Net project these organizations agreed on standards of reporting and database (both textual and spatial) creation and maintenance. The Catchment Activities Management System project created a system that allowed these organizations to record their activities via a web browser and to access the subsequent data in the form of maps or raw data. The key outcome of the two projects is a network of six Regional Data

Servers and the associated administrators that are responsible for collating data, particularly spatial data, from catchment and land care organizations to form regional datasets. The regional datasets in turn were compiled into central statewide datasets.

There was a plan to use the infrastructure developed by the Regional Data Net project to share the flood plain management data with the Catchment Management Authorities and to engage them to update the database in the course of their activities. Currently the flood plain data is distributed to local councils as hard copy maps. Eventually, electronic sharing of data could be done through the Regional Data Net.

The Regional Data Net, Catchment Activities Management System and the data they manage, such as flood plain and other catchment management data, form a component of the corporate SDI of DNRE that is progressively shared with the local communities and agencies. Using the model in Figure 10.2, the above SDI development activities are described diagrammatically in Figure 10.4 under Catchment Management.

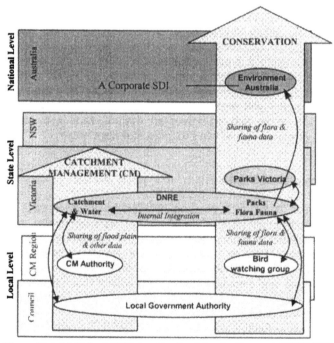

Figure 10.4: Multi-level SDI Interaction Involving the Corporate DNRE SDI

10.3.2 Parks Flora and Fauna

Parks Flora and Fauna is a Program that serves as the custodians of the state's floristic and wildlife databases maintained with contributions from a range of public, private and voluntary individuals, groups and organizations, both local and state-wide, across Victoria. It has formal agreements with the contributors that the data is to be available freely for the conservation of plants and animals in Victoria.

One core business of Parks Flora and Fauna is to make use of its conservation data to help stakeholders across Victoria to plan and prioritise land development projects, and to rehabilitate sites to help conserve threatened or even endangered plant and animal species. This is achieved in different ways including productions of maps of appropriate scale, presentations, expert interpretation of data and making digital data available to stakeholders such as developers, academics, professional consultants and planners.

Coordination by a fully dedicated data asset manager is fundamental to achieving the Program's business objectives. The spatial data, the associated technical expertise and management practices developed by the unit constitute the Parks Flora and Fauna component of the DNRE corporate SDI that is being incorporated into Local and National SDIs.

At a national level, it participated in a Federal program that funded the collation and publishing of Victorian coastal and marine data through the Victorian node of a national network of web sites called the Australian Coastal Atlas. Furthermore, after almost a year of negotiation, it developed a data exchange agreement with its Federal counterpart, Environment Australia, to allow land developers access to Victorian conservation data kept on a proposed purpose built Federal web site.

At a state level, it formulates policies that govern the management of national parks in Victoria in terms of conservation, recreation through Parks Victoria, a strategic partner of DNRE. Again a strategic data sharing agreement has been established to allow exchange of parks conservation and administration data between DNRE and Parks Victoria.

At a local level, it has field staff working with farmers and other land owners, local councils and other agencies to ensure the protection of important plant and animal species. Data is made available to these field staff to facilitate their work. In this regard, Parks Flora and Fauna is currently exploring ways of partnering with local government authorities. The purpose is to make its vast asset of floristic and wildlife data easily accessible to planning departments of local government such as City of Greater Geelong to allow them to better assess development proposals which may have an impact on nature conservation in Victoria. The SDI development activities that Parks Flora and Fauna participated in are illustrated in Figure 10.4 as part of the overall conservation function in Australia.

10.3.3 Department Wide Effort

While Programs in DNRE were engaged in SDI development across Victoria and Australia, DNRE itself was trying to bring the program components of its corporate SDI together. As mentioned earlier, when DNRE was first formed in 1996, these components were pockets of GIS capabilities in various Programs. Owners of these capabilities realised quite early the strategic significance of sharing their spatial data. This was achieved with the creation of a corporate module of infrastructure GIS called the Corporate Geospatial Data Library (CGDL), which existed before 1996 in some DNRE Programs to allow sharing of a limited number of spatial datasets. This formed a key component of DNRE's corporate GIS and SDI. The evolution of the corporate GIS has been documented by Chan and Williamson (1999a, 1999b, 2000). Following a consultant study into the management of GIS in DNRE in 1999 the Geographic Information Unit (GIU) was created to coordinate the development of an integrated corporate GIS. With the day to day up-keeping of the CGDL done through a service level agreement with LIG of Land Victoria, the GIU focussed on establishing an overarching framework to guide its work. This framework included new communication, coordinating and directing mechanisms (Scardamaglia and Chan, 2001).

The framework provided an environment to align GIS development activities by addressing the following issues:

- Raising awareness of and identifying priority business needs;
- Identifying resources requirements, potential funding sources and means of collaboration to secure funding;
- Developing the organizational setting: formal policies, standards, guidelines, protocols and business rules that form to facilitate and govern GIS development and usage;
- Acquire new infrastructure such as browser-based mapping capabilities and develop of new platform for the CGDL to provide greater flexibility and scalability for spatial data management and dissemination.

10.4 SOME PATTERNS OF SDI DEVELOPMENT

The SDI development activities of the City of Greater Geelong (local-state) and DNRE (local-state-national) described above have some similar characteristics. These activities are:

- Multi-organizational;
- Concerned with sharing of spatial data needed by the lead agencies at different levels of government to conduct their business processes;
- Involving new technical infrastructure and some forms of partnership arrangement;
- Expected to be ongoing, with standard processes developed to ensure the currency and integrity of the shared spatial data.

In all cases, the SDI development is a result of interaction among corporate SDIs of respective organizations. Depending on the jurisdiction of each organization, corporate SDIs exist at all levels of the hierarchy of SDI modelled in Figure 10.1. SDI development is the responsibility a full time coordinator or project manager. Depending on where the coordinator is located, in certain cases, the SDI has a home in the Council/DNRE; in others, the Council/DNRE contributed to the development of the SDI. In the cases described in this chapter, SDIs may exist as:

- Individual corporate SDIs at different levels of government,
- Part of the corporate SDI of a lead agency such as DNRE or Environment Australia in Figure 10.4, or
- One of a network of interlinked corporate SDIs involving lead agency as in the case of Australian Coastal Atlas.

10.4.1 Some Good Practices

The experience of the case of the City of Greater Geelong suggests that the success of SDI development activities depends on a number of practices that are equally applicable to the DNRE SDI activities. These practices can be grouped broadly into three categories: *business process improvement, project management and development of standards*.

Under business process improvement the practices, which emphasise on efficiency gain for all agencies, include:

- Recognition of the need to develop interoperable business procedures that capitalise on the SDI;
- The need to understand all operational requirements of participating agencies; in the case of land information, they are the referral authorities that include local councils and the Utilities;
- An appreciation of budget requirements and the need to avoid higher overheads;
- That agencies will participate when automation improves efficiency and service delivery;
- Ensure procedures provide overall efficiency gains without burdening particular areas.

Under project management, there are two good practices. One is to ensure proposals are based on realistic and achievable staged implementation schedules. The other is to identify basic proposals that provide maximum return for minimum expenditure.

While standards are a key component of any SDI, the good practices identified based on the experience in Victoria and Australia are:

- The need for higher than local levels of the SDI to set minimum standards
- All services need to be delivered according to these standards
- Lower level agencies of the SDI should be able to operate competitively using higher standards if they choose.
- Recognition of the potential, wherever applicable, to use legislative powers at State level to control and set minimum standards at the local level.

10.4.2 Two Scales of SDI Development

With corporate SDIs existing at all levels of the SDI hierarchy and serving as the building blocks of SDI development, the cases of the Council and DNRE suggest that SDI development takes place at two related scales, ie., macro- and micro-SDI development. At the macro-scale corporate SDIs interact with one another to form SDIs at different hierarchical levels (Figure 10.4). At the micro-scale pockets of GIS capabilities in an organization are integrated, often initially at the business functional level to form components of the corporate SDI, which are further integrated to form the more holistic corporate SDI.

Macro-SDI development generally takes place vertically within functional areas such as conservation or catchment management. However cross-functional data sharing and SDI development at the micro-scale is generally done in-house within an organization with external data sourced through ad hoc requests or routine purchasing arrangements.

SDI development at the macro- and micro-scales may appear to be different. However early evidence from Geelong City Council and DNRE suggests that there is a common human dimension. Initial observations suggest that dedicated coordinators are needed for SDI development at both scales. However the dedicated coordinator responsible for corporate SDI development will succeed only after, among other things, gaining the cooperation of a network of business managers/data custodians in various functional areas in the organization. In turn, it is this network of business managers/data custodians that will interact with their counterparts in their respective functional areas to form SDIs at macro-scale. The detailed interaction between networks of managers may warrant further investigation.

This overlapping and interacting network of business managers/data custodians suggests that there is a dependency between developments of SDI at the micro- and macro-scales. The better developed two corporate SDIs are, that is at the micro-scale, the more likely that the organizational infrastructure such as supporting policies and business rules, and the technical infrastructure such as the IT backbone and the data servers, are in place for collaboration. As a result, it will be easier for SDI development to take place between the two corporate SDIs at the macro-scale. In return, benefits realised through macro-scale SDI development may encourage senior management to invest further in the organizational and technical infrastructures. This sort of positive feedback cycle is important to the success of SDI development.

The early experiences of SDI developments at micro- and macro-scales suggest that when embarking on a major SDI development effort, it is important that there is a lead organization with a dedicated coordinator. Rather than building the SDI from scratch, it is important for the coordinator to identify potential partnering agencies both inside and outside the lead organization, understand the states of development of the corporate SDIs of these agencies and formulate an engagement strategy. The strategy should use appropriate communication, coordination and guiding mechanisms to initiate and sustaining a positive feedback cycle for SDI development.

10.5 CONCLUSIONS

By examining SDI development activities of the City of Greater Geelong and DNRE, which span local, state and national levels, a number of characteristics can be identified. These activities are:

- Multi-organizational;
- Concerned with sharing of spatial data needed by the lead agencies at different levels of government to conduct their business processes;
- Involve new technical infrastructure and some forms of partnership arrangement;
- Expected to be ongoing, with standard processes developed to ensure the currency and integrity of the shared spatial data.

The same experience also helps to identify a number of practices to ensure the success of SDI development activities. These practices emphasise efficiency gain for all agencies, early delivery of project benefits and means of setting standards. They belong to three groups that are named respectively, business process improvement, project management and standards.

The productional perspective of SDI has been used to describe SDI development outside and inside both the City of Greater Geelong and DNRE. The patterns of development suggest that the building blocks of SDI development at various levels in the SDI hierarchy are corporate SDIs.

SDIs develops at two scales - macro-SDI and micro-SDI. Macro-SDI development applies to the development at the different levels of government, while micro-SDI development to corporate SDI. SDI developments at the two scales are independent.

An important observation is the human dimension of SDI development in which managers involved in micro-SDI development are also likely to be responsible for macro-SDI development. When embarking on a major SDI development project, it is important that there is a lead organization in which a dedicated coordinator is appointed to identify and engage with potential partnering agencies for the project. An engagement strategy should be developed to promote a positive feedback cycle for SDI development in these agencies.

10.6 REFERENCES

Chan, T.O. and Williamson, I.P., 1999a, Spatial data infrastructure management: lessons from corporate GIS development. *Proceedings of AURISA '99*, November 1999, Blue Mountains, NSW, AURISA 99: CD-ROM, Online. <http://www.geom.unimelb.edu.au/research/publications/IPW/ipw_paper31.html>

Chan, T.O. and Williamson, I.P., 1999b, The different identities of GIS and GIS diffusion, *International Journal of Geographical Information Science*, 13: 267-281.

Chan, T.O. and Williamson, I.P., 2000, Long term management of a corporate GIS. *International Journal of Geographic Information Science*, 14:3, pp. 283-303.

Chan, T.O., Feeney, M.E., Rajabifard, A. and Williamson, I.P., 2001, The Dynamic Nature of Spatial Data Infrastructures: a method of descriptive classification. *Geomatica*, Canadian Institute of Geomatics, 55:1, pp. 65-72.

DNRE, 2001, Home page of Department of Natural Resources and Environment, Online. <http://www.nre.vic.gov.au/> (Accessed September 2002).

Jacoby, S., Smith, J., Ting, L. and Williamson I.P., 2002, Developing a Common Spatial Data Infrastructure between State and Local Government - An Australian Case Study. *International Journal of Geographical Information Science*. Vol. 16, No. 4, pp. 305-322.

Rajabifard, A, Chan T O, and Williamson, I.P., 1999, The Nature of Regional Spatial Data Infrastructures, *Proceedings of AURISA '99*, November 1999, Blue Mountains, NSW, AURISA 99: CD-ROM, Online. <http://www.geom.unimelb.edu.au/research/publications/IPW/ipw_paper32.html>

Scardamaglia, D. and Chan, T.O., 2001, NRe-map: implementing a corporate generic web browser-based mapping application to meet the requirements of a

PART FOUR

Supporting Economic, Environmental and Social Objectives

Sustainable Development, the Place for SDIs, and the Potential of E-Governance

Lisa Ting

11.1 INTRODUCTION

One of the challenges in converting the rhetoric of sustainable development into reality is to use the potential of Spatial Data Infrastructures (SDIs) in raising the level of informed dialogue about rights and obligations over land. In order to serve this purpose, by facilitating complex, multi-stakeholder decision-making, SDIs need to operate within a good governance context. From the perspective of sustainable development, this means that there must be frameworks that facilitate the flow of information to relevant stakeholders to enable them to participate in and contribute to informed decision-making about how land and resources are to be used. In this way information supports good governance as well as being an outcome of good governance.

Information and good governance are the two important elements that must work together to form the bases of effective decision-making for sustainable development. There is a synergistic relationship between good governance and information whereby good governance creates a healthy legal, institutional and socio-political-economic framework for information to flow, and the information flow in turn facilitates sound decision-making for good governance.

This chapter will explore the relationship between information and good governance in supporting sustainable development. It will discuss the role of SDIs in delivering the kind of information needed for sustainable development, and will conclude by discussing the potential role of E-governance in enhancing the reach of SDIs and in progressing good governance through wider participatory processes.

11.2 SUSTAINABLE DEVELOPMENT

Since the 1960s, there has been a phenomenon of a rise in civil rights movements at a time of growing mistrust of government and consequent desire to participate in government decision-making. Traditional processes of decision-making in parliamentary democracies began to be affected by civil rights movements as well as a growing awareness of environmental and social issues that reached beyond traditional partisan politics. Environmentalism gradually moved into local, national and regional politics and eventually brought sustainable development onto agendas around the world. A growing phenomenon has been that some communities progressed from being involved in policy-making, to taking a further leap into involvement with the actual policy implementation, which was previously the domain of the administrative arm of government.

Sustainable development demands complex decision-making that weighs up environmental, social and economic consequences of the choices that are made about how resources will be used. Such complex decision-making requires ready access by decision makers and stakeholders to current, relevant and accurate spatial information.

Spatial information and the information technology advances that support its collection, maintenance, analysis, presentation and dissemination are vital tools in the complex decision-making processes of the future, particularly for sustainable development. However, it is good governance that will guarantee equitable access to this information, effective participation from diverse sectors and transparency of process. Figure 11.1 (adapted from Ting and Williamson, 2000) shows the linkages between SDI, information, good governance and sustainable development. Governance is "the manner or function of governing, viewed as rule, influence or regulation with authority. Governance most often refers to the manner of rule of the state, but it may also refer to regulating the proceedings of any organization, such as a business corporation (corporate governance)" (Gleeson and Low, 2000).

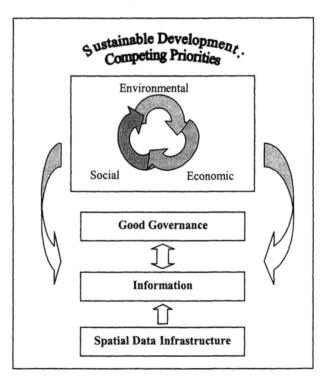

Figure 11.1: SDI, Information and Good Governance Provide a Basis to Manage the Tensions Between Economic, Social and Environmental Imperatives for Sustainable Development

The next sections will explore the relationship between information and good governance in supporting sustainable development.

11.2.1 Information

The dynamic nature of the increasing complexity of rights and restrictions over land are the result of the general trend towards tempering the economic imperative with environmental and social priorities, such as indigenous and women's rights. Reliable information infrastructures are needed to record those rights as well as provide spatial data to facilitate decision-making and conflict resolution.

An example of this is the case of the Amazon forest in Brazil where there was an attempt by commercial interests to sideline the Yanomami Indians' legal rights to vast expanses of the richest land in Amazonia, by moving a Presidential Act that declared the Yanomami area unconstitutional. Vieira noted that the trail of responsibility for the environment and development (and good governance) existed at local, provincial, national and international levels. It was also found that responsibility extended beyond any one department in the administration, to the judiciary as well as the legislature. Building an appropriate database with government and civilian involvement was one important component of the solution put forward by the Ministry for the Environment (Vieira, 1995).

11.2.2 Good Governance

As the Yanomami Indians' example demonstrates, decisions about resource use affect the property rights of any combination of private individuals, corporate entities, community groups and the state, depending on whether one wishes to classify those rights as private or public and whether one is prepared to consider social, environmental and economic impacts.

It has been long-acknowledged that the state itself is not a neutral player. Therefore decision-making supporting sustainable development requires the support of legal, institutional and technological frameworks that enable equitable access to information and transparency of decision-making across government, private sector and the community. The employment of these frameworks is vital to good governance.

Governance is about political decision-making, economic and social planning as well as the way society builds and maintains institutional structures to manage those political, economic and social aspects (Okoth-Ogendo, 1995). Governance is concerned about process as well as outcomes and information is integral to that process.

Informed participation from diverse sectors and transparency of process has been identified as vital to sustainable development. Decentralization of decision-making can assist transparency of process and participatory decision-making and usually requires some policy, regulatory, technical and fiscal support from central government (World Bank, 1997). For example, good governance would encompass the development of standards that would establish ground rules for the exchange and enhanced interoperability of information that would greatly assist in complex decision-making across environmental, social and economic priorities.

SDI contributes to this process through the development of technical policies and standards, access networks, partnerships and through facilitating an environment for the sharing of spatial data. Informed participation requires equity of access to information and the added value of spatial information lies in the fact that it can be identified as relevant to a particular place that is the focus of the dialogue. The role of SDIs in facilitating participative and informed decision-making to support sustainable development is explored below.

11.3 SDIs FOR SUSTAINABLE DEVELOPMENT

The key information principles of sustainable development must be access and equity as well as dissemination in a way that facilitates participation by relevant stakeholders and other interested people. Trends are strengthening towards involving civil society in implementation and service delivery (World Bank, 1999). There is scope for the community to assist in gathering required data and to be trained to understand it enough to participate in consultations. Water catchment committees are good examples of the partnerships that have started to form between government and communities to resolve issues of water quality/quantity that are impacted by land usage. Resultant policies are implemented by the community with assistance from the government where appropriate. Policy implementation requires reliable SDIs to record the social, environmental and economic rights and responsibilities in order to facilitate more sound decision-making and conflict resolution.

From a good governance perspective there are some questions that need to be answered for SDI development generally, as with other spatial data projects. These questions relate directly to the SDI components, and need to be answered in conjunction with stakeholders and relevant community groups as well as the usual technocrats, because their inputs will help to determine:

- What kind of datasets are needed for the project concerned? If the stakeholders and community groups have a range of economic, environmental and social concerns, then what datasets are necessary to address those concerns? On a wider scale, what are the datasets that should be considered fundamental and therefore worthy of resources to develop and maintain for participatory decision-making and implementation?

- Where is the relevant information held? Who are the custodians? How can they work together? (Jacoby *et al.*, 2002) If data is unavailable then is it worth investing its development?

- How can access be established? Should there be categories of people or groups that have access and what rights or responsibilities attach to those categories? Are cooperative mechanisms sufficient? Is there a need for more formal institutional and legislative support and guidance e.g. on matters of privacy and access?

- How can privacy be protected? In what instances would privacy be prioritised above access? Is there a creative middle ground approach to issues such as sacred and secret indigenous sites or the locations of delicate and endangered species?

These principles and questions are explored below by examining the contribution of particular components of SDIs to supporting sustainable development decision-making, and specifically the policies, standards, partnerships, data, access and design of SDIs.

(a) Policies

The information technology revolution has been a catalyst of change. The Geographic Information System (GIS) technologies for data management, manipulation, analysis and integration have had the greatest impact on the spatial information environment, although communications technologies such as the Internet are rapidly becoming the focus of attention for the management of land, and more importantly people's relationship to this process.

SDIs have a central role to play in securing equitable access to spatial information, whilst establishing policies for data development and sharing that retain the privacy of individuals while at the same time involving them. It is when spatial information can be linked with personal information that privacy issues can arise. The International Working Group on Data Protection in Telecommunications (IWGDPT) clarified in Article 2(a) and (c) of Directive 95/46/EC that:

> ...a total scan of all buildings in a city or a country will involve the processing of personal data since much of the information relates to natural persons who are identifiable by factors specific to their physical, economic, cultural and social identity in a data filing system and may be linked directly or indirectly to directories. Therefore the creation of image databases of this kind falls within the scope of national data protection laws in accordance with the EC Data Protection Directive.

> (IWGDPT, 2002)

Whilst reform of legal mechanisms such as privacy laws may be commendable, the speed and reach of information technology, particularly the Internet, and breaches of such laws or guidelines can have much far-reaching (and less traceable) effects than was the case with the print media (Ting and Williamson, 1998). Therefore, thought needs to be given to the drafting and enforcement of legislation that establishes fundamental principles for privacy protection that can provide guidance through technological advances rather than highly prescriptive regulations that will be quickly rendered redundant by such advances. It would be in the interests of good governance and the growth of spatial information industries to be cognisant of and sensitive to privacy matters and by so doing promote the healthy use of spatial information rather than trigger community condemnation and obstruction.

(b) Access

Government is usually the custodian of the largest amounts of spatial and spatially related data. Public access to government-held information is important in order to tap into knowledge beyond government and to maintain accountability for the basis on which policies are being implemented. This is a key role of SDIs. Equitable access to information, especially spatial information, has the potential to facilitate equitable access to dialogue and decision-making. A good example of this was the Wombat State Forest in Victoria, Australia.

In 2001, the local community in the vicinity of the Wombat State Forest publicised their research findings that the Wombat State Forest was not being managed in a sustainable way and demanded to see the GIS data and other information on which the state's management plan was based. Analysis of the data confirmed the community's assessment and was a good example of the potential value in involving the community in the development and administration of policy.

The Delta project in New Brunswick, Canada is another example of the importance of providing public access to information and tools to support participation in decision-making about land and resource use. The Delta project is an initiative that aims to achieve interactive decision-making in physical and virtual meeting rooms on issues in contention about land and resource use. Delta is not an acronym; the name was chosen for its significance as a geographical term. In the Delta project tools such as GIS are used to assist stakeholders towards literally visualizing one another's points of view and reaching agreement with regard to resource use (Davies and McLaughlin, 2000). It is a pilot project and as it progresses it will clarify what institutional and legislative frameworks are required to make it work.

(c) Standards

To conduct dialogue constructively, the range of stakeholders need to be able to access environmental, social and economic information that is interoperable and that can be related to the disputed area as well as its surrounds, such as the water catchment area in which it sits. The range of stakeholders that have the potential to be involved means that they will also bring their own information to the table. Apart from a general need to raise awareness and skills to match the SDI and information infrastructures, interoperability and data standards promoting interoperability of datasets for use and exchange, continues to pose a challenge. Whilst metadata helps users to determine how reliable the data might be for their own purposes, there is still the matter of technical interoperability. Interoperability is a theoretical and technological issue that continues to require resolution (Eagleson *et al.*, 2000, 2002).

(d) The Role of Partnerships

Participatory decision-making and administration requires that relevant government departments set aside their traditional "silo" approaches and instead adopt SDI principles to put resources towards data sharing and common strategies within and between the various tiers or levels of government. For example, in New Brunswick, Canada, the Department of Natural Resources and Energy regional office liaises with the Department of Health over the management of water resources and also land use (through the jurisdiction of local government planning bodies) that affects them (Timms, 2000). Service New Brunswick, which is a property information database established by statute, assists by providing a clearinghouse for information from various government departments about permits and procedures relevant to land use decisions and development so that government may provide a more holistic approach to information sharing between government departments as well as for the public, facilitating involvement from the community.

Canada has already taken steps to institutionalise partnerships with the community for the management of natural resources and data sharing initiatives. Forestry is a well-established area of partnership between government, private sector and the community (MFN, 1999). The forestry license holders are required to cooperate with the government's maintenance of databases on all the public forests in New Brunswick, Canada. Another area where partnerships are developing is concerned with the management of water catchments. The Atlantic Coastal Action Program (ACAP), a Canadian federal initiative, is an example of such support. The coastal and estuarine areas have suffered significant degradation. ACAP started up in 1994 to support five community action groups across the province to collect data, analyse it with relevant support from government and participate in decision-making about those areas (Timms, 2000).

The important role of partnerships in the development of SDIs to support sustainable development is multi faceted and has also been highlighted from different perspectives in previous chapters.

(e) Data

Each country or jurisdiction needs to determine what datasets are necessary and fundamental enough to merit government or private sector attention and support for those datasets to be maintained and made easily accessible to the public through the SDI framework. In Australia, for example, each state has a different policy on what datasets are considered fundamental because each state has different economic, social and environmental policies.

The state of Victoria has what it calls Framework Information (see Section 9.2.4 in Chapter 9) that is information considered fundamental to the development and operation of Victoria's spatial information infrastructure, in that other (business) information cannot be created or maintained without it. Based on that definition, it has eight framework datasets on Geodetic control, Address, Cadastre/property, Transport, Administrative, Elevation, Hydrology, and Imagery (aerial photography and satellite imagery) and from these flow other datasets.

By comparison, the state of Western Australia (WA) defines fundamental datatsets to mean those that are essential to the outcomes of a number of government agencies, and cannot be derived readily from other datasets. Thus WA has 39 fundamental datasets and because mining is a key industry, the dataset on mining tenements was considered fundamental. The state of Tasmania has also applied the definition that fundamental datasets are those that are essential to the outcomes of a number of government agencies, and cannot be derived readily from other datasets. Tasmania has about 50 datasets listed as fundamental.

It is pertinent to note that property datasets are usually the most well-developed of datasets because personal property rights and the economic paradigm have been the main driver of land administration infrastructures until the more recent rise of sustainable development as a priority. The latter led to an additional demand for environmental and social/demographic datasets such as catchment areas, wetlands, conservation areas, heritage sites, aboriginal heritage sites and planning datasets. Sustainable development is also driving the next frontier of dialogue about rights and responsibilities over off-shore areas that will require reliable marine cadastres and other datasets about resource use and conservation that will need to be accessed, analysed and disseminated to support sustainable development decision-making over those marine areas. SDI is the logical framework which can provide the coordination and access to these datasets.

(f) SDI Design and Equity

There is no single formula for determining what spatial information databases are needed and what technology will be needed to store, analyse and disseminate the spatial information within an SDI context. It depends on what priorities and limitations apply in each context. For example, if sustainable development were a priority then there would be a demand for information relevant to each of the competing components of sustainable development, namely environmental, social and economic. Whatever the technology that is applied, steps need to be taken to design governance infrastructures, such as legal and institutional infrastructures, that work to produce a coherent, integrated approach to the collection, maintenance and dissemination of spatial information as well as clarify what access/privacy principles should apply in the design of an SDI. Whatever a country's situation, there are basic requirements such as data sharing policies and metadata standards that can be established and applied early to pave the way to the future.

Developing nations will have limited capacity to maximise use of their existing scarce resources and less capacity to enter the sustainable development decision-making debates in the international arena if they make no effort to think creatively about good governance to facilitate an appropriate regulatory, administrative and educational infrastructure for spatial information. It may well be that for some developing countries, spatial information will need to be collected and developed based on more modest technology but to standards that are sensitive to the technology potential in the future. In doing so, the same awareness of the importance of good governance structures must apply. Technology in itself will not resolve the issue of establishing mechanisms for participatory, integrated decision-making across sectors of society, private interests and government.

As discussed in Section 11.2, the basis of sustainable development is the interrelationship between good governance and information infrastructures (including SDIs) and the accessibility of these to relevant stakeholders across those sectors. The following section will discuss how E-governance could fit into and enhance these infrastrcutures by facilitating improved accessibility to spatial information as well as good governance initiatives.

11.4 THE POTENTIAL OF E-GOVERNANCE

E-government has become widely known as the online delivery of information and services by government through the Internet. Chapter 9 of this book gave the example of the Land Channel website of Victoria, Australia. Canada and Singapore are other leading international examples of comprehensive E-government sites. In 2002, Singapore's eCitizen portal was named winner of the e-Government category in the prestigious Stockholm Challenge in Sweden.

E-governance is about the enhancement of E-government with processes for wider consultation within and between the government, private and community sectors. In simple terms, E-governance builds on E-government. Where E-government has been about delivering information and services from government to the wider community, E-governance would be about creating a more reciprocal relationship that can enhance information flows both ways and improve opportunities for participatory decision-making.

SDIs and related information frameworks are necessary components of E-governance for sustainable development that will assist people to more quickly find common ground, tease out the real issues to be tackled and engage in dialogue more effectively. Sometimes participatory decision-making and administration may result in added expense and time but are more likely to produce lasting results, particularly when supported by the skills, information and processes to work through any conflict. E-government teamed with appropriate SDIs and processes for wider consultation through E-governance, has the potential benefit of dispute avoidance as it provides transparency and accountability on a greater scale than before as to what information (and stakeholder inputs) were relied upon for decision-making.

E-governance would therefore require policy, legal, institutional and technological frameworks that facilitate private sector and the wider community in systematically feeding into that information flow as well as receiving it from government. For example, in Canada, the New Brunswick Conservation Council (NBCC) found that there was no process for government to embrace coastal information gathered by NBCC because of a pre-existing contractual arrangement for the digital version of the spatial information (Coon, 2000). In that case, NBCC had worked with fishermen along the coastline to map data about spawning grounds, and the NBCC wanted to put this data on the base map for the Bay of Fundy with the bathometric lines. However, the Canadian Coastguard had already paid one of the Atlantic Coastal Action Program groups to do the ground-truthing on the shoreline and put in the bathometric lines (Coon, 2000). NBCC had borrowed that baseline information to layer with its own information, but when it came to the digital information their contractor said it belonged to the Coastguard.

The Coastguard in turn would not give or sell it to NBCC because they were concerned about updating and liability issues (Coon, 2000). This simple example illustrates the importance of the contextual policy, legal, institutional and technological frameworks for spatial information to work effectively with and for E-governance.

There are two broad fronts of complex decision-making that will require an E-governance response to link information as well as organizations and people within government, private sector and the community. One is the fact that decision-making for sustainable development requires consideration of up to three competing priorities, namely environmental, social and economic (also known as the 'Triple Bottom Line', as discussed in Chapter 1). The other related front is the increase in the number and diversity of stakeholder interests. This is due in part to the environmental, social and economic aspects of any one issue. It is also due to the pressures of disseminated decision-making, whether resulting from the "rude and crude" methods of getting heard (O'Looney, 1995) or from government restructuring that has put more decision-making (and administration) out into the private sector and community. Put simply, E-governance is about creating the interactive flow of understanding and information between and within the public and private sectors that will institutionalise the dialogue sufficiently to minimise the need for the rude and crude methods of getting heard. A key objective of SDIs should be to facilitate this flow of data and resulting dialogue.

11.5 CONCLUSION

Sustainable development demands complex decision-making that weighs up environmental, social and economic consequences of the choices that are made by a diverse range of stakeholders from government, private and community sectors about how their various rights and responsibilities over land and resources will be exercised. The process of progressing stakeholders with real or potentially competing rights and responsibilities towards a decision requires bases of good governance and information.

The trend towards a more participatory approach to government and governance grew out of a period of civil rights movements and environmentalism. Citizens now not only participate in the consultation stages but also have the option of organising themselves to assist with implementation of policies and monitoring of the results. Good examples of this trend can be found in communities that identify with the water catchment or forest areas in their area. This ability to mobilise themselves is recognised and valued.

Participatory governance requires the support of SDIs that can provide the infrastructure to locate and collate information that is relevant to a particular area. The immediate challenge is to establish the necessary legal, policy, and institutional and technological infrastructures to allow SDIs to develop and support transparency of decision-making within and between the tiers of government, with the private sector and into the wider community. The development, maintenance, analysis and dissemination of spatial information in an effective and equitable manner mean that SDIs need to interact within a context of good governance for sound, inclusive and participatory processes of decision-making. Good governance would encourage the establishment of regulatory and administrative frameworks to set and implement standards for accountability as well as interoperability, and aim for a balance of mechanisms for privacy as well as equitable access through education to build understanding about the opportunities (and limitations) of using spatial information.

E-governance, with an appropriate SDI and good governance framework, has the potential to complement and enhance E-government and also governance in general. E-governance could lend an unprecedented equity in the supply and access to relevant spatial information and enlarge the possibilities of sustainable decision-making by diverse stakeholders with competing economic, environmental and social interests. In short, good governance and spatial information (and associated SDI) have a symbiotic relationship that form the basis of opportunity to systematically tackle sustainable development in a more peaceful, informed and transparent way.

11.6 REFERENCES

Coon, D., 2000, New Brunswick Conservation Council, Canada, Personal communication, 26 October.

Davies, J. L., and McLaughlin, J. D., 2000, Advancing the Participatory Democracy Agenda: The Geomatics Connection, *Geomatica* 54(4): 463-471.

Eagleson, S., Escobar, F. and Williamson, I.P., 2000, Hierarchical Spatial Reasoning Applied to Automated Design of Administrative Boundaries. *URISA*, Orlando, Florida, USA.

Eagleson, S., Escobar, F. and Williamson, I.P., 2002, Hierarchical Spatial Reasoning Theory and GIS Technology Applied to the Automated Delineation of Administrative Boundaries, *Computers, Environment and Urban Systems* 26: 185-200.

Gleeson, B. and Low, N., 2000, *Australian Urban Planning: New Challenges, New Agendas*, (Sydney, Australia: Allen and Unwin).

IWGDPT, 2002, International Working Group on Data Protection in Telecommunnications. Article 2(a) and (c) of Directive 95/46/EC.

Jacoby, S., Smith, J., Ting, L. and Williamson, I.P., 2002, Developing a Common Spatial Data Infrastructure between State and Local Government – An Australian case study, *International Journal of Geographic Information Science*, 16, 305-322.

MFN, 1999, Achieving Sustainable Forest Management through Partnership. Ottawa, Canada, Natural Resources Canada.

Okoth-Ogendo, H.W.O, 1995, Governance and sustainable development in Africa. *Sustainable Development and Good Governance*. K. Ginther, E. Denters and P. J. I. M. d, Waart. Dordrecht, The Netherlands, Martinus Nijhoff Publishers: 105-110.

O'Looney, J., 1995, *Economic Development and Environmental Control: Balancing business and community in an age of NIMBYs and LULUs*. (Westport, Connecticut, USA: Quorum Books).

Timms, J., 2000, Department of Local Government and Environment, New Brunswick, Canada, personal communication, 18 October.

Ting, L. and Williamson, I., 1998, Land Administration, Information Technology and Society. *SIRC '98*: Toward the Next Decade of Spatial Information Research, Dunedin, New Zealand, University of Otago.

Ting, L. and Williamson, I.P., 2000, Spatial Data Infrastructures and Good Governance: Frameworks for Land Administration Reform to Support Sustainable Development, *4th GSDI Conference*, Cape Town, South Africa.

Vieira, S. C., 1995, *Sustainable Development and Good Governance*, Edited by Ginther, K., Denters, E. and Waart, P.J.I.M.d., (Dordrecht, The Netherlands Martinus Nijhof Publishers), pp. 429-440.

World Bank, 1997, *The State in a Changing World*. (Washington D.C., USA: Oxford University Press).

World Bank, 1999, *Entering the 21st Century: World Development Report 1999/2000*, (Washington D.C., USA: Oxford University Press).

CHAPTER TWELVE

SDIs and Decision Support

Mary-Ellen F. Feeney

12.1 INTRODUCTION

One of the key motivations for spatial data infrastructure (SDIs) development is to provide ready access to spatial data to support decision-making. Previous chapters have discussed the importance of this objective in relation to contemporary technical, institutional and social issues, at local through global political and administrative levels.

Internationally the importance of this objective was highlighted in the findings of the United Nations Conference on Environment and Development (UNCED) and was manifest in Agenda 21. The concepts of a *Digital Earth* and of the "Virtual State" (as described in Chapter 1), respond to this Agenda. These concepts emphasise the role of SDIs as essential modern infrastructure, alongside information technology and communication infrastructures, that support a wide range of decision-making environments and recognise the need for decision support to reach a broad spectrum of spatial data users.

This chapter will discuss the challenges and features of SDIs supporting decision environments, decision complexity and the variety of participants in the decision process. Spatial data and tools are central to addressing Agenda 21 action items, because they can be used to understand and integrate social, economic and environmental perspectives, and they can address relationships among places at local, regional, national and global scales. Since the adoption of Agenda 21 in 1992 there has been an increased emphasis on building human and technological capacity to access and use available spatial data to support decision-making. These are important features for unlocking the potential of SDIs and will be reviewed in this chapter in relation to the growing importance of the SDI role in decision support. The chapter concludes with a discussion on how evaluation of decision-support capacity may guide future SDI development and strengthen related data, technological and people activities.

12.2 DECISION SUPPORT FOR SUSTAINABLE DEVELOPMENT

Two decades after the United Nations Conference on the Human Environment (1972) many nations of the world convened to reaffirm its principles and build upon these for sustainable development at the UN Conference on Environment and

Development in Rio de Janeiro. As a result the Rio Declaration on Environment and Development was adopted in 1992. The enthusiasm for this declaration was manifest as Agenda 21 (Agenda 21, 2002), an ambitious plan outlining methods to improve the condition of humans and the environment in many sectors, including the availability, accessibility and applicability of information for decision-making (as outlined in Chapter 40 of the Agenda). The Agenda emphasises that the integration and balance between economic, social and environmental dimensions needed to achieve sustainable development (as discussed in Chapter 11) requires new ways of looking at how we make decisions.

12.2.1 The Role of SDIs

SDI endeavours to provide a basis to achieve the objectives of Chapter 40 of Agenda 21 in relation to spatial information and thus support the overall accomplishment of the Agenda. The aim of SDIs is therefore the ability for people to share data and interact with technology to support their decision-making objectives. To this effect many regional, national and local programs are working to improve access to spatial data, promote its use, and ensure that investment in spatial data collection and management results in an ever-growing, readily available and useable pool of spatial data to support decision-making. The motivation is that the sharing of information (to help resolve difficult economic development, environmental, natural disaster, health, and other problems) can help reduce costs and improve services while laying the groundwork for new information-based industries.

Improved economic, social and environmental decision–making are principal objectives for these investments in SDI development at all political and administrative levels. Therefore SDI developments aim to support the access to and application of data for products and services, but more particularly decision-making. However, whilst the importance of the decision support relationship is identified as inherent to the role of SDI, little has been done to document how SDIs support decision-making and thus how SDI decision support capacity can be evaluated and improved. As a result, it is believed that the potential role of SDIs for spatial decision support is currently underdeveloped, particularly in the application of data and the incorporation of supporting technologies into the decision process.

12.2.2 The Challenges for SDIs

In the opening plenary of the 6th Global SDI Conference, Masser (2002) commented that unless SDIs foster better incorporation of decision support systems, beyond information systems, into the decision process, much spatial data that may support decision-making will be underutilised, as will the potential of the supporting technologies. This concern was highlighted earlier in the month by the World Summit on Sustainable Development (WSSD) in Johannesburg (WSSD,

2002), which was convened to assess global change since the 1992 UNCED and development of Agenda 21. The WSSD implored people to optimise the value of the spatial information for decision-making that would support sustainable development objectives and charged people to "Access, explore and use geographic information by utilizing the technologies of satellite remote sensing, satellite global positioning, mapping and geographic information systems..." (WSSD, 2002).

The WSSD reaffirms the role of information and decision support technologies promoted in Agenda 21. However, what is correspondingly affirmed is that the increasing importance of spatial information for decision-making and its use in a wide range of applications, demand new mechanisms and efforts for improving its access, sharing, integration and use. It further challenges that the range of advances in the information, communication and decision support technologies utilising these applications and data be reachable to everyone (as discussed by Borrero, 2002).

There is currently an unequal distribution of information, tools, and models among countries and continents. Although ambitious programs are being carried out in order to reduce the gap, advance production of seamless spatial data, and gain more equal development of SDI around the world, even distribution of capabilities is not likely in the near future, nor is it necessarily the desirable condition. It may be more desirable to share resources than to develop or install duplicate resources in all places. Thus inefficient and costly duplication of effort could be avoided and scarce resources could be applied to build better alternatives (models, data, hardware, software, communications...etc.) which could be shared through SDI to build distributed decision support capabilities.

Therefore, by developing understanding about the role of SDIs for spatial decision support, and the decision support capacity of different SDI frameworks, it is believed that SDI development may be guided towards better meeting some of the challenges offered by the decision environment and technological and user capabilities.

12.3 SPATIAL DECISION SUPPORT AND SDIs

Decision support capability refers to the mechanisms for users (decision makers and stakeholders) to access applicable data and technologies to support their decision-making. Given that spatial data is one of the most critical elements that underpin decision-making for 'many disciplines, SDI represents a significant potential to increase the range and coordination of decision support mechanisms for spatial data use. Decision support capability is a crucial function of SDI, to realize the maximum value and use of the spatial data and the policies established to provide common and timely bases upon which to frame decisions.

Decision support refers to the automation, modelling and/or analysis that enables information to be shaped from data which is useful to decision-making and enables improvement in the decision process. Decision support can be used to structure, filter and integrate information, model information where gaps occur in data, generate alternative solution scenarios as well as weight these according to priorities, and importantly enable group as well as distributed participation in decision-making. Applications and technologies contributing to decision support may be integrated into systems.

A Decision Support System (DSS) can be generally defined as "an interactive, computer-based tool or collection of tools that uses information and models to improve both the process and the outcomes of decision-making" (Lessard and Gunther, 1999). DSS may cover a spectrum for which the extremes are roughly marked at one end by a focus on decision support and at the other end on systems (Keen, 1987). This is as much because the technology and applications that DSS draw on constantly change and there is no independent or idiosyncratic technical base for it; as new tools become available and suitable new types of DSS will be built (Shim *et al.* (2002) provide a contemporary review). DSS have evolved by marrying the pragmatics of how to work with decision-makers, with a focus on computer tools that are useable. This has resulted in the emergence of Spatial DSS (SDSS) as a result of combining the capabilities provided by DSS with those from geographical information technologies, such as GIS.

SDSSdiffer in several important ways from more traditional DSS's, as a result of the type and context of the problems they address: they apply to entire communities, public lands, or publicly owned natural resources; they must attempt to bring together the disparate values and objectives of multiple stakeholders; and they need to incorporate multiple sources of expertise (e.g., hydrology, economics, biology, and engineering). Increasing interest in SDSS draws into contrast, traditional DSS which have generally focussed on private sector resources, single decision makers, and a narrower base of expertise.

This evolution over the past decades has marked a transition for DSS from systems typically used by one or a small number of specialists to those that are capable of reflecting broad social objectives. The focus has shifted from narrowly defined private sector objectives using proprietary databases to one that encourages collaborative decision-making about the places we live and work. Citizens and community groups are increasingly demanding a voice in these decisions, and developers are responding.

Part of responding to the challenge is that despite there being a range of tools and systems available, or under development, to support decision-making, no one system will provide the broad range of capabilities required by decision-makers and decision stakeholders. Nor will one agency or developer have the resources or mission to develop and maintain the range of tools needed by the decision-maker. These considerations, in conjunction with the current and evolving technology, fixed and shrinking budgets, and increasingly complex challenges of resource management combine to provide an opportunity for expanded cooperation in accessing and developing interoperable decision support tools and systems (Gunther, 1999).

The role SDI may play in this endeavour is thus becoming of increasing importance to enabling people access to and supporting their participation in the decision environment, be it through supporting user's data needs, or helping them to visualise and analyse a problem using available applications and technologies, such as SDSS. The nature of the decision support role for the decision environment will be the subject of discussion in the next section.

12.4 SUPPORTING THE DECISION ENVIRONMENT

There are two drivers influencing the development of the spatial data environment for decision-making. The first relates to the people environment, not least that there is a growing need for governments and businesses to improve their decision-making and increase their efficiency with the help of proper spatial analysis (Gore 1998). The second driver is the advent of cheap, powerful information and communications technology which facilitate the more effective handling of large quantities of spatial data. As these forces converge their nexus is a decision environment influenced by the data, technological and people components of SDIs.

Therefore the decision environment as it relates to SDIs is the result of the interrelationship of the technological, data and people components classified by Rajabifard *et al.* (2002) and discussed in Chapter 2. That is to say, the decision environment is shaped by the people environment of users (decision-makers and stakeholders in a decision process) and producers of data and any policies that regulate the relationship between people and data. The decision environment is also shaped by what data is available to be used in decision-making, its quality, cost, scale, coverage, completeness, whether it is accessible and how it may be applied to the decision or problem. Finally the decision environment is shaped by the technological environment, entailing the standards and technologies enabling data to be discovered, exchanged, integrated visualised and actually applied to the decision process. The interrelationship of these three environments and the impact on the decision environment are shown diagrammatically in Figure 12.1 and will be discussed in further detail in the following sections.

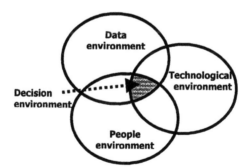

Figure 12.1: Describing the Decision Environment

12.4.1 The Data Environment

SDIs are tools to provide an environment within which people and technology interact to foster activities for using, managing, sharing and producing spatial data. However, to empower SDI framework implementation and optimization in the spatial information industry, technical and institutional issues of access to useable data need to be addressed. There is need for spatial information and to decrease the gap in availability, quality, standardization and accessibility of data between countries. People need to be able to access accurate, exchangeable, consistent and up-to-date spatial data resources to make informed decisions and to implement spatial initiatives. This process relies on the availability, accessibility and applicability of data to support spatial decision-making.

(a) Availability
Problem formulation involves scanning the environment for data relevant to finding the answer(s) to the proposed decision. It requires that data be available for collecting and integrating, to form information that will support problem formulation and evaluation of decision alternatives. For instance, one of the challenges in employing models for decision support is the availability of data from across various data warehouses within an organization.

Availability consequently refers to many forms of information sources which might be available to the decision maker through Intranets, Internet-based databases, catalogues, directories, clearinghouses and other online services. Information sources may include spatial data, metadata, schemas, symbols, feature type catalogues, place name gazetteers, scientific literature, knowledge bases, expertise, operational studies and others (Nebert, 2002; Cleaves, 1999).

Availability via the Internet, networks and other communication and information technologies has the advantage of improving the quality of data by gaining user feedback (and updates), promoting common information bases for decisions, as well as extending datasets' application range beyond the purposes for which they were originally created. This enhances the investment in spatial data collection and management and is the basis of the SDI concept.

(b) Accessibility
The information gathered from that which is available during problem formulation needs to be processed and examined for clues leading to the best decision, and is thus reliant on access to data and technologies enabling the analysis of possible courses of action, generation of possible solutions and testing of solutions for feasibility (Cleaves, 1999).

Common access to data and technologies for decision support promotes an environment for the sharing of spatial information as well as encouraging the integration of different datasets to enhance a decision maker's understanding of a problem and trade-offs associated with its resolution. It further encourages collaboration for decision-making. Access to data may be gained via Intranets, Internet Portals, atlases, clearinghouses, Research and/or Service Centres (such as the geospatial data service centres discussed by Groot and McLaughlin (2000)).

SDI also has the potential to facilitate access to and the exchange of derived datasets, models, and other data services that not only increase the capacity of the spatial data market, but also the ability to support different parts of the decision process. Participants with different levels of spatial data processing skills are able to access the data directly, or through technologies able to visualise, model, analyse and/or summarise spatial data.

Ultimately, the variety of mechanisms accessible to enable the presentation and application of spatial data for decision-making activities makes the decision-support capabilities of an SDI more flexible to the evolution of the spatial data market and adaptable to the changes presented by users and new technologies.

(c) Applicability

The process of accumulating data is in itself insufficient to assess and manage the complex process of sustainable development and its broad implications for the environmental, health and social issues that confront policy makers and citizens (NAS, 2002). These decisions often involve compromises and trade-offs, which will require integrating and displaying data in an understandable form and providing ways to chose among alternative solutions. This requires that a decision maker be able to build relationships between different types of data, merge multiple layers into synthetic information, weigh outcomes from potentially competing alternatives and forecast.

Whilst inroads have been made in promoting data availability, the shortfall in many current SDI frameworks occurs in the promotion of the applicability of data to support decision-making beyond information discovery and visualization, through to analysis and modelling. Technologies and applications which may need to be employed to use the data include user interfaces, models, web mapping and data symbolization, overlay, integration and transformation techniques. These enable the application of datasets produced by different standards and specifications to be used together to greater effect to support decision-making.

However, despite the range of geographic information technologies and applications available to augment the use of spatial data, Environmental Information Systems-Africa (EIS-Africa) concluded from a review of information initiatives in five countries in Africa that few application-oriented examples demonstrated advanced analysis of geographic information (NAS, 2002). Reasons cited include:

- Projects are oriented toward data production and updating rather than useage or application;
- Focus on technical issues instead of data management in support of the decision-making process;
- Lack of inclusion of stakeholders in information networks in several countries undermining incorporation of new knowledge and data applications being produced daily in universities, government agencies, and by private enterprises.

Therefore, despite the range of tools available that are able to enhance the application of spatial data for decision support, much of the potential applicability of spatial data is being underutilized. To overcome this barrier, models, other tools, data, and information must be readily available, easily accessible and integrateable with the results being in the hands of the users when and where they are needed, to support their decision-making. Only when this vision is realized will some of the shortfalls in spatial data support for decision-making be remedied.

Figure 12.2, illustrates the relative support of SDIs for spatial data use in decision-making. The vertical axis, namely the reach of support to decision makers, describes the relative number of datasets available, the number of these datasets that are readily accessible to users, and the number of datasets that are standardized and may be utilised with a variety of technologies and are thus easily applicable to the decision-making process. The horizontal access describes the relative increase in the number of ways decision-making may be supported if data are not only available but accessible by a number of different means and can then be applied using the huge variety of spatial technologies that exist. This axis therefore refers to the range of mechanisms supporting the spatial data environment.

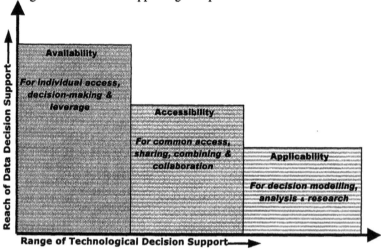

Figure 12.2: Relative Support for Spatial Data Use in Decision-Making

12.4.2 The Technological Environment

The technical nature of the digital spatial data environment requires that SDIs are able to continue to adapt to the rapidity with which technology develops and the changing relationship (including people's rights, restrictions and responsibilities) between people and data.

This suggests an integrated SDI cannot be composed of spatial data (and value-added services) alone, but instead involves other important issues regarding interoperability, policy implementation (for access, pricing, licensing, security, privacy) and access networks, which constitute some of SDIs core technological components.

For example, the rise of the Web as a common platform extends the capabilities of decision support tools to a very large number of users. The fact that a standard Web browser can be used as the user interface/dialogue means that organizations can introduce new data and technologies at their sites at relatively low cost. A Web browser user interface allows access to data and implementation of decision support technology with little user training. Therefore the potential exists for Web-based SDI to increase productivity and profitability, and speed the decision-making process in a distributed decision-making environment. Through increased support for decision-making, reduced cost and reduced support needs Web-based SDI can significantly improve organizations' use of their existing infrastructures.

Another trend influencing the technological environment for decision support is the standardization of Web-based technologies for interface design, for instance for model-based decision support software (Shim *et al.*, 2002). Shim et al. (2002) report that most major decision support software developers now have web sites and offer downloading trial software for further exploration. Alternatively, by using an Application Software Provider (ASP) model of delivery for decision support functionality, customers may rent the software on a per-use basis from an ASP who hosts the decision support application and provides secure access over the Internet for software purchase and installation (Shim *et al.*, 2002). It seems likely that mobile tools, mobile e-services and wireless Internet protocols will mark the next major set of influences of the technological environment.

According to Chapter 2, anyone (data users through producers) wishing to access datasets must utilise the technological components of an SDI. What is important to remember is that SDIs exist to support the access to and application of data for products, services, but particularly decision-making. The strength of the infrastructure therefore comes from its ability to support other infrastructure and a range of technologies to facilitate these activities (see for example Chan *et al.*, 2001), in particular the ability for people to share data and interact with technology to support their decision-making objectives.

How the technological environment interfaces with the spatial data and people (user-producer) environments to support decision-making may be depicted in the following framework adapted from Nebert (2002) in Figure 12.3.

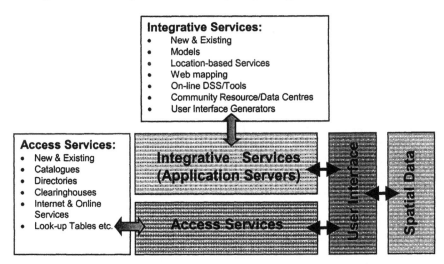

Figure 12.3: Framework Illustrating the Relationship Between Technological SDI Components and Decision Support for Accessing Data and Technologies

The framework illustrates the means for SDI to support decision-making through facilitating user access to data and decision support technologies and services. This enables decision support to reach a broad spectrum of spatial data users and a wide range of decision-making environments, using some of the technological innovations discussed above. The tools and services associated with such a framework may promote equity of access to decision support through SDI by decision stakeholders, by people with different levels of experience with geographic information and to forums within which many stakeholders involved in an issue can collaborate.

12.4.3 The People Environment

People are the key to transaction processing and decision-making. All decisions require data and as data becomes more sensitive, human issues of data sharing, security, accuracy and access forge the need for more defined relationships between people and data. The rights, restrictions and responsibilities influencing the relationship of people to data become increasingly complex, through compelling and often competing issues of social, environmental and economic management.

As the SDI concept evolves the current emphasis for people accessing data directly will become increasingly transparent to data users, as access is increasingly sought to 'simplified answers', or data shaped (integrated, analysed, modelled) through integrative services and technologies like decision support systems, rather than in its raw form. The transition to this model of 'applied' data access and use, will require the gradual harmonising of data and information services allowing eventual seamless integration of systems and datasets. It will also require that SDI develop mechanisms to facilitate the availability and access to integrated data and interoperable technologies. As the amount of data available increases, people will need assistance to structure, filter, visualise and integrate data, model information where gaps occur in data, generate alternative solution scenarios as well as weight these according to priorities, and importantly enable the use of different forms of information, from people's experience, specialised knowledge, or incorporating social, environmental and economic factors.

The potential for SDI to facilitate this complex and evolving relationship between people and data, as well as in technology uptake and utilization, is great, however it will require that the human capital be developed so that the people environment may complement the developing data and technological environments. This may be achieved through training and skills programs, specialised education, exposure to technology demonstrator projects and applications and improving the ease with which spatial data may be discovered, accessed and used. Potential mechanisms for expanding SDI capabilities in such directions are summarised in Table 12.1. Human capital may be developed by these means as part of a program of capacity building and in association with an enabling technical framework as discussed in the former section.

Table 12.1: Potential Mechanisms for Expanding SDI Capabilities to Support the Decision Environment (Adapted from Cleaves, 1999: 22-24)

Data	Potential Technological Mechanisms
Availability	Decision Structuring tools that emphasise information search and value. Examples include network analyses, decision tree models and confidence rating schemes that allow the communication about the relative strength and importance of various information types and sources.
	Scenario development aids that help decision makers rebuild and circulate alternative scenarios for strategic planning. These could be tied into Internet-based planning data and projections along with other ongoing databases. They represent cumulative resources that are often unavailable for reuse.
	Smart databases that codify, catalogue and provide access to agency decision experiences. Summaries of different types of decisions would serve as a guide for future directions.
	Tools for eliciting expert judgment that reveal and manage biases, represent causal pathways that drive expert predictions, represent ambiguity and uncertainty, and facilitate group judgements.
Accessibility	Guides to jurisdiction-relevant interagency experts, organised by specialty, experience and contact information, to support the assessment of the state of knowledge in particular problem domains and a resource for accessing and assembling expert panels and advice.
	Online links to other efforts to solve similar problems, enabling real-time learning advantages and avoidance of ineffective alternatives.
	Technology Tool Boxes in the public domain employing modularity and interoperability in the design of decision support tools that enable iterative decision support design as the problem unfolds.
Applicability	Problem visualization aids, including spatial displays of data and a wide variety of options for graphic representations of abstract concepts.
	Prompts for eliciting and documenting selection rationale that would enable the systematic checking of multiple attributes and objectives. This enables accountability and broad participation and review of the decision process.
	Uncertainty representation schemes that communicate the nature of uncertainty and risk in datasets and models, through to ranges of expected scenario outcomes/consequences.
	Trade-off analysis models, which require precise description of what is being given up in the selection of one alternative over another, and documents the responsibilities of the trade-off and who will bear them.
	Chat boxes and bulletin boards that allow analysis teams and decision makers throughout an organization to exchange views on ongoing analyses and decisions.

Capacity building, as defined by Georgiadou (2001), may refer to the provision of foundation data, metadata standards, clearinghouse functionalities and a facilitating environment for decentralising GIS application in manageable application domains within the SDI concept. Capacity building enables the productivity of the relationship between people and data (as well as technology) to be supported and improved, which therefore increases the economy of the relationship. Capacity building supports the prospects for continuing and strong productivity growth between people and data by enhancing the investment in human capital – knowledge, skills and ideas - through education, research and development.

Economists only became aware of the value of the relationship between investments in human capital and productivity in the early 1990s with the advent of "new growth theory", however, the implications of this theory for the development of the relationships between data, people and technologies in the context of SDIs is quite significant. Dowrick (2001) argues that investment in human capital, unlike physical capital such as machines and buildings, is less likely to be subject to diminishing returns and brings us closer to the factors that drive technological advance. The latter supports the principle that the more knowledgeable and skilled the workforce the more rapid the adoption of new technologies is permitted. This is because of several key attributes which are the product of investment in human capital or capacity building: .

Non-rivalry – unlike physical capital that can generally only be used exclusively by one or a group of people (once one person is using a machine no other person can), knowledge or ideas can be used simultaneously by many people. Therefore, unlike a machine, once an idea has been announced to the world it can simultaneously be used to develop a wide range of applications.

Complementarity – the more people invest in human capital, the more the numbers increase amongst which the skill/knowledge may be exercised. In this way, an investment in acquiring a particular skill or set of knowledge actually increases the benefit of those who have already invested in the skill/knowledge.

These qualities of human capital, when developed through capacity building and harnessing the technological and data environments, provide an important key to unlocking the potential of SDIs to support decision-making.

12.5 THE DEVELOPING DECISION SUPPORT STATUS OF SDIs

SDI is developed to support ready access to spatial information to support decision-making processes at different scales for multiple purposes. SDI is achieved through the coordinated actions of organizations (people) that promote awareness and implementation of complementary policies, common standards, and broad access networks through effective institutional mechanisms. This section discusses the development of SDI frameworks that not only support data products but also address the inclusion of technologies and services in order to support a broader range of spatial decision-making processes and the role of people in these processes.

The role of spatial technologies in the operations of SDI has been recognised as essential to meeting the needs of the multi-disciplinary and multi-participant environments that characterise decision-making for sustainable development (GSDI, 2001). SDI cannot exist as a means in itself – it is essential the infrastructure support the development of spatial data products, services and the needs of diverse decision-making environments.

Government agencies are the custodians of large amounts of spatial data useful to the community. Recent technological developments have democratised the ways communities' access information and knowledge (Barker *et al.*, 1999). The demand has grown for more flexible service delivery from government, including access to tools, spatial information and the skills to interpret the information. Recent work has proposed ways by which government can interact with community and industry to improve the flow of information (Barker *et al.*, 1999), particularly spatial information, between data custodians and users. Government has an important role to play.

There are many existing methods to distribute data and knowledge to communities to assist spatial decision-making. Ideally, an SDI provides the facilities for stakeholders to share and exchange information, so the process is not simply one of spatial data dissemination. Such facilities range from individuals including consultants, government representatives and data brokers, who have access to data, GIS and other spatial technologies and interpretative skills, through to online services delivered through Web sites either created and maintained by communities and centred on the business and information needs of that community, or government web sites. Resource centres for communities are a further example where a suitable combination of integrated datasets, internet tools, web sites, data-provider extranets and facilitators/consultants are selected to help users with their information needs. These mechanisms are part of SDI frameworks being developed to support access to data and technologies for structuring spatial data for decision-making.

Atlases and directories provide the means for text and often spatial searches for data to satisfy data discovery in the decision process. These access services as yet fall short of providing access to data which is downloadable for analysis and modelling or to derived data products where this analysis has already been done by value-adding resellers, or where models could be found to contribute to the design phase of the decision process.

The capability for directories to provide more than unidirectional links to other sources of data and accept the addition of independent data, provide the foundations to support a competitive data market and clearinghouse development for access and two-way (download-upload) data exchange.

Online services, the community resource centres and decision support technologies have a different emphasis to the atlases and directories. These mechanisms have greater capability to extend data availability and discovery in the decision process to making tools, information, and the skills with which to interpret data, more accessible, as well as offering broader support to more of the decision process. In terms of decision support technologies and systems, this refers more to extending the interpretative capabilities to those without the analytical or modelling resources or experience, or alternatively extending the datasets derived from such analyses.

12.6 DECISION SUPPORT IN THE FUTURE OF SDIs

Improved economic, social and environmental decision–making are principal objectives for investing in the development of SDI at all political and administrative levels and is promoted by improving the availability, accessibility, and applicability of spatial information for decision-making.   o

The need to build SDI that supports a broad range of decision-making and is user-friendly for a variety of decision makers underpins local, national, international and global cooperation for problem-solving that transcends borders, where data as well as models and tools can be accessed and used to help spatial decision support.

Yet, the ability of SDI to support decision-making processes and a varied user environment is dependent on the current and developing capabilities of SDIs to support spatial data and derived data products, models and decision support technologies, through the mechanisms developed within SDI frameworks. These represent the interface of the people, data and technological environments influencing decision support capability for SDIs.

Future directions for SDI development will continue to require greater framework development to meet the challenges of diverse decision-making environments and spatial data interpretation. Issues such as urban renewal, forest management, native title administration, coastal economic zone management, defence, drought relief and land care cannot be addressed without available, accessible and applicable spatial data that support the decision process, and more particularly decision support technologies. This will require that greater consideration be given to the range of mechanisms available to support decision processes, provide interpretation, analysis and flexible applicability of spatial data to decision-making in SDIs, as well as the perception of the infrastructure as purely a data facility.

The interaction of the spatial data users and suppliers and any value-adding agents in between, drive the development of any SDI. These present significant influences on the changing spatial data relationships within the context of SDI jurisdictions. This chapter looks at how the availability, accessibility and applicability of spatial data are being fostered alongside the development of technological and human capability to enhance the decision environment and thus support the changing data-people-technology relationships within the context of decision supportive SDIs.

12.7 REFERENCES

Agenda 21, 2002, Agenda 21. Online.
    <http://www.un.org/esa/sustdev/agenda21text.htm> (Accessed November 20).
Barker, A., Fry, W., Hardman, J., Peter, R. and Nelson, K., 1999, Model for community resource centres to empower the community. Discussion Paper, Department of Natural Resources (Project DNRS9819), Queensland, Australia. 43pp.

Borrero, S., 2002, The GSDI Association: State of the Art. Report of the GSDI Steering Committee, *6th GSDI Conference*, Budapest, Hungary, 16-19 September. Online. <http://www.gsdi.org> (Accessed 15 December).

Chan, T.O., Feeney, M., Rajabifard, A. and Williamson, I.P., 2001, The Dynamic Nature of Spatial Data Infrastructures: A Method of Descriptive Classification. *Geomatica* 55(1):65-72

Cleaves, D. 1999 Supporting the Decision Process: What can we hope for and expect from DSS's? Report on the Decision Support Systems Workshop, Denver, Colorado February 18-20, 1998. US Department for the Interior and US Geological Survey Open file Report 99-351, 1999, pp. 19-24.

Dowrick, S., 2001, Investing in the Knowledge Economy: Implications for Australian Economic Growth. Australian National University, Canberra, Australia.

Georgiadou, Y., 2001, Capacity Building Aspects for a Geospatial Data Infrastructure. *5th GSDI Conference*, 21-25 May, Cartagena de Indias Colombia.

Gore, A., 1998, The digital earth: understanding our planet in the 21st century. *The Australian Surveyor*, 43(2), 89-91.

Groot, R. and McLaughlin, J., 2000, *Geospatial Data Infrastructure: concepts, cases and good practice.* (Oxford University Press), New York, 286 p.

GSDI, 2001, Resolutions of the *5th GSD Conference*, Cartagena, Colombia, May 21-24, 2001. Online. <http://www.gsdi.org/docs/240501.htm> (Accessed May 2002).

Gunther, T., 1999, The Interagency Group on Decision Support. Proceedings of the Decision Support Systems Workshop. Denver, Colorado, 18-20 February 1998. U.S. Department of the Interior and U.S. Geological Survey. USGS Open-File Report 99-351, 1999.

Keen, P.W.G., 1987, Decision Support Systems: The Next Decade, *Decision Support Systems* Vol. 3, pp. 253-265.

Lessard, G. and Gunther, T. (Eds), 1999, Introduction. In Report of Decision Support Systems Workshop, Denver, Colorado, February 18-20, 1998. U.S. Department of the Interior and U.S. Geological Survey. USGS Open-File Report 99-351, 1999, pp. 1-5

Masser, I., 2002, Opening Plenary *6th GSDI Conference*, Budapest, Hungary, 16-19 September. Online. <http://www.gsdi.org> (Accessed 15 December 2002).

NAS, 2002, *Down to Earth – Geographic Information for Sustainable Development in Africa*, National Academy of Sciences. (Washington, USA: National Academy Press), Online. <http://www.nap.edu> (Accessed December 2002).

Nebert, D., 2002, Global Spatial Data Infrastructure Technical Working Group Status. Report of the Technical Working Group for the GSDI Steering Committee, *6th GSDI Conference*, Budapest, Hungary, 16-19 September. Online. <http://www.gsdi.org> (Accessed 15 December 2002).

Rajabifard, A. Feeney, M. and Williamson, I.P., 2002, Future Directions for SDI Development. *International Journal of Applied Earth Observation and Geoinformation*, Vol. 4, No. 1, pp. 11-22.

Shim, J.P., Warkentin, M., Courtney, J.F., Power, D.J., Sharda, R. and Carlsson, C., 2002, Past, Present, and Future of Decision Support Technology. *Decision Support Systems*, Vol. 33, pp. 111-126

WSSD 2002, Plan of Implementation, Means of implementation, World Summit on Sustainable Development, Johannesburg, 2-11 September. Online. <http://www.johannesburgsummit.org> (Accessed November 20).

CHAPTER THIRTEEN

Financing SDI Development: Examining Alternative Funding Models

Garfield Giff and David Coleman

13.1 INTRODUCTION

Previous chapters have introduced the concept and hierarchy of a Spatial Data Infrastructure (SDI) and reviewed SDI initiatives from a Global to a Local perspective. However, associated with the implementation of these initiatives are important technical, economic, institutional and socio-political issues. An SDI typically crosses different political and organizational boundaries, and ensuring its ultimate success depends on employing the diverse collective skills of people of different organizational and environmental cultures. Understanding these issues will be essential to the successful implementation and maintenance of an SDI.

In the field of Geomatics and its related professions different definitions have been used for the term SDI (Rhind, 1997; Coleman and McLaughlin, 1998; Nebert, 2001; and Rajabifard and Williamson, 2001). In this chapter, reference to an SDI does not include National Mapping Agencies (NMA), private/public owned spatial databases or individual Geographic Information Systems (GIS) alone. Again, it refers to the larger integration of datasets, policies, standards, institutional arrangements, human resources and the technical issues necessary to facilitate the dissemination of current and well defined spatial data throughout the information society.

While all these elements can be considered components of an SDI, the mix is different from country to country and even among the respective SDI initiatives of different jurisdictions within a given country. Further, this mix will change as a program (and its user base) evolves from early data collections through more advanced information management and application stages (McLaughlin, 1991). Other factors affecting the mix include the following:

- In a particular jurisdiction, the relative emphasis placed on basic data collection vs follow on data discovery, access, visualization and application;
- The natural role and relationship between various SDIs operating at the national, state/provincial and local levels;
- The presence and relative degree of influence of the various datasets (e.g. cadastral data, topographic data and road networks) within the SDI effort;
- The existence and function of the private sector in development and delivery of spatial data products and services to both targeted end-users and the mass market; and

- The influence of natural emergencies, wartime situation or catastrophes which may require a significant sustained response in term of new information, technologies and other spatial data applications.

Since the mix of SDI components and emphases differs from place to place, it follows that there will be no single funding approach that will meet the needs of all countries. Rather, a mix of financing models may be required in any given jurisdiction – and that mix will certainly change over time.

With this in mind, the aim of this chapter is to explore and encourage additional interest in the economic issues associated with SDI implementation especially the area of funding. After a brief review of the economic issues associated with SDI implementation, the chapter presents an in-depth study of funding models for SDI development in different implementation environments.

13.2 THE ECONOMIC ISSUES OF SDI IMPLEMENTATION

Infrastructure economics may be viewed as the science which sorts techniques of quantitative analysis used for selecting preferable alternative(s) from several technically viable ones (Fraser *et al.*, 2000). In infrastructure economics cost and benefits analysis forms the primary area of focus. However, for an infrastructure to be implemented efficiently in the information society other areas of infrastructure economics – structured funding models and the relationship between infrastructure economics and the inherent complexity of today's society – must also be taken into consideration.

Like any other infrastructure, an SDI should provide the underlying framework for easy access to spatial data. It should be effective, efficient, easy to use and transparent to the users. The users should be aware that an SDI exists and expect it to be there always to facilitate them in their quest for spatial data but never be over shadowed by its present (ECA, 2001). Since an SDI is similar to general infrastructure it will have similar economic characteristics of general infrastructure. This is evident from the number of research and actual benefit cost analyses carried out on the implementation of SDIs. These benefit cost analyses are absolutely necessary since, the implementation of an SDI will require significant capital investment that will have to be justified in terms of returns on investment (financial and/or social benefits). However, associated with the implementation of an SDI are a number of other economic issues that must be addressed by coordinating agencies.

The economic issues associated with the implementation of an SDI spans across both macro and micro-economic levels. The implementation issues at the macro-economic level are closely associated with the institutional and socio-political issues and will not be covered here. Although both levels do provide tremendous challenges for SDI coordinating agencies, the challenges at the micro-level are the more intriguing and less addressed of the two (Beerens and de Vries, 2001; Coleman and Giff, 2001).

At the micro-level, some of the more challenging issues that SDI coordinating agencies must address in order to ensure successful implementation and maintenance of their SDIs are:

a. The creation of a Business Plan -The main objectives of which are to secure timely financial and political support for the implementation and maintenance of the SDI. A typical business plan should contain the following main components (CIE, 2000):
    -   The Nature of Spatial Information/Data and the demand for it
    -   The Stakeholders
    -   Benefit Cost Analysis (difficulty in identifying intangible cost and externalities)
    -   Return on Investment/Net Public Benefits
    -   Risk Management and Analysis
b. The Structuring of Pricing Policies
    -   Who should pay
    -   What price should be paid
    -   Licensing fees for value added provider
c. The Marketing of the Products of an SDI
d. Funding Models - How will the implementation and maintenance of the SDI be funded?

Attempts have been made by both SDI coordinating agencies and academics to address the above economic issues, resulting in implementation policies and publications on these issues (Beerens and de Vries, 2001; Groot, 2001; Rhind, 2000). However, the concept of having funding models for SDI implementation and maintenance is the least addressed of the micro-level economic issues and therefore, will be the focus of this chapter.

13.3 THE CONCEPT OF SDI FUNDING MODELS

The main function of an SDI funding model is to act as a guideline to SDI coordinating agencies on how to formalize, structure, present and source financing for the implementation and maintenance of an SDI. The funding models introduced in this chapter are only conceptual since the implementation environment (economic climate, political climate, legislation, etc.) of individual SDIs may differ thus, requiring adjustment to the models. However, conceptual funding models can become very important to SDI development since they can provide SDI coordinating agencies with guidelines to such questions as:

*   Where and how to seek out funds?
*   What are the relationships amongst the different funding components?
*   How best to present the funding arrangement to Governments and Financial Institutions?
*   How funds should be structured to facilitate efficient implementation?
*   Over what period will the funds be disbursed? and
*   What are the effects of funding on pricing policies?

The answers to the above questions are important to SDI implementation in any country but even more in emerging nations nations who are in the process of developing free market economies based on openness and competition (Edwards, 1995). Such nations are usually affected by very limited financial resources, poor capital markets and inadequate political structures (IIPF, 2001). These and other factors will make infrastructure financing - on its own a formidable task - an even more complex problem in these countries. Emerging nations and nations in transition cannot afford to totally finance their SDIs from their local budgets thus, additional funds must be obtained externally. Long-term capital financing models have the potential of becoming an important tool for assisting SDI coordinating agencies of these nations in sourcing, structuring and formalizing funding for SDI implementation and maintenance.

Although the concept of SDI funding models is important to SDI implementation within the information society, the development/selection of the appropriate funding model(s) will differ among developed nations, nations in transition and emerging nations due to the nature of the implementation environment. In the developed world, the implementation environment normally consists of a vibrant economic climate of which the geomatics information sector provides on average 0.5% of GNP (Tveitdal, 1999). This and other favourable factors of the implementation environment of the developed world allow the funding models to be developed with strong public and private sector components. As well, developed nations are now working on the second generation of SDIs, which are now beyond the status of marginal cost providers and thus more favourable to commercialization, for example, the creation of value-added products and services (Giff and Coleman, 2002).

In contrast to that of developed nations, the implementation environments of emerging nations and nations in transit vary from having sustainable to very poor economic climates. In these economies, the contribution of the geomatics information sector to a much lower GNP is on average a mere 0.1% (Tveitdal, 1999). SDI developments are also in the infancy stage here, so most of the funding, policies, and operational decisions are focused on developing and implementing the basic databases or putting datasets in place. In such cases, the "SDI" may essentially be a euphemism for a national mapping program. Other features of the implementation environments of these nations that will make the funding models differ from those of the developed nations are:

a. The Governments of these nations are usually poor thus, the dependent of external agencies to finance infrastructure development;
b. Spatial data activities are normally developed on a project basis with no funds allotted for continuation or maintenance;
c. A lack of coordination amongst projects funded by external agencies (de Montalvo, 2001);
d. The lack of institutional coordination and the awareness of the usage of spatial data (Giff, 2002);
e. Unstable social and political climate;
f. The lack of efficient utility infrastructure, technology, Internet providers and trained professionals to support an SDI (Ezigbalike *et al.*, 2000); and
g. The existence of legislation and other barriers that limit the role of the private sector in the collection and dissemination of spatial data.

The above list offers some of the key issues that make the selection/development of funding models for SDIs in emerging and transition nations differ from that of developed nations. While many of these issues also exist in developed nations, it is the level of existence that affects SDI implementation. Issues affecting the selection of funding model(s) will also vary within the group classified as "emerging nations and nations in transition", since the implementation environment (economic, political, social, cultural climate, etc) within this group tend to have distinctive variations.

13.3.1 Funding Models of the First Generation of SDIs

The majority of today's existing SDIs evolved from the programs and/or strategic support of National Mapping Agencies (NMAs). A significant proportion of their funding was derived from the budgets of these NMAs and the remainder was found through specially-funded projects and private sector investment to a lesser extent (Coleman and Giff, 2002). The first generation of SDIs evolved from the NMAs without the development of long term financing mechanisms. In an attempt to structure the funding of the first generation of SDIs, Rhind (2000) concluded that there are at least four different models today that could be applied to SDI funding. They are:

a. Government Funding (Funds derived from taxation);
b. Private Sector Funding (Derived from fees charged to customers);
c. Public Sector Funding (Derived from fees charged to customers); and
d. The Indirect Method (Funds derived from advertising, sponsorship and other indirect methods).

The applications of these models to current SDI initiatives vary from 'stand alone' to a combination of one or more or all of the above models.

A review of general infrastructure and Internet financing examples suggests that the first model was and still is the most widely used. In the past, government made the most significant contribution to infrastructure development; for example, the US Highway Network (Sorensen, 1999). However, since governments are now moving away from investing heavily in infrastructure, alternative sources must be sought to fill the void and thus, the need for new models.

Income generated from the usage of the infrastructure ranked next (e.g. income from fix-lines and cellular networks). In the context of an SDI the sale of spatial data is on the increase and is viewed by many as one such source of alternative funding to government. To this end, NMAs around the world (e.g. Europe- Ordnance Survey and The National Geodetic Mapping Agency of Portugal and Asia-Pacific- Land Information New Zealand) are in the process of implementing policies aimed at making themselves more self-sufficient through income generated from the sale or provision of spatial data. Success stories can be seen in some European Countries, for example the Ordnance Survey of Great Britain (Ordnance Survey, 1999; Bing, 1998; de Jong, 1998; Kok and van Loenen, 2000).

These organizations are now aiming at producing and marketing their goods (spatial data/information) and services more efficiently. In his review on infrastructure funding, Crandall (1996) concluded that an increase in the supply of affordable spatial information/data in the format required by the consumers may in the long-term result in an increase in demand. This increase in demand may lead to increased revenue — thus the possibility of funding an SDI through fees derived by Statutory Bodies (Government owned companies) and the private sector.

In some countries – and especially in programs at the state, county and local levels -proceeds from land market activities carried out by one or more government agencies (i.e., real property registration, land titling, and valuation transactions) may provide important financial support to SDI efforts (please see Williamson *et. al*, 1998 for an Australian perspective and Bevin, 2002 for a New Zealand perspective on this matter). In these cases, if a host organization is established as a special operating agency able to retain any revenue it generates, then income from customer fees, subscriptions, and levies may be rolled back into the organization rather than being simply collected and stored in the government's general revenue accounts. Service New Brunswick in Canada and PSMA Australia Limited are two good examples of organizational models where important and significant revenues from land market activities are retained and applied towards their respective program's operations and goals (SNB, 2002) (please see Chapters 7 and 8).

Finally, Crandall's review also covered the indirect method and concluded that this method of generating revenue is on the rise but the environment is still risky. That is, the expected returns from advertising in the spatial data environment are low since advertisers are not yet convinced of the volume of traffic a spatial data/information site will command. Thus, an influx of investment in this area is not expected in the near future. However, the authors are more optimistic than Crandall. This optimism is based on success of companies such as MapQuest™ and Microsoft's TerraServer™ to a lesser extent in this area. These companies are presently generating steady revenue from advertising on their spatial data/information related websites.

Although there are some success stories, the latter two models are in their infancy stage and are yet to be tested on a broad scale in a spatial data environment. However, the likelihood of using these models along with other models seems plausible.

13.4 ALTERNATIVE FUNDING MODELS

Clearly, central governments have acted as the major financiers of infrastructure development to date. However, in order to produce more efficient, leaner and transparent governments with low public deficits, governments world-wide began embarking on public sector reform programs through the 1990's (Campos and Pradhan, 1997). A product of such reforms is leaner budgets and resultant cutbacks in infrastructure financing. In order to fill the void left by government, either alternative funding models must be developed or alternatively, persuasive arguments for the return of government financing must be structured by SDI coordinating agencies.

A number of alternative methodologies may be used to design funding models for SDI implementation and maintenance. These methodologies will differ based on conditions within a given jurisdiction, for example, economic condition, the level of implementation and government policies (Giff and Coleman, 2002). In this respect, an SDI is similar to any other infrastructure providing public good with positive externalities. This notion of an SDI as a provider of public goods is one of the key designing factors in the proposed models.

13.4.1 Infrastructure Classification

This section will briefly review the two main categories preferred by most economists in discussing infrastructure financing and attempt to position SDI in these categories. This classification will then be used as the basis for the selection of a methodology(s) for the development of funding models. The two main categories under which SDIs can be best analysed in an economic sense are:

SDI as a Classic Infrastructure: where capacity or usage is not as important as providing a public good. Public Good may be defined as that which is both non-rival and non-excludable. A non-rival good is that which "...one person's consumption does not diminish the amount available to other people", and non-excludable is the concept where "... one person cannot exclude another person from consuming the good in question" (Yevdokimov, 2000). Spatial information, the product of an SDI clearly satisfies the above conditions and thus, the justification of placing an SDI in this category.

SDI treated as a Network Infrastructure: with each section or component of the SDI viewed as a node of the network (Figure 13.1). In this category, the capacity of the network is of utmost importance. The success of the network and thus, the level of investment/new investment in the network will be determined by the capacity the network handles over a fix period of time. Therefore, in order to secure funds for this type of network, it would be necessary to demonstrate that the network has or will have the capabilities to operate at a capacity that will generate satisfactory returns on investment.

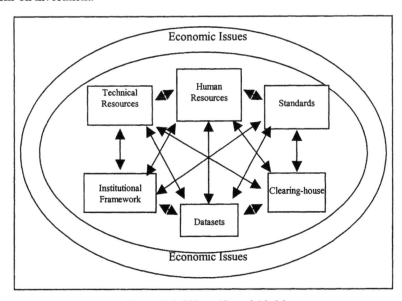

Figure 13.1: SDI as a Network Model

(a) Natural Monopoly

Another strategy is to define an SDI in terms of the social benefits it generates. For example, governments can be persuaded to invest more heavily if SDIs are placed in a category that identifies them with maximum social benefits. In order to strengthen the arguments for government investment in SDIs, an SDI could be viewed as a natural monopoly instead of a pure monopoly. This classification is justified since most information industries tend to become natural monopolies (Yevdokimov, 2000). "A natural monopoly is an industry characterized by economies of scale sufficiently large that one firm can mostly efficiently supply the entire market" (Yevdokimov, 2000). That is, a natural monopoly is an industry in which the advantages of large scale production makes it possible for a single firm to supply the entire market output at a much lower average cost than a larger number of firms producing smaller quantities. The ownership of a natural monopoly is similar to that of a pure monopoly in that it a can be government owned and regulated or privately owned with some government regulations or owned by both government and private sector. However, unlike a pure monopoly a natural monopoly arises "naturally" without the use of barriers to entry into the market. Therefore, the main difference between a natural monopoly and a pure monopoly is that, for a pure monopoly to exist there must be some barrier(s) preventing the entry and survival of a competing firm.

If an SDI is seen as a pure monopoly, then it would be the responsibility of government to encourage its divestment possibly through more private sector participation. Pure monopolies should be discouraged since:

- they tend to support smaller networks and charge higher prices (Economides and Himmelberg, 1995); and also
- the pure monopolist's goal of profit maximization usually leads to lower production levels and higher cost to consumers (Economides, 1996).

A natural monopoly classification of an SDI will have the following strong arguments for government investment in its development, implementation and maintenance. They are:

- It can be proven that the demand for the products of natural monopolies normally reflects significant social benefits (Economides, 1996; Yevdokimov, 2000), for example, they provide consumers with a single uniform system, while paying approximately the same price as in a competitive system (de la Vega, 2000); and
- The interaction amongst the demand, the externalities of a network and the monopoly's marginal cost normally produces maximum benefits to the society as well as the lowest prices and total production cost (Yevdokimov, 2000).

Thus, in summary, classifying an SDI as a natural monopolistic network provides a strong reason for governments' financial and political support since, it is a proven economic theory that natural monopolies provide social benefits (Economides, 1996; Economides, 1993; Yevdokimov, 2000). This is in part due to the fact that they produce goods and services at a lower cost. However, it should be noted that natural monopolies also have their drawbacks in that they suffer from the same inefficiencies as pure monopolies and will reflect very low if not zero economic profit in the long term.

13.4.2 Funding Models for SDIs Classified as Classic Infrastructures/Natural Monopolies

If an SDI is classified as a classic infrastructure/a natural monopoly then its should be possible to treat the financing of the SDI similar to that of any other infrastructure which is of national importance in terms of economical and social benefits. Thus, it should be possible to develop SDI funding models using analogies and lessons learnt from the financing of other infrastructures such as highway networks, railroads, telecommunication and sewage system to name a few. Possible SDI funding models developed out of the study of general infrastructure financing include the following:

a. Direct government funding – SDIs that fall into this category as mentioned previously should have strong arguments for total government funding. An example of an SDI created from centralized government funding is the Portuguese National SDI (de Montalvo, 2001).

b. The application of incentives such as matching ratios to stimulate investment – Under this type of arrangement the federal government would match (according to the specified ratio) the amount of funds invested into the SDI by a state. This type of venture would encourage states to seek out investment for their SDIs so that they can access government funds. This type of model could be well suited for Federated Government structures, for example Australia where individual states, provinces or territories are required to develop their own portion of an SDI.

c. Special Taxation – The use of taxation for the financing of an SDI can be in the form of positive or negative taxation or a combination of both methods. Positive taxation can be Tax Incentives (i.e. the waiving of taxes on goods/profit) while negative taxation can be in the form of Tax Increment Financing (i.e. the imposing of a special tax on goods/services/properties). Examples of tax increment financing can be seen in the imposition of a special tax on telephone services to support E-911 programs in North America and in Indiana special property taxes were levied on selected parcels to finance infrastructure development within the specific area where the parcels were located (DeBoer *et al.*, 1993). Another special tax worth mentioning is a levy on spatial data related activities (e.g. user fees, transaction fees, subscription fees and levies on property transactions). These incomes would then be channelled to a special fund which would then be used to finance SDI implementation. For example in New Zealand to assist in the implementation and maintenance of the Land-Online strategy a special tax was levied on all land transfer.

d. The establishment of special banks or financial institutions to underwrite low interest loans for the investment in SDIs. These institutions can have a similar structure to Agriculture Credit Banks established in Europe to support the growth of agriculture. The Federal Geographic Data Committee (FGDC) is presently carrying out research in this direction (Urban Logic, 2000).

e. The issuing of medium and long term tax-free bonds specially targeted at (for example) large public and private spatial data user and spatial data software developers. Since the issuing of bonds is very dependent on market conditions, research in present and future market conditions should be undertaken before applying this option.

f. SDI funded through partnerships –The most popular model is to have a Public/Private Sector partnership for one or more components of the SDI. An example of this type of initiative is Teranet Inc™. Teranet Inc™ was established through an equal partnership between the government of Ontario Canada and Teramira Holdings Inc. However, other partnerships amongst different levels of the Public Sector and International Agencies are also possible.

g. Limited-recourse Structures – Over the past thirty years, the private sector's contribution to public infrastructure financing has increased via the usage of limited-recourse structures. In this technique, the private sector will undertake the construction, financing, operating and maintenance of the infrastructure for a limited concession period (Buljevich and Park, 1999). During the term of concession, the private sector is able to collect returns on their investment by charging a fee to the users of the infrastructure. The more commonly used limited-recourse structures are: build, operate and transfer (BOT); build, own and operate (BOO); build, own, operate and transfer (BOOT); and build, lease and transfer (BLT). The most popular limited-recourse structure financed project to date is the Eurotunnel (Buljevich and Park, 1999). Other successful projects are the 407 Highway in Ontario, Canada (Fraser *et al.,* 2000) and the Aguas Argentinas water distribution system in Argentina (Buljevich and Park, 1999).

h. In some cases depending on the implementation environment the models proposed above might fall short of raising the complete capital investment required for an SDI. A solution to this problem could be a combination of the models listed above. Combining the models would depend on government structure, financial markets and the political climate to name a few.

Treating an SDI as a "classic infrastructure" implies that governments will assume the lead role in implementing and managing the SDI in a given jurisdiction. However, this definitely does not exclude private sector participation. Quite the contrary, depending on contracting policies and industrial capacity in a given jurisdiction, private companies may play a major role in terms of contract data collection, software and system development, and outsourcing of data management and distribution activities. However, under this classification, ultimate responsibility for effective infrastructure creation and maintenance would remain with government.

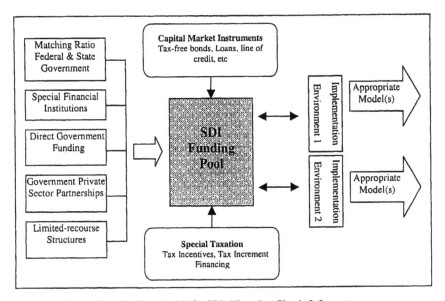

Figure13. 2: Funding Models for SDIs Viewed as Classic Infrastructures

If the SDI is classified as an essential part of a nation's capital infrastructure producing goods for the public benefits, then the above models should be favourable alternative methods for obtaining capital financing. These models can be collected together to create a pool of SDI funding models (Figure 13.2). From this pool, one or more models can be selected for SDI financing based on the environment in which the SDI is to be implemented. The models depicted in Figure13.2 are applicable to SDI implementation in general but would be most suited for SDI implementation at the national and state level.

13.4.3 Funding Models for SDIs Classified as Network Infrastructures

The main goal of an SDI in this category is to operate at a self-sustainable level or even at a profitable level. Therefore, the role of government here may be reduced to that of coordinator/regulator and a partner in the development of the SDI. An SDI in this category should be developed through private sector participation (inclusive of non-profit organizations and academic institutions) and/or by a public-private partnership. Capital investment in the SDI should then be provided by either the private sector and/or by government and private sector contributions. The nature of the investment provided by the different sector will vary across the different components of an SDI. The private sector will invest primarily in the creation of software, system development, value-added products, management and marketing strategies, and other technologies needed to support an SDI. On the other hand, the public sector's main focus is on data collection and policies. The public-private partnership is the approach taken by both the United States of America and Canada (FGDC, 1997; Labonte *et al.*, 1998; and Masser, 1998). The challenges in this category lay not only in formulating funding models but also in developing "common criteria for spatial data infrastructure investments, align annual public and private budget cycles more effectively, and pool and leverage spatial data investment" (OMB, 2001).

Capital financing models in this category will largely depend on the structure of the organizations of the different partners and the nature of the economic climate of the implementation environment. Also, some basic level of government funding will be required to offset at least a specified percentage of implementation cost. The success of these models will greatly depend on whether or not there is an active private sector and a vibrant capital market in the implementation environment. Models that may provide incremental or supplementary funding of SDIs in this category include:

a. The creation of a consortium or incorporated association to manage and generate funds for SDI implementation. The incorporation of the above type of organizations will facilitate the following:

    -   The issuing of shares in the organization on the stock exchange or through private subscriptions (Urban Logic, 2000).
    -   Large users of spatial data can be asked to pay a membership fee to these organizations (Urban Logic, 2000).
    -   The solicitation of contributions from the individual partners, which should be considered as capital investment into the consortium.
    -   Access to capital market for financial assistance such as revolving loans and other similar debt structures.

a. The financing of an SDI through the sale of spatial data by government and private sector partners – The application of this model can be seen at the Ordnance Survey of Great Britain and Service New Brunswick in Canada.

b. Project Finance – A variety of variables has forced the world financial markets to seek out new financial innovations to cope with the changing market conditions, for example, high interest rates, fluctuation in interest rates and changes in policies of financial institutions (please see Buljevich and Park (1999) and Pollio (1999) for a more comprehensive listing). One such new innovation used in infrastructure financing is Project Finance. The term Project Finance used in this paper may be defined as "limited recourse loans, where repayment depends uniquely upon the cash flow generated by a single, self-liquidating investment" (Pollio, 1999). Two examples of infrastructure projects financed through this model are: The Clover power plant in Pennsylvania and the second runway of the Eldorado Airport in Colombia (Pollio, 1999). The application of this model to SDI implementation would require SDI coordinating agencies to prove that SDIs will generate adequate returns on investment. An efficient tool to illustrate the benefits of an SDI and its potential returns on investment is a good business plan.

c. Private Sector non-cash contribution – Under this model the private sector may invest in the implementation of an SDI by providing goods and services that may be used in the implementation/maintenance of component(s) of an SDI. The provision of service can take on one or more of the following forms, database management, managing clearing house or portals and research to facilitate more efficient methods of data sharing to name a few. While provision of goods can be in the form of data, computer hardware and software, and other relevant technology. An example of this type of initiative can be seen in the Open GIS Consortium's (OGC) contribution to SDI implementation through the creation of standards to facilitate interoperability.

d. Responses to declared emergencies, special projects funding and/or alignment with central/state government financed special initiatives – This model offers SDI coordinating agencies the possibility of:

-   Ensuring that data is collected in a manner suitable for sharing.
-   Advising on the implementation of local GISs generated from the project (ensuring they support interoperability).
-   Accessing funds to implement SDI components that support the project(s) goals.

Examples of special projects which SDI coordinating agencies could possible align with in order to access funds and/or goods and services are: Homeland Security programs through the developed world e.g. in the United States Homeland Security is a $38 billion (US) program of which the access to spatial data is seen as a key issue (ESRI, 2001; Lantz, 2002). Another example is the United States' Federal Communications Commission's Enhanced 911 mandate. The E911 mandate requires that for all calls made to the emergency 911 line from mobile phones, the mobile phone (and hence its user) must be located to within 125m (FCC, 1999), which requires spatial information for its support (Senate Committee, 2002).

a. SDI funded through partnerships – A number of different combinations of partnerships are available for financing SDIs in this category. Examples of available partnerships are:

-   Government and private sector partnerships (e.g. The Nationaal Clearinghouse Geo-Informatie (NCGI) in the Netherlands).
-   Government partnerships with community organizations (e.g. with environmental bodies, forestry, tourism and other community organizations). Community groups can contribute to SDI development through the sharing of data/information they have collected and/or through the provision of services.
-   Private sector partnerships with community organizations.
-   Government, private sector and community organization(s) partnerships.

a. The final model to consider is a combination of one or more of the models listed above. The combination can take on different formats and can include all of the models depending on the implementation environment (Figure 13.3).

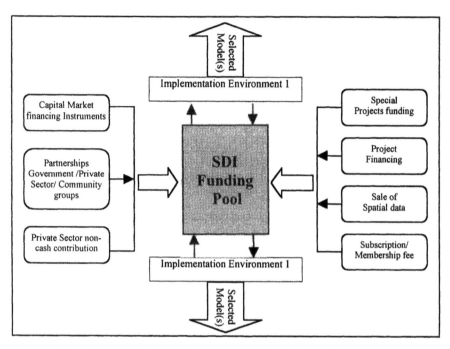

Figure 13.3: Funding Models for SDIs Classified as Network Infrastructures

The above list does not totally exhaust all the possible techniques available for funding an SDI but instead proposes the most significant models. When used on their own or in some combination, the above models would make a significant contribution to the implementation of a SDI. Although the funding pool in Figure 13.3 was designed for SDIs classified as network infrastructures they are also highly suited for SDI implementation at the local level. There are also other procedures that can be combined with the proposed models to increase the access to additional funding for SDI implementation (Please see Urban Logic, 2000). However these steps on their own will not make a significant contribution and thus, will not be discussed by this chapter.

13.4.4 Government's Function in this Category

To date, governments throughout the information society are the major financiers of SDI implementation and maintenance. However, there are other significant roles governments can play in order to increase the general level of investment in SDIs. These non-financial roles are in the form of restructuring of spatial data related organizations, legislation to support the usage of spatial data, restructuring the business of government and increasing the awareness of the usage of spatial data to name a few. An example of how a government can enhance financial contributions to the development of an SDI (in both categories introduced in this chapter) is through the implementation of procedures to remove the barriers that prevent the commercial aspect of an SDI to be successful. Some of the steps a government can take to remove these barriers are:

a.   Give tax breaks to investors in the consortium and also to the partners of the consortium;
b.   Monitor and correct the economic problems associated with this type of infrastructure;
c.   Foster the sharing of data within the Public Sector;
d.   The modernization and restructuring of organizations providing the framework data
e.   Improving the laws associated with copyrights and database protection issues;
f.   Address the issues affecting government and private data pricing and data licensing (Urban Logic, 2000); and
g.   Address the legal issues affecting data transmission (Bandwidth and licensing).

13.5 CUSTOMIZING THE ALTERNATIVE FUNDING MODELS FOR EMERGING NATIONS

SDI implementation in emerging nations will differ from that of developed countries due mainly to differences in their prevailing economic and political climates. Therefore, it follows that the funding models for SDI implementation in emerging nations will also be different from those of developed countries.

The models discussed in the previous paragraphs rely heavily on the assumption that a vibrant economy and a stable political climate exist and thus, may not be applicable in emerging nations. These models may not be applicable because the economies of most emerging nations are very weak if not stagnant. Also, government-funded models may not be applicable in emerging nations since the struggling economies of emerging nations may not be capable of financially sustaining the implementation of an SDI through government budget. The scarce budget resources of these nations are most likely to be invested in more tangible urgent projects, especially if these projects are likely to return success in the short term (de Montalvo, 2001). This is because governments normally have the tendency to invest in projects that will demonstrate benefits during their term in office.

It is evident from the state of the economies of emerging nations that the funding models proposed earlier would not be applicable without significant modification. The models proposed in Sections 4.2 and 4.3 are either dependent on a vibrant economy or a government structure that can afford to and is willing to invest in long term SDI development. Emerging nations cannot afford to support SDI implementation from the budgets' of their governments and their economies are not vibrant enough to motivate a private sector or a private sector/government funded implementation. Therefore, funding must be sought externally and are usually obtained from donor agencies.

Supporting infrastructure development in emerging nations is not a new concept to donor agencies (World Bank, 1997). However, what is new is the concept of an SDI, which must be sold to the donor community if funding is to be secured. Also these agencies must be sold on the evolving concept of fostering partnerships with the private sector to facilitate SDI implementation. The donor community is familiar with the need for creating government partnership in financing infrastructure development but the concept of partnership with the private sector is a new and necessary one in today's changing economy.

Receiving donations from these agencies is not the total answer to an SDI program manager's prayer since these donations normally have conditions attached. For example, there is normally a limited life span placed on the project plus the restriction of limited funds (de Montalvo, 2001). Therefore, there is a necessity to formulate funding models, which will sustain the SDI in the long term. Funding models that might be suitable for emerging nations are:

a. Partnerships – The creation of partnerships amongst local and international organizations with interest in SDI implementation. Examples of possible partnerships are:

    -   Government and Donor Agencies partnerships;
    -   Donor Agencies and private sector partnerships;
    -   Donor Agencies, government and private sector partnerships; and
    -   Partnerships with International Private Sector (e.g. local-international private sector partnerships, international private sector-government partnerships and local private sector-international private sector-government partnerships.

b. The creation of a donor pool. That is, a partnership amongst different donor agencies with each agency responsible for different aspects of the SDI (de Montalvo, 2001). This donor pool should be organized in such a manner that it will ensure there is sufficient funds to sustain the SDI until it becomes self-sufficient or other methods of funding are secured (ECA, 2001).

c. Matching ratios with the Private Sector (local and international). Tax incentives can be used to support cash or as a substitute for the scarce resources (DeBoer *et al.*, 1993).

d. Private Sector non-cash contribution (see Section 13.4.3).

e. The establishment of special banks or financial institutions to underwrite low interest loans for the investment in SDIs. This can be done in conjunction with international lending agencies.

f. Project financing (see Section 13.4.3).

g. Special projects funding and/or aligning with central/state government/donor community financed special initiatives (e.g. environmental management project or land reform projects, see Section 13.4.3).

h. Tied Aid Financing – Funds are tied to purchases from donor country(s) and/or organization(s) providing the funds.

i. Special Taxation (see Section 13.4.2).

A number of the above models might not be able on their own to finance the implementation of an SDI. However, two or more could be used in combination to finance component(s) of an SDI and therefore, should be included in the development of a business plan. The development of a business plan is an important aspect of financing an SDI. A good business plan will facilitate an SDI program manager in the selling of the concept of an SDI to the donor community, the private sector and decision-makers within governments (CIE, 2000). It will be an absolute necessity for the application of a number of the models proposed in Figure 13.4.

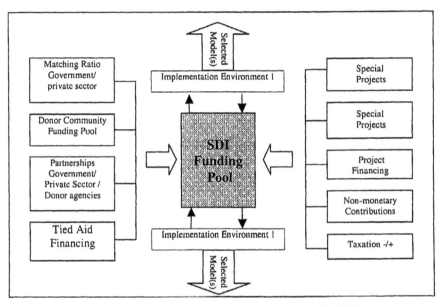

Figure 13.4: Funding Models for SDI Implementation in Emerging Nations

13.6 DISCUSSION

The models developed in Section 13.5 for SDI implementation in emerging nations still require further customization based on additional knowledge of the environment of emerging nations. These models will adopt elements from the two categories of models described in Sections 13.4.2 and 13.4.3 and customize these elements to suit the political, socio-economic and cultural characteristics which collectively define the 'general environment' of emerging nations. The three categories of models would then overlap with each category having models applicable in either environment. The models for emerging nations although having their bases in the other two categories will not be very applicable in the other categories since they will be specifically customized for the environment of emerging nations. However, to achieve this feat, more information about the environments of emerging nations must be researched. Information such as the level of utility and communication infrastructure available, the economic, social and environmental differences between urban and rural areas, the differences between the digital divide in the rural and urban areas, the status of the national economy and political climate and the type of capital markets available. Other external factors that should be considered are the current and future policies of donor agencies on SDI development, the financial status of donor agencies and the policies of the developed world on infrastructure development in emerging nations. With this type of information on hand, it should be possible to customize the existing and proposed models to be more applicable to specific emerging nations.

13.6.1 Future Directions

The models proposed in Sections 13.4.2, 13.4.3 and 13.5 contained a mix of financing instruments and organizational structures. This mix evolved since, not only are new innovative financing techniques required for the implementation of the next generations of SDIs but also the restructuring of spatial data related organizations. Restructuring and the formalization of new organizations will facilitate SDI coordinating agencies in accessing non-traditional financing techniques.

To begin clarifying the relationships within this mix, Table 13.1 presents a preliminary matrix of financing instruments compared with possible structures of SDI host organizations. This matrix can then be used to evaluate the likelihood of employing one or more particular financing instruments within a given organizational structure in order to finance all or part of an SDI implementation. Preliminary analysis of the table suggests that the most favourable organizational structures for SDI implementation are formal partnerships. The analysis also indicated that a number of the financing instruments (e.g. matching ratio, data sales, non-cash contribution and commercial loans to name a few) were well suited for SDI implementation.

Also vital to the successful implementation of SDI funding models will be the development of further metrics to evaluate the models and measure their performance in actual implementation environments. The usage of simulation software by SDI program coordinators to analyse the models and determine how they will react to changes in the implementation environment will also contribute to the effectiveness of the model(s). This is even more important if the models are to be customized for implementation in emerging nations and nations in transition.

Table 13.1 The Application of Financial Instruments to Organize SDI Implementation

Financing Instruments	Single Organizations			Informal Partnerships				Formal Partnerships/Incorporated associations		
	Single Govt. Org.	Private Sector Companies	Donor Agencies	Different Level of Govt.	Public/ Private	Govt./ Community org.	Donor Agencies/ Private	Public	Public Private	Private
Direct Govt. Financing	Y	P	N	Y	Y	Y	P	Y	Y	P
Special Banks	Y	Y	P	Y	Y	Y	P	Y	Y	Y
Bonds	Y	Y	P	Y	Y	N	P	Y	Y	Y
Matching ratio	Y	Y	Y	Y	Y	Y	Y	Y	Y	Y
Limited-recourse structures	N	Y	P	N	Y	P	Y	P	Y	Y
Special taxation	Y	P	N	Y	P	N	N	Y	P	N
Shares	N	Y	N	N	N	N	N	P	Y	Y
Membership fees	N	N	P	N	N	N	Y	P	Y	Y
Commercial Loans	P	Y	Y	P	Y	P	Y	Y	Y	Y
Data Sales	Y	Y	P	Y	Y	Y	Y	Y	Y	Y
Project Financing	P	Y	P	N	P	N	P	Y	Y	Y
Special projects	P	Y	N	Y	Y	Y	P	Y	Y	Y
Tied Aid	Y	P	Y	Y	Y	P	Y	Y	Y	Y
Non-cash contribution	Y	Y	P	Y	Y	Y	P	Y	Y	Y
Rating for SDI implementation	Fair	Fair	Potential	Fair	Fair	Potential	Potential	Fair	Good	Good

Key :
Y- Yes;
N – No; and
P – Has potential

13.7 CONCLUSION

Three categories of SDI funding models were presented by this chapter. These categories were based on the classification of an SDI (in an economic sense) and the implementation environment. Funding models for SDIs classified in the Classic Infrastructure/Natural Monopoly category were designed with an emphasis on sourcing government related/managed funds and private sector investment to a lesser extent. Alternatively, the funding models for SDIs classified as Network

Infrastructure were developed mainly on the concept of private sector funding with government playing a minor role when compared to the previous category.

The third category of models was proposed based on the unstable economic environment of emerging nations. The main designing consideration of this category was the inability of emerging nations to secure funding for SDIs from their governments and local capital markets. These models ignored other significant variables associated with the environments of emerging nations since including them requires further research.

This chapter has reinforced the importance of having in place funding models for the efficient implementation of an SDI. These models will serve as a guide to SDI coordinating agencies on how to identify and secure the finance necessary for the implementation of their SDIs. This concept is even more important to emerging nations because of their very limited ability to generate financial resources for infrastructure development. With SDI initiatives on the increase and the allotted funding for the first generation of SDIs now running out, it is clear that creative new approaches to financing must be identified and assessed.

13.8 REFERENCES

Bevin, T., 2002, e-Survey and Title in New Zealand-Land Online, *AURISA and Institution of Surveyors Conference*, Adelaide, Australia. 25 – 30 November 2002.

Bing, J., 1998, Commercialization of Geographic Information in Europe, Proceedings of the seminar Free accessibility of geo-information in the Netherlands, the United States and the European Community October 2, 1998, Online. <http://www.eurogi.org/geoinfo/publications/commerce.html> (Accessed January 2003).

Beerens, S. and de Vries, W., 2001, Economic, Financial and Capacity Aspects of National Geospatial Data Infrastructure, In GISdevelopment.net, Online. <http://www.itc.nl/library/Papers/0006.pdf> (Accessed December 2002).

Buljevich, E. and Park, Y., 1999, *Project Financing and The International Financial Markets*. (Boston: Kluwer Academic Publishers).

Campos, J.E. and Sanjay, P., 1997, Evaluating Public Expenditure Management Systems, *The Journal of Policy Analysis and Management*, Vol. 16 pp 423-445.

CIE, 2000, Scoping the business of SDI development, Centre for International Economics Online.
<http://www.gsdi.org/docs/capetown/businesscase/scoping.pdf>(Accessed December 2002).

Coleman, D. and Giff, G., 2001, Financing Spatial Data Infrastructure Development: Towards Alternative Funding Models, *International Symposium on SDI*, Melbourne Australia Nov. 2001.

Coleman, D.J. and McLaughlin, J.D., 1998, Defining Global Geospatial Data Infrastructure (GGDI): Components, Stakeholders And Interfaces, *Geomatica*, Vol. 52, No. 2, pp. 129-143.

Crandall, R.W., 1996, Funding the National Information Infrastructure: Advertising, Subscription and Usage Charges. In The Unpredictable Certainty:

Information Infrastructure Through 2000. White Paper of the NII, Chapter 18. National Press, Washington. Online. <http://stills.nap.edu/html/whitepapers/ch-18.html> (Accessed December1 2002).

DeBoer, L., McNamaraa K., and Gebremedhin T., 1993, Tax Increment Financing: An Infrastructure Funding Option in Indiana, Cooperative Extension Services, Purdue University, IN.

ECA, 2001, The Future Orientation of Geoinformation Activities in Africa, An ECA Position Paper, The United Nations Economic Commission for Africa Online. <http://www.uneca.org/disd/geoinfo/FutureGIAfrica.PDF> (Accessed December 2002).

Economides, N., 1996, The Economics of Network, *International Journal of Industrial Organization*, Vol. 14, No. 2.

Economides, N., 1993, A monopolist's Incentive to Invite Competitors to Enter in Telecommunication Services, In *Global Telecommunication Services and Technological Changes*, Edited by G. Pogorel. (Amsterdam: Elsevier).

Economides, N and Himmelberg C., 1995, Critical Mass and Network Size with Application to US Fax Market, Discussion Paper No. EC-95-11, Stern School of Business, N.Y.U. mimeo.

Edwards, S., 1995, Crisis and reform in Latin America: from despair to hope, World Bank, Washington D.C.

ESRI, 2001, GIS for Homeland Security" An ESRI® White Paper, November 2001, Online. <http://www.esri.com/library/whitepapers/pdfs/homeland_security_wp.pdf> (Accessed December 2002).

Ezigbalike, E., Selebalo Q., Faiz S., and Zhou S., 2000, Spatial Data Infrastructures: Is Africa Ready?, *4th GSDI Conference*, Cape Town, South Africa, March 13-15, 2000, Online. <http://www.gsdi.org/docs/capetown/ezig.rtf> (Accessed December 2002).

FGDC, 1997, A Strategy for the NSDI. FGDC Publication, Reston, VA 20192. April 1997. Online. <http://www.fgdc.gov/nsdi/strategy/strategy.html> (Accessed December 2002).

FCC, 1999, Third Report and Order, Federal Communications Commission, Online. <http://www.fcc.gov/Bureaus/Wireless/Orders/1999/fcc99245.pdf> (Accessed July 2000).

Fraser, N., Bernhardt, I., Jewkes, E. and Tajima, M., 2000, *Engineering Economics in Canada*. (Scarborough, Ontario: Prentice Hall).

Giff, G., 2002, A Critical Review of the GSDI Cookbook from the Viewpoint of SDI Implementation in Emerging Nations, *Geomatica*, Vol. 56, No. 3, pp. 246 - 250.

Giff, G. and Coleman, D., 2002, Funding Models for SDI Implementation: from Local to Global, *6th GSDI Conference*, Budapest, Hungary.

Groot, R., 2001, Economic Issues in the Evolution of National Geospatial Data Infrastructure, *7th UN Regional Cartographic Conference for the Americas*, NY, USA.

IIPF, 2001, Project Financing in Developing Countries, The Institute of International Project Finance Online. <http://www.economics uni linz.ac.at/IIPF2001/ > (Accessed July 200...

de Jong, J., 1998, Access to Geo-information in the Netherlands; a Policy Review, Proceedings of the seminar Free accessibility of geo-information in the Netherlands, the United States and the European Community October 2, 1998, Online.   <http://www.euronet.nl/users/ravi/proceed210.html#1>   (Accessed January 2003).

Kok, B.C. and van Loenen, 2000, Policy Issues Affecting Spatial Data Infrastructure Development in the Netherlands, Online.
<http://www.spatial.maine.edu/~onsrud/gsdi/Netherlands.pdf>   (Accessed December 2002).

Labonte, J., Corey,M. and Evangelatos T., 1998, Canadian Geospatial Data Infrastructure (CGDI)- Geospatial Information for the Knowledge Economy, *Geomatica*, Vol. 52, No. 2, pp. 194 -200.

Lantz, B. 2002, How Location Intelligence is Paramount to Protecting Our Nation, Directions Magazine, November 12, 2002, Online.
<http://www.directionsmag.com/article.php?article_id=267> (Accessed December 2002).

Masser, I., 1998, The first Generation of National Geographic Information Strategies, 3rd GSDI Conference on SDI, Canberra, Australia. Online. <http://www.gsdi.org/docs/canberra/masser.html> (Accessed December 2002).

McLaughlin, J.D., 1991, Towards National Spatial Data Infrastructure, *the 1991 Canadian Conference on GIS*, Ottawa, Canada, pp. 1-5. Canadian Institute of Geomatics, Ottawa, Canada. March.

de Montalvo, U. W., 2001, Outreach and Capacity Building, In Developing Spatial Data Infrastructures: The SDI Cookbook, Ed. D. Nebert, Online.
< http://www.gsdi.org/pubs/cookbook/cookbook0515.pdf>   (Accessed December 2002).

Nebert, D., 2001, The Cookbook Approach, In Developing Spatial Data Infrastructures: The SDI Cookbook, Ed. D. Nebert. Online.
< http://www.gsdi.org/pubs/cookbook/cookbook0515.pdf>   (Accessed December 2002).

OMB, 2001, OMB Information Initiative, Office of Management and Budget, Published in the FGDC Newsletter, Vol. 5, No. 1. Online.
<http://www.fgdc.gov/publications/documents/geninfo/Spring01NL.pdf>
(Accessed December 2002).

Ordnance Survey, 1999, Joined-up geography for the new millennium, Ordnance Survey Information Paper, Southampton, UK. Online.
<http://www.ordnancesurvey.gov.uk/literatu/infopapr/1999/pap1399.htm>
(Accessed December 2002).

Pollio, G., 1999, *International Project Analysis and Financing,* (Michigan: University of Michigan Press).

Rajabifard, A. and Williamson I.P., 2001, Spatial Data Infrastructures: Concept, SDI Hierarchy and Future Directions, *Geomatic '80 Conference*, Tehran, Iran, 29April-3 May, pp28-37.

Rhind, D., 1997, Implementing a Global Geospatial Data Infrastructure, *2nd GSDI Conference*, Chapel Hill, North Carolina, USA, 19-21 October,1998. Online. <http://www.gsdi.org/docs/ggdiwp2b.html> (Accessed December 2002).

Rhind, D., 2000, Funding an NGDI, In *Geospatial Data Infrastructure Concepts, Cases and Good Practice*, Edited by Groot, R. and McLaughlin, J., (New York, NY: Oxford University Press), Pp39-55.

Senate Committee, 2002, House Bill Report E2SSB 6034, Online. <http://www.leg.wa.gov/pub/billinfo/2001-02/senate/6025-6049/6034-s2_hbr.pdf> (Accessed December 2002).

SNB, 2002, Service New Brunswick Business Plan 2002-2005, Service New Brunswick Queen's Printer, Province of New Brunswick, Fredericton, N.B., Canada. April 1,2002.

Sorensen, M., 1999, Institutional Linkages for National and Regional GIS-Management Issues, Opportunities and Challenges, *Ordnance Survey Conference*, July 1999, Cambridge, UK.

Tveitdal, S., 1999, Economics of EIS. Environment Information Systems in Sub-Saharan Africa (EIS-SSA), Publication, May 1999, Pretoria, Republic of South Africa. Online. <http://www.grida.no/eis-ssa/products/econom/index.htm> (Accessed January 2003).

Urban Logic, 2000, Financing the NSDI: National Spatial Data Infrastructure. Online. <http://www.fgdc.gov/whatsnew/whatsnew.html#financing> (Accessed December 2002).

de la Vega, B., 2000, Why software is a natural monopoly and some repercussions, Online. <http://www.fecund.org/bd/monopoly.html> (Accessed December1, 2002).

Williamson, I.P., Chan, T.O. and Effenberg, W.W., 1998, Development of Spatial Data Infrastructures – Lessons Learned from the Australian Digital Cadastral Databases, *Geomatica*, Vol. 52, No. 2, 177-187.

Worldbank, 1997, The Private Sector's Role in Infrastructure Development, World Bank Publication, 1997. Online. <http://www.worldbank.org/html/extdr/backgrd/idrd/infrastr.htm>. (Accessed April 2000).

Yevdokimov, Y., 2000, The Economics of Information Networks, *In Economics for Engineers*, Custom Made Edition for the University of New Brunswick, Edited by Bade, R., Parkin, M. and Lyons. B., (Canada: Person Canada Publishing), pp. 303-317.

CHAPTER FOURTEEN

Developing Evaluation and Performance Indicators for SDIs

Daniel Steudler

14.1 INTRODUCTION

This chapter aims to introduce the role and value of evaluation and performance indicators for Spatial Data Infrastructures (SDIs). Evaluation involves assessing the strengths and weaknesses of programs, policies, personnel, products, and organizations to improve their effectiveness. It is about finding answers to questions such as 'are we doing the right thing' and 'are we doing things right'. These are prominent questions for SDI, the development of which has been very dynamic over the last decade and has involved a lot of learning from other national or local initiatives.

The comparison and evaluation of SDIs can help to better understand the issues, to find best practice for certain tasks, and to improve the system as a whole. Evaluating and comparing public and private administration systems can be significant in terms of improvements to processes and institutional structures. The application of these principles to the development of SDIs will therefore come to play a crucial role in the management of our land information and that pertaining to the administration of our societies.

The field of land administration is one where evaluation principles are being developed, with much of these principles relevant to SDIs. Land administration systems are essential parts of countries' national infrastructures (UN-FIG, 1999). They are concerned with the administration of land and land resources and therefore also with land-related, spatial data.

Spatial data are required for the management and location of land issues, land resources and other land related phenomena. Within national administrations, spatial data are often acquired and maintained by different organizations, resulting in problems such as datasets not being compatible with each other and data not being shared across organizations, leading to inefficiencies and duplication of efforts. The common objectives of the different organizations have resulted in the development of the SDI concept at different political and administrative levels in regard to the facilitation and coordination of the exchange and sharing of spatial data between stakeholders (Rajabifard *et al.*, 2002).

The commonalities between SDIs and the objectives of efficient and effective land administration systems provide strong grounds for the derivation of evaluation and performance indicators for SDIs from land administration principles. To achieve this, the chapter first identifies key components of land administration systems and SDIs. It then develops a general evaluation framework, which can be applied to SDI and its different components before it draws some conclusions.

14.2 LAND ADMINISTRATION SYSTEMS AND THE ROLE OF SDIs

The UN-ECE (1996) defines land administration as "the processes of determining, recording and disseminating information about the tenure, value and use of land when implementing land management policies. It is considered to include land registration, cadastral surveying and mapping, fiscal, legal and multi-purpose cadastres and land information systems".

Dale and McLaughlin (1999) define land administration as "the process of regulating land and property development and the use and conservation of the land, the gathering of revenues from the land through sales, leasing, and taxation, and the resolving of conflicts concerning the ownership and use of the land." They continue that the basic building block in any land administration system is the cadastral parcel and that land administration functions can be divided into four functions: juridical, regulatory, fiscal, and information management. The first three functions are traditionally organised around three sets of organizations while the latter, information management is integral to all the other three components.

Along with the progress in information technology, the information management function has been developed considerably over the last few decades, when there have been many efforts to establish information systems dealing with land information based on the cadastral parcel. Within national administrations, spatial data are however often acquired and maintained by different organizations, resulting in problems such as datasets not being compatible with each other and data not being shared across organizations leading to inefficiencies and duplications of effort.

SDI is an initiative attempting to overcome these shortcomings and to create an environment in which all stakeholders in spatial data can co-operate and interact with technology to better achieve their objectives at different political and administrative levels. SDIs have become important in determining the way in which spatial data are used in an organization, a nation, different regions and the world. By reducing duplication and facilitating integration and development of new and innovative business applications, SDIs can produce significant human and resource savings and returns.

Regardless of the fact that different interest groups view SDIs differently, researchers have identified a number of core components that are common to all SDI implementations (Coleman and McLaughlin, 1998; Rajabifard *et al.*, 2002). As discussed in Chapter 2, these components are: *people, access networks, policy,* technical *standards* and *datasets* (Figure 2.2). All components are strongly related to each other resulting in an interrelated infrastructure.

While land administration systems are foremost concerned with supporting the management of land issues – ownership, use, value – and land resources, the focus of SDI is mainly on the data and information about the land. As such, SDI is the underlying infrastructure for operating land information systems (Dale and McLaughlin, 1999) which by themselves are underpinning the land administration process.

It is this interaction between SDIs and land administration systems that is crucial for both parts. The interaction can be investigated and understanding fostered by searching for 'best practice' and for evaluation methods looking at those specific issues through qualitative and quantitative indicators describing the relationships between them.

14.3 EVALUATION AND A FRAMEWORK FOR EVALUATION

Evaluation is mainly concerned with questions such as: are we doing the right thing, are we doing things right, what lessons can we draw from experiences, and what can we learn from similar situations. Such questions are an integral part of steering and management tasks within programs and projects and can be formulated and partly be answered by means of an evaluation (SDC, 2000). The objectives of an evaluation can be to verify the impacts, the objectives or the efficiency of a project or a system, to find answers to specific questions associated with the project or system context, to prepare information for reporting, or to draw lessons for future phases.

An important decision that has to be taken beforehand relates to how the evaluation has to be carried out. For the purpose of better being able to handle and understand large projects or systems, they have to be broken down and divided into comprehensible subclasses. In a World Bank Seminar about "Public Sector Performance – The Critical Role of Evaluation", Baird (1998) emphasised four elements that are central in how to evaluate the performance of an organization or system. They are:

a) well-defined *OBJECTIVES* (to know where to go to):
   - define the targets for the whole system;
   - might involve historical and social aspects, the cultural heritage as well as the political, legal, and economic basis;
b) clear *STRATEGY* (to know how to get there):
   - defines the way forward to reach and satisfy the objectives (institutions, organizations, finances, activities);
c) *OUTCOMES* and monitorable *INDICATORS* (to know if on track):
   - outcomes are the results of the activities arising from the objectives and strategies;
   - indicators must be monitorable and relevant for feedback to objectives and strategies;
d) *ASSESSMENT OF PERFORMANCE* (to gain input for improvements):
   - the process which takes the outcomes and indicators into account in order to evaluate and review the objectives and strategies on a regular basis;
   - looks at the performance and reliability of the system and how the initial objectives and strategies are satisfied.

These four evaluation elements must be thought of as a cyclical process, allowing a regular assessment of the performance and a review of the initial objectives and strategies. The review cycle can for example be annually for the strategies while the objectives might be reviewed only every four years (Figure 14.1).

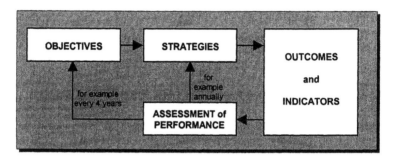

Figure 14.1: Basic Evaluation Elements and Cycle of Assessment

In order to fill the elements with content, they would also have to be brought in context with the relevant stakeholders. For that purpose, the organizational pyramid with the three organizational levels can provide a valuable basis. Any organization is structured into different divisions, subdivisions and sometimes even external units, each with separate functions. Regardless of the organization, the three levels of the organizational pyramid can in general be distinguished, representing the different organizational tasks and responsibilities. The three levels are the policy level, the management level, and the operational level.

The organizational levels can be correlated with the evaluation elements introduced in Figure 14.1 as well as with the stakeholders. The policy level can be related with defining the objectives, for which the government or the executive board is responsible. The management level includes the definition of the strategy, for which the administration or management of the organization is responsible. The operations required for the outcomes are handled in the operational level for which the operational units are responsible (Figure 14.2).

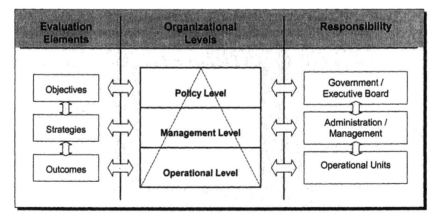

Figure 14.2: The Rlation Between Evaluation Elements and Organizational Levels

The organizational levels provide the basis for defining the actual fields or areas of evaluation. For evaluating an administration system as a whole, however, another two areas would need to be considered as well. Firstly, the *assessment of performance* area that is looking at how the whole system performs and how objectives and strategies are satisfied. Secondly, there are *other influencing factors* that have an impact on all three organizational levels. These are factors such as human resources, capacity building, or technology that all influence the organizational levels in one way or another and need to be addressed as well (Figure 14.3).

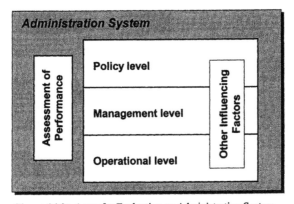

Figure 14.3: Areas for Evaluating an Administration System

These evaluation areas provide the basis for the evaluation framework in which all areas are evaluated separately, although with a holistic perspective and respecting the overall purpose of the system. For the evaluation, the areas need to be broken down into smaller units, which are supported by performance indicators, measuring the performance of key variables such as quality, time, and cost. The evaluation of the areas and indicators can then be done on the basis of predefined "good practice" criteria, which are representing a presumed "ideal" system. The criteria of this ideal system are to be based on the actual objectives and strategies of the system, on the results of previous lesson-learning and comparison projects, or ideally on both.

Table 14.1 illustrates a generalised evaluation framework where the evaluation areas are further expanded with possible aspects, indicators and good practice criteria.

Table 14.1: Evaluation Framework with Possible Aspects, Indicators and Good Practice for Each Area

Area	Possible Aspects	Possible Indicators	Good Practice
Policy Level	• objectives and tasks of the system • historic, legal, social, cultural background • equity in social and economic terms • viability of system (economical, social)	• list of objectives and tasks • legal and historic indicators • social indicators • economic indicators (expenses, incomes, fees, costs)	• system is well defined by objectives and tasks • system responds to needs of society • system is equitable for all • system is economically viable
Management Level	• structural definition of system • strategic targets • institutional and organizational arrangements • cooperation and communication between institutions • involvement of private sector	• definitions and characteristics of system • list of strategic targets • list of institutions and their responsibilities and strategies • links between institutions (legal, organizational, technical) • number of contracts with private sector	• structure of system is useful and clearly defined • strategies are appropriate to reach and satisfy objectives • involved institutions have each clearly defined tasks and cooperate and communicate well with each other • private sector is involved
Operational Level	• outcomes • technical specifications • implementation	• products for clients • technical indicators • implementation factors	• products respond to objectives • technical specifications and implementations are appropriate to strategic needs
Influencing Factors (Human Resources, Capacity Building, Technology)	• Human Resources (personnel, training) • capacity building • professional association • technical developments	• number of personnel, eduction • continuing eduction (seminars, etc.) • number of universities and students • is there a professional association (y/n) • new technologies on the market	• appropriate number of personnel in relation to task and population • continuing eduction on a regular basis • appropriate number of universities and students • professional association takes active role • new technologies are evaluated on a continuing basis
Assessment of Performance	• review of objectives and strategies • performance and reliability of system • user satisfaction	• review of objectives and strategies (y/n) • turnover, time to deliver, number of errors • review of user satisfaction (y/n)	• regular review process • system is efficient and effective • system delivers in time and with few errors • appropriate, fast and reliable service to clients

14.4 EVALUATION OF SDIs

Masser (1998) used an analytical framework to compare first generation National SDIs. The framework considered the main criteria of geographic and historic context, main data providers, institutional context, and national geographic information strategy elements.

The evaluation framework that was developed in the previous section, however, attempts to take a more comprehensive approach and to also consider issues such as the different stakeholders in the organizational pyramid, and the recurring and regular review of the objectives and strategies through performance assessment. If an SDI is evaluated through the general evaluation framework developed (Table 14.1), the SDI components (discussed in Section 14.2 and illustrated in Figure 2.2) can be mapped into the evaluation areas mentioned in Figure 14.3. The *policy* component obviously can be associated with the policy level, the *standards* component with the management level, while the *access network* and *data* components are attributed to the operational level. The *access network* component may have to be considered in both management and operational levels given the varying maturity of SDI developments established over the last decade. The *people* component has an influence on all three organizational levels and is therefore associated with the other influencing factors area. The result is shown in Figure 14.4.

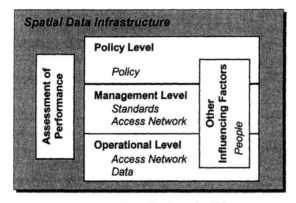

Figure 14.4: Evaluation Areas for SDI

If the evaluation areas for SDIs are further expanded according to the suggested framework in Table 14.1, each area would need to be specified in terms of ASPECTS, INDICATORS, and GOOD PRACTICE. For SDI, the evaluation framework may consider the following aspects:

(a) Policy Level
Policy: One aspect to be considered for the *policy* component is the geographic, historic, and social context of the country. A second aspect is how the government handles the overall policy regarding the collection, dissemination, and legal protection of spatial data, for example issues such as intellectual property rights, privacy issues, and pricing. Indicators might be the existence of a government

policy regarding the mentioned issues, and how the issues are dealt with. Good practice is when the government has taken actions for an SDI and when issues have been handled in a comprehensive and satisfying way in relation to the geographic, historic and social context of the country.

(b) Management Level

Standards: The evaluation of the *standards* component includes how the government administration is dealing with organizational arrangements for the coordination of spatial data. It may include the assessment of government agencies involved in providing spatial data for land titles, for large- and small-scale mapping, and for socio-economic statistics. The evaluation has to consider standardization issues like the definition of core datasets, data modelling practices, and interoperability at the national level.

Indicators for the management level might be a list and the size of government agencies involved in spatial data, their size and activities, and how they communicate and cooperate with each other. In order to permit comparisons with other countries, indicators might point out the definitions of the core datasets, the data modelling techniques used for defining spatial datasets, and the standardization decisions for the access networks.

Access Networks: The evaluation of the *access networks* component may include issues like the definition of data summaries, formats of available data, delivery mechanisms for the data, whether access will have associated costs, and whether data-access privileges will be defined for different user groups.

Indicators might point out access pricing, access delivery mechanisms and procedures, whether access is defined by privileges or is open to all users, as well as whether there are inter-institutional links for data access, or value-adding arrangements established with the private sector.

(c) Operational Level

Access Network: The responsibility for the operational level is with the government's operational units that have to make things happen in terms of access network and data provision. The *access network* component is to be evaluated by considering the type of available network and its capacity and reliability. Indicators might be the data volume and response time, and good practice would be when the network can handle a large data volume reliable with short response time.

Data: The *data* component can be evaluated by assessing the data models of the spatial datasets of the different agencies, the creation of a national core dataset, the data formats, data capture methods, data maintenance, and data quality and accuracy. Good practice might be when data are defined in clear and transparent ways (content, quality, accuracy), so that they can easily and readily be shared among the different agencies and users.

(d) Influencing Factors

People: The evaluation of the *people* or human resources component has to take the three groups into account that have been identified relevant in the SDI context: end-users, data integrators respectively value adders, and data providers. The evaluation will have to assess the situation within these three groups in terms of personnel, opportunities for training and capacity building, and market situation for spatial data. Good practice will be when end-users are easily and readily getting the data product that they are looking for, when integrators can operate and prosper in favourable market situations, and when data providers are able to deliver the data in efficient and effective ways.

(e) Assessment of Performance
This aspect has not much been addressed in SDI research papers so far, but is equally important for the overall assessment of national infrastructures. It might include the review of objectives, strategies, performance, and reliability of the system as well as user satisfaction. Indicators can be the adoption of SDI principles, use and diffusion of spatial data, and user satisfaction surveys. Good practice can be considered as when all SDI principles are adopted, when there is large use and diffusion of spatial datasets, and when users indicate satisfaction about the products and services offered.

Table 14.2: Possible Indicators for Evaluating SDIs

Area	Possible Indicators
Policy Level *– Policy*	• existence of a government policy for SDI • handling of intellectual property rights, privacy issues, pricing • objectives for acquisition and use of spatial data
Management Level *– Standards*	• standardization arrangements for data dissemination and access network • institutional arrangements of agencies involved in providing spatial data • organizational arrangements for coordination of spatial data • definition of core datasets • data modelling • interoperability
Management Level *– Access Network*	• access pricing • delivery mechanism and procedure • access privileges • value-adding arrangements
Operational Level *– Access Network*	• type of network • data volume • response time
Operational Level *– Data*	• data format • data capture method • definition of core datasets • data maintenance • data quality and accuracy
Other Influencing Factors *– People*	• number of organizations and people involved • opportunities for training • market situation for data providers, data integrators, and end-users
Performance Assessment	• degree of satisfying the objectives and strategies • user satisfaction • diffusion and use of spatial data and information • turnover and reliability

The areas and possible indicators suggested in Table 14.2 are only a general framework for evaluating SDIs but are nonetheless useful for providing a first-order evaluation of an SDI and for eliciting valuable indicators. An example of the indicators that can emerge from such an application is summarised as a strength-weakness-opportunity-threat (SWOT) matrix in Table 14.3, based on a state-level SDI analysis in Australia. The five evaluation areas were reviewed for the State SDIs with the main findings being fed into the SWOT matrix. The insights provided from this first-order evaluation indicate the value that may be derived from more in-depth applications of the evaluation areas and further development of the indicators and criteria specific to the SDI being evaluated. It must be emphasised that the areas and possible indicators suggested in Table 14.2 are only a general framework for evaluating SDIs and would require further development to optimise the benefit of an evaluation.

Table 14.3: SWOT Matrix Summarizing a General (first-order) Evaluation of a State SDI

Strengths:	Weaknesses:
• Comprehensive review of land information strategy takes place on a regular basis	• Strategy does not consider the cadastral issues to their full merits
• One government department is responsible for spatial data, which is favourable for strong leadership and decision-making	• No promotion of data modelling and interoperability and hence freedom of systems and methods
• Strong academic sector	• No independent board which could promote and coordinate spatial information
• Good cooperation between public-private-academic sectors	
Opportunities:	**Threats:**
• Vision of spatial information being crucial for good governance	• Not being able to bring the diverging interest groups together
• Strengthen political support	• Losing political support

14.5 CONCLUSIONS

This chapter sets out a broad strategy for evaluating SDIs. An evaluation framework has been suggested, based on an approach originally developed for evaluating land administration systems. This framework attempts to accommodate the well-recognized SDI components namely people, access network, policy, standards and data, which may be considered as the main evaluation areas within the suggested framework. However, the main evaluation areas need to be complemented by the additional evaluation area of *performance assessment*, which evaluates the progress towards objectives and strategies that were initially defined. In this respect, the framework mainly assesses the effectiveness (are we doing the right thing?) and efficiency (are we doing it right?) of SDIs.

There is a lot in common between land administration systems and SDIs, especially at the state level where cadastres are a main component of both. Therefore, while the SDI evaluation strategy is still evolving, much can be learned from systems being developed for evaluating land administration, where a number of benefits have been identified such as:

- standardized benchmarking provides an unbiased way of comparing systems;
- standardized benchmarking procedures can improve productivity, efficiency, and performance;
- cross-jurisdiction or cross-country comparisons can help better understand one's own system;
- benchmarking and evaluation can help identify categories of processes and systems;
- they provide a basis for comparisons over time;
- they provide help to demonstrate strengths and weaknesses.

The main innovation that the suggested framework provides is the incorporation of *performance assessment* as an evaluation area. The field of SDI is still under development and the body of knowledge still growing, yet there is already considerable attention given to the field's development, although not specifically relating to its quantification and qualification. Therefore, for the further development of SDI evaluation, greater emphasis must be placed on the recurring and regular review of objectives and strategies.

The most important benefit from evaluating and comparing SDIs with each other will be the lessons learnt and identification of good practices. Performance indicators measuring the performance of key variables will provide the basis for this. The broad framework presented in this chapter is strongly related to land administration but suggests a way forward for SDI evaluation.

14.6 REFERENCES

Baird, M., 1998, The Role of Evaluation. In *Public Sector Performance – The Critical Role of Evaluation.* Selected Proceedings from a World Bank Seminar. Editor Keith Mackay, World Bank Operations Evaluation Department, Evaluation Capacity Development, Washington D.C., April, p. 7-12.

Coleman, D. J. and McLaughlin, J.D., 1998, Defining Global Geospatial Data Infrastructure (GGDI): Components, Stakeholders and Interfaces. *Geomatica,* 52:2, pp. 129-143.

Dale, P. and McLaughlin, J.D., 1999, *Land Administration Systems.* (Oxford: Oxford University Press), 169 p.

Masser, I., 1998, *Governments and Geographic Information,* (London: Taylor & Francis), 121 p.

Rajabifard, A. and Williamson, I.P., 2001, Spatial Data Infrastructures: Concept, SDI Hierarchy and Future Directions. *Geomatics'80 Conference,* Iran, Online. <http://www.geom.unimelb.edu.au/research/publications/IPW/4_01Raj_Iran.pdf> (Accessed October 2002).

Rajabifard, A., Feeney, M. and Williamson, I.P., 2002, Future Directions for the Development of Spatial Data Infrastructure. *Journal of the International Institute for Aerospace Survey and Earth Sciences,* ITC, Vol. 4, No. 1, pp. 11-22.

SDC, 2000, *External Evaluation - Part 1*. Working Instruments for planning, evaluation, monitoring and transference into Action (PEMT). Swiss Agency for Development and Cooperation, 30p., June.

UN-ECE, 1996, *Land Administration Guidelines*. Meeting of Officials on Land Administration, UN Economic Commission for Europe. ECE/HBP/96 Sales No. E.96.II.E.7, 111 p, Online.
<http://www.unece.org/env/hs/wpla/docs/guidelines/lag.html> (Accessed Oct. 2002).

UN-FIG, 1999, The Bathurst Declaration on Land Administration for Sustainable Development. Report from the UN-FIG Workshop on Land Tenure and Cadastral Infrastructures for Sustainable Development, Bathurst, NSW, Australia, 18-22 October, Online.
<http://www.fig.net/figtree/pub/figpub/pub21/figpub21.htm> (Accessed October 2002).

PART FIVE

Technical Dimension

CHAPTER FIFTEEN

Administrative Boundary Design in Support of SDI Objectives

Serryn Eagleson and Francisco Escobar

15.1 INTRODUCTION

Health, wealth and population distributions are all examples of spatial data commonly referenced to administrative boundaries. In fact, there are few areas of the economy and environment that do not rely either directly or indirectly on the integration of data attached to administrative boundaries for planning, maintaining or rationalising activities. Conceptually, as outlined in Chapter 2, an SDI incorporates the technology, policies, standards and human resources necessary to facilitate the integration of administrative boundary data. In practice, however, the fragmentation of administrative boundaries is a serious problem that restricts the integration and potential benefits of spatial data. As outlined by Flowerdew and Green (1994), situations frequently arise where the analyst wants to compare a variable that is available for one set of administrative units with a variable that is only obtainable for a different incompatible set.

A number of organizations have realised the advantages of using administrative boundaries for the collection and collation of data. For example, once the administrative boundaries are established, the data is easy collected and efficient to store. Even in light of technological advancements, other forms of spatial data, such as address point and line data are still relatively expensive to produce, difficult to manipulate and require large amounts of memory to store (Rajabifard and Williamson, 2001). Many organizations are thus using established polygon-based administrative boundaries as a base for the collection and collation of spatial data. As we move into an era of spatial decision-making, there is recognition amongst the users that current technical issues relating to the non-coterminous alignment of administrative-boundaries need to be addressed.

Imagine someone has just obtained the census data detailing the population distribution attached to boundary set A. They are interested in planning a new healthcare facility. To determine the best location for this facility, the person needs to cross-analyse the census data with health statistics that are reported on boundary set B. Due to the incompatible boundary systems used by the agencies, though, it is not possible to accurately and efficiently cross-analyse the health and demographic data. Consequently, the user must rely on their own judgement, to compare the datasets and decide the most logical position for the new centre.

The objective of this chapter is to highlight future directions of administrative boundary design, delineation and dissemination that meet the needs of stakeholders within the SDI framework. To achieve this objective, the chapter proposes the development of an administrative boundary hierarchy to facilitate the design, delineation and dissemination of administrative boundaries in support of SDI objectives.

15.2 A DEFINITION OF THE SPATIAL-HIERARCHY PROBLEM

Historically, countries have divided social, economic and political responsibilities amongst a variety of agencies. In turn, these agencies have established independent administrative, planning and political boundaries that rarely coincide (Robinson and Zubrow, 1997; Huxhold, 1991). Figure 15.1, illustrates an abstract view of the current situation. Each agency establishes a differently sized or shaped spatial unit, based on their individual — and often unique — requirements, using the land parcel (in most cases) as the bottom layer. In turn, each agency aggregates these boundaries in a hierarchical fashion to cover the state. Data integration is possible within each agency; however, under this current system additional methods such as data interpolation must be employed to facilitate cross-analysis between agencies.

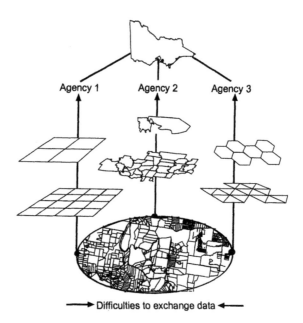

Figure 15.1   An abstract Illustration of the Various Boundary Layers that Exist in Victoria
(Adapted from Eagleson *et al.*, 2002a)

Essentially, the spatial-hierarchy problem has occurred because, in the beginning, individual organizations hand-drafted the majority of boundaries on paper maps. With advances in technology, these hand-drafted maps have been digitised for incorporation into GIS, a technology for which they have not been adequately designed. In an effort to improve data integration between non-coterminous administrative boundary layers, a number of methods have been developed to enhance data integration. As detailed below, surface modelling, data interpolation, derived boundaries and data re-aggregation are techniques developed to facilitate data integration between non-coterminous boundary units.

15.2.1 The First Solution: Surface Modelling

Within a GIS, administrative boundaries are traditionally defined by (x, y) coordinates, and these coordinates are joined by lines, forming closed polygons. To overcome the problem of data integration between two non-coterminous polygon layers, Martin and Bracken (1991), and Bracken, (1994) have developed raster-based models to integrate the originally polygon-based data. Using their model, variables attached to administrative boundary polygons, such as census districts or postcodes, are referenced to the polygon centroid and converted to point data. Various techniques are then used to map this data onto a raster-based density surface, thus allowing the data to be easily represented and integrated with other raster-based datasets.

Although the raster-based model does facilitate data integration and exchange, limitations do exist. For example, the transfer of data between data structures inevitably causes errors in the accuracy of the data. Additionally, as highlighted by Morphet (1993), boundaries themselves can often add valuable information in analysis; therefore, it is not always sensible to exclude them from the data analysis.

15.2.2 The Second Solution: Data Interpolation

The problem of cross-analysing data between two boundary systems can be restated as the problem of deriving data for one set of boundaries given the relevant data for another set. Techniques that are able to complete this process of data transfer between boundary units are commonly known as *areal interpolation techniques* (Flowerdew and Green, 1994). Areal interpolation often requires complicated mathematical algorithms for the transfer of attribute data between non-coterminous boundary systems (Goodchild *et al.*, 1993; Martin 1998; Trinidad and Crawford, 1996). Although the interpolation process appears to provide an approximate solution to the problem, many assumptions are made in the process. One, often invalid, assumption is that the distributions of values in the original source map are constant (Goodchild *et al.*, 1993).

In an effort to increase the accuracy of interpolation, and minimise the number of assumptions, supplementary data such as road networks, land-use maps, satellite imagery, road networks and administrative boundaries are often used as "controls" for the interpolation process. Although areal interpolation techniques are valuable for providing a basis for analysis not currently possible with a single boundary layer, the errors and assumptions inherent in the techniques can lead to a less than optimum solution (Eagleson *et al.*, 2002b).

15.2.3 The Third Solution: Derived Boundaries

In an attempt to make data readily usable, some organizations have created derived boundaries. Derived boundaries are formed through the re-aggregation of agency boundaries that approximately nest within more publicly recognizable administrative units. One example is the derived postcodes generated in Australia by the Australian Bureau of Statistics (ABS). For operational reasons, the Australia Post postcode boundaries do not necessarily match the ABS census collector district (CCD) boundaries. In recognition of the separate functions undertaken by these agencies, the ABS aggregates CCDs to approximate the Australia Post postcode boundaries, producing ABS derived postal areas. Discrepancies between the boundaries of these two postal zones can easily arise since the two systems are not coordinated. The derived postal areas may be quite different from the actual postcode boundaries, both in terms of shape and area. Figure 15.2, illustrates the problem. The two sets of spatial entities (postal zones) are, nevertheless, given the same identifier by the agencies, consequently leading to the misinterpretation of data by users. A discussion on this issue can be found in Jones *et al.* (2003). If users remain uninformed about the origin of the data boundaries, subsequent decisions will not be well supported. The use of these derived boundaries can lead to confusion between agencies using the data when differences between derived postcodes and postcodes cannot be clearly identified by the user.

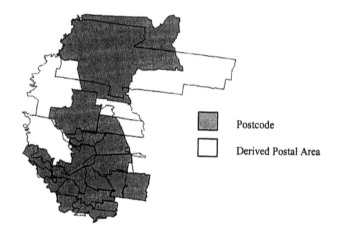

■ Postcode

☐ Derived Postal Area

Figure 15.2: An Illustration of the Difference Between Derived Postcode Boundaries and Actual Postcode Boundaries in the North West Melbourne Health Division

15.2.4 The Fourth Solution: Re-Aggregation

A fourth method for the dissemination of datasets across incompatible boundary regions is the re-aggregation of point data and polygon data. The re-aggregation of point data requires data to be stored at a parcel level and aggregated to a different spatial unit at any time. Although the process of aggregation accurately solves the problem, other problems exist.

First, this solution is not viable in Australia and many other countries, primarily due to stringent laws protecting confidentiality. For instance, once household data is collected by the ABS, it must be aggregated to the CCD boundaries (approximately 220 households) and the individual household data destroyed (ABS, 1998). If confidentiality is not guaranteed, it is probable that people will not complete census forms truthfully, degrading the accuracy and reliability of census information for planning purposes.

Second, a large quantity of storage space is required to store data associated with individual land parcels, and each re-aggregation of data to new boundaries would be extremely time consuming and costly. Additionally, problems such as differencing exist when data is aggregated to a number of different boundaries. As Duke-Williams and Rees (1998) explain, if polygons containing confidential information are overlapping, in some circumstances it may be possible to subtract one set of polygons from the other to obtain statistics for subthreshold areas, thus breaching confidentiality.

The re-aggregation of polygon data involves the re-aggregation of existing units into new boundaries more suitable for specific analysis techniques. For example, Openshaw (1977) devised the automated-zone-design program (AZP) for investigating the modifiable-area-unit problem (MAUP). With the introduction of new technology, digital data and improved algorithms during the 1990s, AZP was further refined and extended forming the zone-design system (ZDES) (Openshaw and Rao, 1995; Openshaw and Alvanides, 1999). These zone-design systems allow the analyst the freedom to start with data at one scale and then re-aggregate it to create a new set of regions designed to be suitable for a specific purpose, independent of the collection boundaries used (Openshaw and Rao, 1995). If these initial boundaries are not designed as layers within a hierarchy, however, the problem of data integration between overlapping polygons remains. Although research has been conducted into the cross-analysis of boundary-referenced data, the problem of incompatible boundary design is still a major concern for spatial analysts around the world. These concerns are largely due to the limited accuracy and specialist skills that may be required to operate the technical solutions. The issue of technical skills is one of the problems limiting the diffusion of GIS in a number of applications such as social-service planning as highlighted by Hugo (1997):

> In GIS, as in all technology, there is a real danger that the elite will gain control of it and that access among the mass community will remain limited. This must be guarded against especially since the technology and methodology of GIS, as in other areas involving computers, is becoming cheaper and more user-friendly and not necessitating years of training to interface with and use.

The core components of SDI — in relation to administrative boundaries — requires further development to address the issues of data integration between non-coterminous administrative boundaries and to empower the SDI framework to facilitate an optimum level of analysis in the spatial-information industry.

15.3 ADMINISTRATIVE BOUNDARIES WITHIN SDI

The spatial industry has experienced a transition phase from being data-poor, especially in terms of spatial data, to one that is now comparatively data rich. However, the means of organising, managing and using data to which there is now access have not kept pace with the need to make informed decisions and the technology now available (Openshaw 1998; UCGIS 2000). In order to meet the future needs of spatial-information analysts, institutional initiatives must be developed to address the different aspects of administrative boundary integration, sharing and management within an SDI (Feeney *et al.*, 2002).

It is proposed that well structured SDIs can reduce data duplication and facilitate data integration across administrative boundary systems and through time. The following section of this chapter addresses each of the five SDI components discussed in Chapter 2 and highlights their role in coordinating administrative boundary data integration.

15.3.1 Access

Improved technology and the greater penetration of GIS into government, business and society has produced a driving need for access to reliable and accurate spatial data (Nairn and Holland, 2001). Due to economics, culture and laws governing the extent of disclosure of spatial information, however, it is often impossible for spatial-information analysts to gain access to the data they require.

Administrative boundaries fulfil a niche within the spatial data market. They are relatively inexpensive to produce, meet privacy standards and provide spatial analysts with a plethora of information. Postcodes are a prime example of administrative boundaries within the SDI "...with postcodes you can locate people and see the hows, where's and whys of markets, customers and prospects, competitors, prices, suppliers, routes and profits. Postcodes neatly define convenient demographic zones and are familiar to everyone" (Geoscience Australia, 2001). As the potential of data analysis based on administrative boundaries is realised, policy related to data-access issues — such as pricing, copyright and licensing along with technical data standards — needs to be firmly established.

15.3.2 People

The interaction between the users of spatial data, data suppliers and any value-adding agents in between them drives the development of any SDI (Chan *et al.*, 2001; Rajabifard *et al.*, 2000). Considering the important and dynamic interaction between people and data, to develop effective SDIs, it is important to consider the changing nature of communities and their needs, which, in return, requires different standards and sets of administrative boundary data

In general, users of administrative boundary data are far more experienced and aware than previously and have increasingly demanding and more diverse expectations (Openshaw *et al.*, 1998). As a result, there is an increasing need to deliver administrative boundaries that meet the needs of users. One problem creating confusion amongst spatial-information analysts is the attempt by some organizations to aggregate their data to boundaries that are representatives of existing publicly recognizable units, such as the derived boundaries discussed in section 15.3.

15.3.3 Data

Boundaries are no longer just mechanisms through which order can be created and maintained. They can also act as a spatial device through which improved economic, social and environmental decision-making can take place (Marquart and Crumley, 1987). In order for this process to occur effectively, however, aside from the general data requirements relating to the content, quality, condition and completeness of spatial dataset, the issues of confidentiality and the modifiable-area-unit problem (MAUP) present two problems specific to the development of data attached to administrative boundary polygons.

(a) Confidentiality

The use of personal information within GIS arouses the conflict between societies' demand for increasingly accurate information and individuals' rights to preserve their privacy (Escobar *et al.*, 2001). The vast majority of social databases have grown from information collected from individuals and groups. The importance of maintaining confidentiality in the use of these databases is imperative to both the individuals and the public standing of the agencies involved in the data collection. As many social applications rely heavily on client-group confidence and the cooperation of community groups operating in the field, the development of improved inter-agency data exchange must be accompanied by effective procedures that protect individual confidentiality (ABS, 1998).

(b) The Modifiable-Area-Unit Problem (MAUP)

The MAUP is a form of ecological fallacy associated with the aggregation of individual data into areal units for spatial analysis (Fotheringham and Wong, 1991). An example of the process is census data, which is collected from every household but released only at census boundaries. When the values are averaged through the process of aggregation, variability in the dataset is lost, and values of statistics computed at different boundary resolutions will be different. This is called the *scale effect*. In addition to the scale effect, the analyst gets different results depending on how the spatial aggregation occurs. The MAUP is integral to the display of demographic data as the information relayed through mapping and statistics is a product of the size, shape and scale of the administrative boundaries used in the data-aggregation process. As outlined by Openshaw *et al.* (1998), in the past, the MAUP has been largely ignored by administrative agencies, with analysts unable to alter the boundaries provided to them. As a result, new developments are required to enable spatial analysts the freedom to design new output areas for the analysis of spatial data at a range of scales and aggregations, whilst preserving confidentiality.

15.3.4 Technical Standards

Technical standards are essential for efficient sharing of products and to provide information about spatial data. Technical standards are designed to simplify access and data quality and integration. Currently, the SDI policies, in general, have been designed to govern reference systems, data models, data dictionaries, data quality, data transfer and metadata. One area that needs to be fully developed is that of technical standards relating to the design, update, maintenance, consistency and cartographic representation of administrative boundaries. The Kansas Geospatial Jurisdictional and Administrative Boundaries Standard is an example of a set of standards that have been developed to facilitate the maintenance, representation and dissemination of boundary information so that it can be more easily integrated with other spatial structures (DASC, 1999).

The role of the standards is to provide 'best practise guidelines' to ensure that all maps, boundary descriptions, district names, and digital representations are complete, current and correct. As digital administrative boundary maps become commonplace, they will be used more frequently by a wide number of people. It is expected that these people will be using the data on a daily basis and will require more frequent updates of the data (DASC, 1999).

One technical issue related to standards that is highlighted in this chapter is the design criteria for new political and administrative boundaries. One initiative, that has been undertaken within Victoria, Australia by the authors is the reorganization of administrative boundaries into a coordinated hierarchy based on hierarchical-spatial-reasoning (HSR) theory.

(a) Hierarchical Spatial Reasoning (HSR) Applied to Administrative Boundaries

It is proposed that the reorganization of administrative-agency boundaries within a common, hierarchical spatial framework will enhance data integration and analysis methods. Figure 15.3 illustrates the proposed solution. Through the application of HSR theory, the spatial boundaries of different agencies are organised in a coordinated hierarchical system (Car, 1997). Data exchange and aggregation is possible within, and amongst, individual agencies, providing aggregated data at all levels. Currently, hierarchical properties are used in an array of different disciplines to break complex problems into subproblems that can be solved in an effective manner (Timpf and Frank, 1997). Although spatial hierarchies are designed using the same principles — to break complex tasks into subtasks or areas — relationships between levels within the hierarchies are complex (refer to Chapter 2). Section 15.4.4.2 details the structural complexities involved in the creation of a coordinated spatial-hierarchy model.

(b) The Structure of Administrative Boundaries

Structurally, within a GIS, administrative boundaries are considered as objects in a layer, such that each layer contains the same type of boundaries interacting in the same way among themselves (Car, 1997). The layers differ only in the degree of detail; therefore, to establish each layer in a hierarchy a set of rules are required. These rules must consider the boundary layer from both a functional and analytical perspective. Arguably, one of the most complex problems to overcome is the lack of clear business rules and constraints governing the design and shape of administrative boundaries. To be successful, it is imperative that common criteria can be established for the design of coordinated administrative boundaries. Figure 15.3 illustrates an example of a spatial hierarchy. The cadastre forms the base layer because the smallest administrative unit stored in the system determines the most detailed boundary system available (Volta and Egenhofer, 1993), and the cadastre is one of the most important infrastructure layers available (Dale and McLaughlin, 1988).

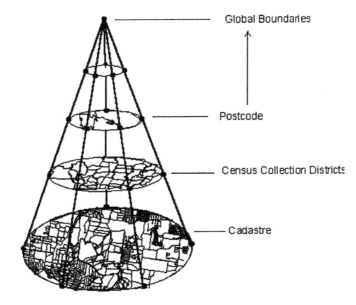

Figure 15.3:   Future hierarchically organised administrative structures
(Adapted from Eagleson *et al.,* 2002b)

The development of a coordinated spatial-hierarchy is intended to provide a framework in which agencies are able to construct administrative boundaries based on a common spatial layer, in this instance the cadastre. These boundaries are then aggregated to form new administrative units that meet the needs of more than one agency. If required, it is also possible for spatial-information analysts to create synthetic boundaries based on the core boundaries within the hierarchy. These synthetic boundaries allow the analyst freedom to examine alternative scenarios, whist preserving the confidentiality of individuals.

15.3.5 Policy

It has been established that exchanging, sharing and integrating spatial data based on administrative boundaries from various sources has become increasingly important. As described above, however, little policy governing the design and delineation of administrative boundaries exists, with emphasis predominately, focussed on technical aspects of boundary design (Eagleson *et al.*, 2002a; 2002b). Nevertheless, it has been proven that it is technically possible to develop a hierarchy of boundary units based on the criteria of two agencies. It is therefore important to develop policy that will further support these technological advancements. This will, in turn, facilitate the sharing and exchange of information between the public and the private sectors.

It must be recognised, though, that developing a policy alone cannot ensure the free flow of information from one organization to another unless institutional issues are addressed. In order to begin addressing these issues, there is a need to better understand the complex nature of SDIs and their ability to facilitate the implementation of new methods for designing administrative boundaries in the future.

To further promote the coordinated design of administrative boundaries, it is proposed that incentives for agencies to participate in the hierarchical design framework need to be established. These incentives may include the accreditation of agencies establishing boundaries within the spatial-hierarchy and/or benchmarking administrative boundary hierarchies to assess the comparative effectiveness of the system in facilitating data integration and exchange. As detailed below in Section 15.4.6 there are a number of recommendations that could improve SDIs and, consequently, improve the integration and exchange of data attached to administrative boundary systems.

15.3.6 Summary

The role of administrative boundaries has changed from that of an era of analogue mapping by individual agencies to the realised need for a coordinated boundary system incorporating the requirements of many SDI stakeholders. Additionally, technology is, to a certain degree, driving the way agencies do business. For example, the Internet has been suggested as a future tool to conduct censuses (Mobbs, 1998). If this form of collection is realised then the boundary delineation criteria set for establishing boundaries to represent this data will no longer need to consider the distance and time taken by census collectors; therefore, the method established for boundary design will need to be flexible and dynamic, taking into account the technology-related changes of the future.

Table 15.1 summarises the components of SDI and the mechanisms required to guide the design, delineation and dissemination of administrative boundaries and polygon-based data into the future.

Table 15.1: The Role of SDI and Recommendations to Facilitate the Use of
Administrative Boundary Data

	Role of an SDI	Recommendations
Access Networks	Provide users with mechanisms to access administrative boundary data.	• Improve data availability and ongoing assessment of requirements. • Provide a range of data products at different file sizes to facilitate a range of user needs.
People	Develop partnerships between administrative boundary users and the agencies establishing administrative boundaries.	• Educate spatial-data users. • Promote the benefits of spatial data amongst potential users. • Develop mechanisms to assess the requirements of users.
Technical Standards	Provide standards for the design, delineation and dissemination of administrative boundaries.	• Establish criteria for boundary delineation. • Establish methods for automated boundary delineation. • Derive metadata standards specific to administrative boundaries. • Provide guidelines for the cartographic representation of boundaries. • Improve mechanism for updating boundaries and providing notification of changes made.
Data	Provide standards for data attached to administrative boundaries. Reduce the cost of data production and dissemination.	• Facilitate the development of complete and up-to-date data beneficial for a range of applications. • Reduce duplication of datasets.
Policy	Facilitate the design of policy for the coordinated design, delineation and dissemination of administrative boundaries and associated metadata.	• Make ongoing assessment of requirements. • Provide guidelines to data custodians. • Delineate technology and methods. • Access and disseminate methods established. • Provide incentives to participate; i.e. accreditation, benchmarking and standards. • Provide mechanisms for research into the refinement of administrative boundary data.

15.4 CONCLUSION

Administrative boundaries are a product of both the era and the constraints of the individual agencies for which they were developed. This chapter demonstrates the significance of administrative boundaries within the SDI framework. Additionally, the chapter highlights one of the most prevalent problems currently limiting the use of data within a number of GIS applications: the spatial-hierarchy problem. In response to this problem, a number of technical developments have been made in the areas of surface modelling, interpolation, derived boundaries, and the re-aggregation of point and polygon data. These developments have contributed to a better understanding of the problem and nature of incompatible boundaries however, as the spatial industry expands, more accurate solutions are required. The research summarised in this chapter has demonstrated that the reorganization of boundaries into a coordinated spatial hierarchy is possible. However, as stated previously in this chapter developing a technical solution alone cannot ensure the development of a hierarchy of administrative boundaries until an organised SDI infrastructure is in place (see Chapter 2).

As SDI develops as a mechanism facilitating the transfer and access of spatial data to a wide array of data users the structuring of administrative boundaries in a coordinated manner will become increasingly important. This chapter has focussed specifically on the role and developments necessary to incorporate the unique properties of administrative boundaries within the SDI.

15.5 REFERENCES

ABS., 1998, *Australian Standard Geographical Classification (ASGC)*, The Australian Bureau of Statistics, Canberra, 12160.0.

Bracken, I., 1994, A Surface Model Approach to the Representation of Population-Related Social Indicators, In *Spatial analysis and GIS*, Edited by Fotheringham, S. and Rogerson, P., (Bristol: Taylor & Francis), pp. 247-260.

Car, A., 1997, Hierarchical Spatial Reasoning: Theoretical Consideration and its Application to Modeling Wayfinding, *PhD thesis*, Department of Geoinformation, Technical University Vienna.

Chan, T.O., Feeney, M., Rajabifard, A. and Williamson, I.P., 2001, The Dynamic Nature of Spatial Data Infrastructures: A Method of Descriptive Classification. *Geomatica* 55(1): 65-72.

DASC, 1999, *Kansas Geospatial Data Standards: Jurisdictional and Administrative Boundaries*, State of Kansas Data Access and Support Center (DASC), Kansas.

Dale, P. and McLaughlin, J., 1988, *Land Information Management,* (New York: Oxford University Press).

Duke-Williams, O. and Rees, 1998, Can census offices publish statistical for more than one small area geography? An analysis of the differencing problem in statistical disclosure, *International Journal Geographical Information Science,* vol. 12, no. 6, pp. 579-605.

Eagleson, S., Escobar, F. and Williamson, I. P., 2002a, Hierarchical spatial reasoning theory and GIS technology applied to the automated delineation of administrative boundaries. *Computers, Environment and Urban Systems*, no. 26, pp. 185-200.

Eagleson, S., Escobar, F. and Williamson, I. P., 2002b, Automating the Administrative Boundary Design Process using Hierarchical Spatial Reasoning Theory and GIS, *International Journal of Geographical Information Science,* (in press).

Escobar, F., Green, J., Waters, E. and Williamson, I., 2001, Geographic Information Systems for the Public Health: A proposal for a research agenda, In

Geography and Medicine. Geomed'99, Edited by Flahault, A., Toubiana, L. and Valleron, A., (Paris: Elsevier), pp. 139-148.

Feeney, M.E.F, Williamson, I.P. and Bishop, I.D., 2002, The Decision Support Role of Institutional Mechanisms in Spatial Data Infrastructure. *Cartography Journal* 3, No 2, pp 22-37.

Flowerdew, R. and Green, M., 1994, Area interpolation and types of data, in *Spatial Analysis and GIS*, Edited by Fotheringham, S. and Rogerson, P., (London: Taylor and Francis), pp. 121- 145.

Fotheringham, A. and Wong, D., 1991, The Modifiable Area Unit Problem, *Environment and Planning A,* 23, pp. 1025-1044.

Geoscience Australia, 2001, *Administrative boundaries*, Online. <http://www.auslig.gov.au/products/digidat/admin.htm> (Accessed March 2002).

Goodchild, M., Anselin and Deichmann, U., 1993, A framework for the areal interpolation of socioeconomic data, *Environment and Planning A,* vol. 25, pp. 383-397.

Hugo, G., 1997, Putting People back into the Planning Process: The changing role of Geographical Information Systems, Online. <http://www.gisca.adelaide.edu.au/gisca/hugo_paper.html>   (Accessed   May 2002).

Huxhold, W., 1991, *An introduction to urban Geographic Information Systems,* (Oxford: Oxford University Press).

Jones, S. D., Eagleson, S., Escobar, F. and Hunter, G. J., 2003, Lost in the mail: The inherent errors of mapping Australia Post postcodes to ABS derived postal areas, *Australian Geographical Studies (In press).*

Martin, D. and Bracken, I., 1991, Techniques for modelling population-related raster databases, *Environment and Planing A,* 23, pp. 1069-1075.

Martin, D., 1998, Optimizing census geography: the separation of collection and output geographies, *International Journal of Geographical Information Science,* vol. 12, no. 7, pp. 673-685.

Marquardt, W. R. and Crumley, C. L., 1987, Theoretical issues in the analysis of spatial planning, In *Regional dynamics: Burgyndian landscapes in historical perspective*, Edited by Crumley, C. L. and Marquardt, W. H., (San Diego: Academic Press), pp. 1-17.

Mobbs, J., 1998, Australia comes to its census, *Proceedings of FIG 7*, Melbourne Australia, p. 14.

Morphet, C., 1993, The mapping of small area census data - a consideration of the effects of enumeration district boundaries, *Environment and Planning A,* vol. 25, 9, pp. 1267-1277.

Nairn, A. and Holland, P., 2001, *The NGDI of Australia:* Achievements and challenges from federal perspective*s*, Online. <http://www.gisdevelopment.net/application/gii/global/giigp0003.htm> (Accessed October 2001).

Openshaw, S., 1998, Toward a more computationally minded scientific human geography, *Environment and Planning A,* vol. 30, pp. 317-332.

Openshaw, S., 1977, An optimal zoning approach to the study of spatially aggregated data, In *Spatial Representation and Spatial Interaction*, Edited by Masser, I. & Brown, P., (Leiden: Martinus Nijhoff), pp. 96-113.

Openshaw, S. and Alvanides, S., 1999, Applying geocomputation to the analysis of spatial distributions. In *Geographic Information Systems: Principles and Technical Issue*, Edited by Longley, P., Goodchild, M., Maguire, D. and Rhind, D., (New York: John Wiley and Sons).

Openshaw, S., Alvanides, S. and Whalley, S., 1998, Some further experiments with designing output areas for the 2001 UK census, Online.
<http://www.geog.soton.ac.uk/research/oa2001/resources.htm>   (Accessed November 2000).

Openshaw, S. and Rao, L., 1995, Algorithms for reengineering 1991 Census geography, *Environment and Planning A.*, no. 27, pp. 425-446.

Rajabifard, A., Escobar, F. and Williamson, I. P., 2000, Hierarchical spatial reasoning applied to Spatial Data Infrastructure, *Cartography*, vol. 29, No. 2, pp. 41-50.

Rajabifard, A. and Williamson, I. P., 2001, Spatial Data Infrastructures: Concept, SDI hierarchy and future directions, *Geomatics'80 Conference*, Tehran, Iran. Online.
<http://www.geom.unimelb.edu.au/research/publications/IPW/4_01Raj_Iran.pdf> (Accessed October 2002).

Robinson, J.M. and Zubrow, E., 1997, Restoring continuity: Exploration of techniques for reconstructing the spatial distribution underlying polygonized data, *International Journal of Geographical Information Science*, vol. 11, no. 7, pp. 633-648.

Timpf, S. and Frank, A. U., 1997, Using hierarchical spatial data structures for hierarchical spatial reasoning. In *Spatial Information Theory - A Theoretical Basis for GIS (International Conference COSIT'97)*, Edited by Hirtle, Stephen C. and Frank, Andrew U. Lecture Notes in Computer Science 1329, (Berlin-Heidelberg: Springer-Verlag).

Trinidad, G. and Crawford, J., 1996, A tool for transferring attributes across thematic maps, *AM/FM*, pp. 127-137.

UCGIS, 2000, *Spatial analysis in a GIS environment*, Online.
<http://www.ucgis.org/research_white/anal.html> (Accessed October 2001).

Volta, G., and Egenhofer, M., 1993, Interaction with GIS Attribute Data Based on Categorical Coverages. In *European Conference on Spatial Information Theory*, Edited by A. Frank and I. Campari, (Italy: Marciana Marina).

CHAPTER SIXTEEN

SDI and Location Based Wireless Applications

Jessica Smith and Allison Kealy

16.1 INTRODUCTION

Current SDI design is mainly focused on networked access to datasets. However, technological advancements are providing spatial information producers and users with new data access opportunities. The convergence of wireless communication techniques, for instance with positioning technology and network computing are providing the mechanisms for a range of facilities and applications known as Location Based Services (LBS). LBS often rely on a range of common spatial fundamental datasets. However, at this stage no coordinated effort has been made to use SDI to facilitate access and retrieval of parts of these common datasets. In order to take advantage of the technological developments enabling new data access opportunities, there is a need to augment the existing SDI model. Augmentation is proposed through the revision and expansion of current infrastructure element definitions. The revised elements should encompass a high degree of flexibility so that they can continue to adapt to the wide variety of applications that are expected to arise from these and future technological convergences.

With this in mind, the objective of this chapter is to demonstrate the need for the development of a framework to enable SDI to incorporate new applications for data use and access. This is achieved through a review and evaluation of existing access models, polices and standards, the proposal of new opportunities and the infrastructure requirement for the key components of SDIs (data, access network, policies, standards and people) in relation to wireless applications that use spatial information. Based on this evaluation, recommendations and future directions are suggested to facilitate SDI development for all levels in the SDI hierarchy.

16.2 SDI AS A FOUNDATION FOR LOCATION BASED SERVICES

Since the concept of an infrastructure to support data access and management in the 1980s and the inception of the term Spatial Data Infrastructure in 1989 (Tosta, 1999), much research and practical experience has been gained regarding SDIs both nationally and internationally. Intended to facilitate the collection, management, access, delivery and utilization of geographic data, SDIs require participation and co-operation between both government and private sector organizations. The data sharing environment that is made possible by an SDI has the potential to promote widespread use of available datasets and thus increase business opportunities for the geographic information industry as a whole.

Irrespective of the fact that stakeholders from various disciplines view SDIs differently, researchers have identified a number of core components that are common to all SDI implementations (Coleman and McLaughlin, 1998; Rajabifard *et al.*, 2000, 2002). As shown in Figure 16.1 (and previously in Figure 2.2) these components are: people, access networks, policy, technical standards and datasets. Each component is strongly related to the other four components resulting in a cohesive infrastructure.

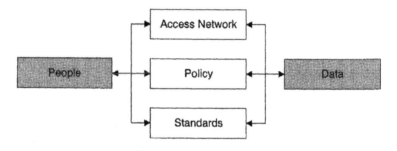

Figure 16.1: SDI Components (Rajabifard *et al.*, 2002)

It is these five core components that need to accommodate the recent opportunities presented by the wireless communication field in relation to data access and dissemination. Mobile phones and handheld computing devices (such as personal digital assistants) with wireless connectivity capabilities are examples of the new platforms that are enabling wireless Internet access. This nomadic computing and communications paradigm as described by Leiner *et al.* (2000) reverses the geographic anonymity of the traditional, fixed line Internet with location now playing a significant role in the access and dissemination of information.

Whilst the mobile Internet is promoted as a facility that can be operated anywhere, anytime (PricewaterhouseCoopers, 2001), these location and time 'attributes' will actually be populated for each access made by a user. Using this spatial and temporal knowledge, services can be tailored to present appropriate and relevant information to users. It is these sorts of services that reinforce geography and a sense of place to mobile subscribers and are referred to as Location Based Services. LBS have evolved from the convergence of network computing, wireless telecommunications and positioning technologies as well as in response to regulatory initiatives such as the United States' Federal Communications Commission's Enhanced 911 mandate. The E911 mandate requires that for all calls made to the emergency 911 line from mobile phones, the mobile phone (and hence its user) must be located to within 125m (FCC, 1999). Techniques using enhanced mobile phone handsets and the telecommunications network to determine the position of mobile phones have been the focus of many research initiatives (Drane and Rizos, 1998; ETSI, 2000; Hayes, 2000). The technologies supporting LBS are well established in their own right, but will continue to evolve. As a result, LBS applications will also evolve over time and will provide improved services to users.

Location based services have been forecast as a driver for the mobile Internet and the basis of many new applications (Blonz and McCarthy, 1998; McCabe, 1999; Hayes, 2000). To achieve their predicted success, these applications will have to contribute to a user's personal mobility by providing substantial benefit to users whilst they are away from locations of fixed Internet access. The potential to integrate real-time data with other location based information is one such advantage of this communication method that is likely to significantly enhance a mobile user's experience.

Location based services rely on an infrastructure that is capable of performing spatial queries and transmitting this information to mobile devices (with limited screen dimensions, screen colour depth, power restrictions and computing power) in an appropriate form. Many of the LBS developed to date rely on similar datasets (for example transportation networks, cadastral information and administrative boundaries in the form of postcodes, suburbs, etc.). Whilst much of this data must be obtained from various sources, it is not the core business functionality of an LBS developer to update and maintain such datasets. Issues of data access and reuse promoted by SDI initiatives suggest that these principles could be applied to the LBS domain; a corporate level SDI for LBS could be established that would facilitate LBS application development, and as a result, enhanced data access to a wide range of users. LBS developers could then develop their applications to access the appropriate data through the infrastructure and be guaranteed of its quality and completeness. For this to become a reality however, the specific infrastructure requirements necessary for ensuring that SDIs can support this form of data delivery need to be determined. This need has been identified by the Open Location Services initiative, a sub-group of the Open GIS Consortium:

> Spatial connectivity is a primary, universal construct for business planning and modelling, service development and deployment, network provisioning and operation and customer satisfaction. Location application services are of universal industry significance and depend upon the availability of relevant spatial information infrastructures in forms useful for small devices.
>
> (Open GIS Consortium, 2000).

16.3 AUGMENTING THE SDI MODEL

The strength of the SDI lies in the interconnected and cohesive nature of the five components of access network, standards, policy, people and data. Whilst these components could be investigated in isolation, they have all been recognised in a range of SDI implementations of varying scales. Therefore determining the specific requirements of this infrastructure in the context of LBS will require an examination of all five of the components. Some of the issues specific to the wireless environment are detailed in the following sections along with some proposals for detail enhancement or augmentation to the SDI model.

16.3.1 Access Network

The access network component of an SDI is critical from a technical perspective to facilitate the use of data by people. Unlike a fixed line Internet environment, the wireless environment poses many limitations on the transmission of data. Most prominent is the issue of bandwidth.

Bandwidth is the width of a band of electromagnetic frequencies of a signal, essentially the amount of frequency spectrum that a single electronic signal occupies (Held, 2000), and is generally expressed in terms of the speed at which data can flow on a particular transmission path – bits per second (bps). Fixed line modems and Local Area Network Ethernet connections with bandwidths of 56kbps and 10Mbps respectively offer much more scope for data transmission when compared to the most prevalent wireless telecommunications system, the GSM (Global System for Mobile Communications) network which offers a bandwidth of only 9.6kbps.

Despite developments to improve the bandwidth available over wireless networks (such as High Speed Circuit-Switch Data, the General Packet Radio Service and 3G services – which require new telecommunications infrastructure investment), it is unlikely that they will ever be able to match the capacity of fixed line connections and therefore restrictions on the amount of data transmitted to mobile users should be imposed. This will not only ensure that the wireless network does not operate close to capacity at all times, but will also enhance the user's experience by providing prompt responses. Using location as an information filter can also help to restrict data flow. Rather than sending a mobile user the complete yellow pages directory entries for restaurants, a small section of the entries need only be returned – those that list restaurants in the user's current area. This is also a more useable and useful solution for users.

Whilst increasing bandwidth may be feasible from a network perspective, it introduces another problem. In order to process more bandwidth, more power is required. Handset form factors limit the size of batteries and hence the available power of mobile devices, therefore wireless data solutions must be able to overcome these network limitations to still deliver a satisfactory experience to the user. A balance must be found between bandwidth and the power required to process the bandwidth.

In terms of data transmission, the current standard is the Wireless Application Protocol (WAP). WAP is designed to complement existing wireless standards, but does not specify how data should be transmitted over the air interface. Rather, WAP 'sits on top' of existing bearer channel standards (including Short Message Service (SMS), Circuit Switched Data (CSD), Unstructured Supplementary Services Data (USSD) or General Packet Radio Service (GPRS)) (Buckingham, 2000). This means that any bearer standard can be used with the WAP protocols to implement product solutions (MobileInfo.com, 2001). WAP is also optimised for the limited functionality and display capabilities of current mobile terminals.

Whilst most of these observations relate to the end user's experience, access issues are also of concern to integrators and data providers. Integrators need to ensure that the data they are enabling access to is appropriate for the intended purpose, and presented in a useful form to end users. Additionally, services that the integrator develops need to provide seamless data access (alleviating the user from understanding the underlying infrastructure) and be both responsive and sympathetic to the network mechanism through which the access is being provided (particularly in the case of wireless networks with limited bandwidth). In the process of developing applications, integrators must have appropriate access mechanisms to the underlying data – a catalog or repository of datasets available and suitable for LBS could be established. Data providers, whose data is useful for LBS services and conforms to appropriate standards for LBS, should advertise their data through such a catalog service to facilitate the sharing and reuse of datasets.

16.3.2 Policies

The policy and administrative component of the SDI definition is critical for the construction, maintenance, access and application of standards and datasets for an SDI implementation. In general, policies and guidelines are required for SDI that incorporate:

- Spatial data access and pricing;
- Spatial data transfer;
- Custodianship;
- Metadata; and
- Standards.

Whilst these policy and administrative arrangement categories may be appropriate for SDIs that support fixed line data access, they also need to be appropriately structured to accommodate wireless data transfer. Additionally wireless communication poses some privacy concerns and thus privacy policy or regulations should also be a part of an LBS SDI policy definition.

Throughout its history, the nature of the spatial information industry, and the use of spatial technologies have raised concerns regarding personal information privacy, intellectual property rights of geographic information, liability in the use of geographic datasets, public access to government geographic datasets and the sale of geographic information by government agencies (refer to Onsrud *et al*, 1994; Cho, 1995; Clarke, 2001). The evolving domain of location based services that is often reliant on position determination of a mobile device (and hence its user), is also raising concerns amongst subscribers. Irrespective of these emerging applications, tracking is inherent in existing telecommunications networks, albeit at a coarse level that is related to cell sizes (which can range from less than 1km to approximately 20km). This location monitoring is required in order to direct incoming calls to the appropriate device (through the relevant cell's base station). The movement of mobile users can be inferred from temporal analysis of these tracking records. Whereas the traditional telephony model identifies a location through a telephone number, mobile telephone numbers now identify a user (the location of whom can be determined from network records). Whilst these records are not typically available to the general public, telecommunication companies have recently become aware of the benefits associated with analysing this data (Clarke, 2001).

Location based services extend the position information that can be obtained about a mobile device at a particular time, by using this position variable to provide specialised content to users. While currently LBS in themselves do not pose any further infringements of privacy, they are increasing awareness of the ability to pinpoint a user through their mobile device. As improved location techniques are devised and positioning mechanisms are implemented in mobile devices, users are likely to become more concerned with this range of services and the use of data that is kept about them.

As a result, policies must be put in place to ensure the appropriate use of personal and location data captured via telecommunications. In Australia, these policies must be inline with the new *Privacy Amendment (Private Sector) Act 2000* to the *Privacy Act 1998* (Cth) which aims to 'regulate the way many private sector organizations collect, use, keep secure and disclose personal information' (Office of the Federal Privacy Commissioner, 2001). The policy component in an augmented SDI should also specify guidelines on pricing, data standards and personalization. Even though the details of these policy issues will vary for each location based service implementation, there should be some generic principles that could be recommended at the infrastructure level.

16.3.3 Standards

To ensure interoperability amongst the datasets and access mechanisms defined by an SDI, standards are essential. Standards can be applied at many different levels within an SDI. In terms of data, Australia's former national mapping organization the Australian Land Information Group, AUSLIG (2001) identify that standards are required 'in reference systems, data models, data dictionaries, data quality, data transfer and metadata'.

Even though the fields from which SDI and LBS have evolved are well established, the SDI and LBS areas themselves are still relatively young and as a result many standards are still under development. Organizations such as the International Standards Organization Technical Committee on International Geographic Information Standards and the Open GIS Consortium are researching standards related issues for location based services from a global perspective. (Refer to www.isotc211.org and www.opengis.org respectively). Similarly on the SDI front, many countries are exploring standards issues in association with the development and implementation of their National SDIs (Hissong 1999; Nairn 2000; Arias 2001; Hayes 2001; Viergever 2001). There are also a number of regional and jurisdictional initiatives, particularly in Europe, investigating the impact and importance of standards on SDI implementations.

From a wireless communications perspective, telecommunications network standards will pose some restrictions on the quantity and format of data that can be delivered wirelessly. Whilst the spatial information industry will have little, if any, input into how these networks should be structured, an understanding of the network capabilities and limitations is essential in order to develop appropriate spatial applications that rely on this medium. As mentioned above bandwidth limitations will require minimised query and response data flow from users to the data repositories. Standard interfaces, such as those proposed by the OGC (refer to Open GIS Consortium, 2001), will facilitate this.

Standards on data quality need to be specified for data that can be considered suitable for location based services. Whilst it is critical that all spatial information used in decision-making be of appropriate quality for its intended use, the expanded user base encouraged by location based services raises the need for unique quality standard requirements. Users of LBS will typically be accessing the service on a mobile device with limited screen size and resolution. It is likely that they will be simultaneously carrying out tasks whilst using the service, and will commonly be travelling by foot. Not necessarily being familiar with spatial information, users could have difficulties executing navigation and orientation instructions as portrayed by LBS. Standard presentation formats could assist in the cognition of this information by users and hence the usability of systems. Standards for positional and attribute accuracy of data would also be of use considering the criticality of this information to pedestrians.

16.3.4 Data

Interoperability is a key consideration of both the standards and data component of an SDI. Data within an SDI should be compatible in terms of format, reference system, projection, resolution and quality.

A major challenge for the wireless dissemination of spatial data is that of quantity and quality. As discussed previously, wireless communication networks offer restricted bandwidth compared with fixed line networks. Spatial databases can be quite large (in terms of geographic extent, level of detail stored and hence quantity of storage space required).

The nature of mobility will mean that mobile users will not require access to large portions of data, rather they will be interested in small data portions that are relevant to their current position. Additionally, the format and presentation of this information must be carefully structured. Whether current data models are appropriate to meet these challenges is yet to be determined.

Considering that many LBS applications rely on the same sort of data (e.g. transport networks and address information) which LBS service providers may not wish to maintain themselves, the establishment of an SDI containing these datasets could be feasible. LBS providers could then develop their applications to use data via the SDI, but could also supplement the fundamental sets with their own data specifically suited to their application. This would ensure data quality of the fundamental datasets would be maintained, as it would remain a responsibility of the data custodians.

Data quality is an important issue for mobile users who demand information in real time. Users, who typically will not be trained in the spatial sciences, will not be interested in trying to assess the quality of information that they are receiving. Rather, they will expect the information to be of an appropriate quality standard for their purposes, and will be unlikely to subsequently use the service if they discover otherwise. Guptill and Morrison (1995) describe data quality as consisting of: lineage, positional accuracy, attribute accuracy, logical consistency, completeness and temporal accuracy. These elements have been incorporated into the framework guidelines and framework dataset compliance auditing criteria of many nations. For example Somers (1997) documents the guidelines for the US NSDI and Nairn (2000) the ASDI guidelines.

Whilst issues of content, extent, custodianship, format, metadata, standards and access can all help to ensure quality data, of prime concern to a mobile user are the issues of spatial and attribute accuracy, currency and logical consistency. Without some standard or guarantee for these elements, location based services would offer little value to end users.

16.3.5 People

Through increased use and awareness of spatial data, a dramatic growth has occurred in the user base of spatial information. With the proliferation of online web mapping, and navigation/direction information, an increasing number of people are using GIS and spatial information often without even realising. The provision of spatial information to handheld devices will continue to increase the spatial information user base substantially.

The people involved in a location based service SDI can be categorised into three groups: (i) data providers; (ii) integrators - LBS developers or service providers; and (iii) end users. Data producers collect and create the spatial databases for the SDI. The LBS developers design and implement LBS applications that use this data and disseminate it using wireless communication mechanisms to users' mobile devices. Finally the end users are the people who use the LBS applications on their mobile devices. They will not necessarily be trained in the use of spatial information, but should be able to use the service to solve their spatial problems, and assist them with their decision-making.

Wireless location based services will help to promote the use of spatial information by a broad proportion of the community. Whilst capacity building initiatives encompassing a range of education and training programmes would be useful in enabling the wider community to become spatially aware and literate, it should not be a prerequisite for location based service use. Particularly in the early stages of these services, it is the responsibility of data providers and integrators to develop intuitive services that can be used by the general community irrespective of their spatial ability and knowledge. As society does become more technically savvy, spatially aware and more demanding of services available through mobile devices, more detailed and enhanced services are likely to be required.

16.4 FRAMEWORK TO FACILITATE WIRELESS APPLICATIONS

16.4.1 SDI Requirements

As noted above, the specifics of wireless dissemination of spatial data to a diverse user group presents many challenges. As a result, a number of requirements have been identified for an SDI in this environment (refer to Figure 16.2). It should be noted that the requirements identified in Figure 16.2 and described below do not form a definitive list and other requirements may be found to exist for each SDI component.

The access component is required to provide an efficient request-response turnaround for the service and be capable of meeting the demands of multiple simultaneous users without degrading service speed significantly. Whilst these issues are relevant for all SDI, these issues are regarded more highly by users in the wireless environment. Associated with the access component are the issues of information presentation and cognition. Information should be presented in a way that can be quickly and easily read and understood given that mobile users are likely to be completing other tasks simultaneously. Interaction methods (including personalization features) also form part of the access component's requirements.

Generic policy requirements are more difficult to define than the requirements for the access component, as they are likely to vary more widely for each LBS implementation. Four broad requirements have been identified; firstly guidelines on the knowledge of user's location and/or activities, and use of this information are required. Secondly, provisions for informing users of the validity, accuracy and currency of the information presented are required. Thirdly, unlike other SDI implementations, the decision-making process undertaken by a mobile user must be completed rapidly, and even in some cases dynamically; by allowing users to customise services, this may be able to occur more rapidly. Finally, pricing models will need to be determined on a case by case basis. In some instances a subscription model may be appropriate, in others a pay per use model may suffice.

The standards component requires seamless integration and interoperability between datasets. As is the case with other SDI this may include data of varying resolution and granularity. Most critically, expressions of data quality (including currency, and precision) are required.

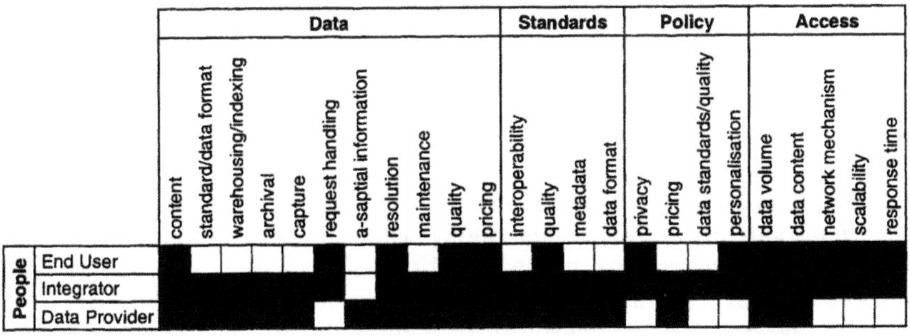

Figure 16.2: SDI Requirements for the Diverse User environment, Categorised by People

16.4.2 User Environment

To date, the SDI user environment has been predominantly focussed on the public sector. Increasingly the private sector are realising the benefits that can be obtained by the data collaboration and sharing practices encouraged by SDI.

The three people categories identified earlier (namely end users, data integrators and data providers) describe the three actors and their corresponding roles within an SDI. Typically the public sector will play the role of data provider, and this would be no different in an LBS SDI. For example, the base datasets of road network and cadastre are the responsibility of local and state governments respectively. However, this does not exclude the private sector from the role of data provider. In fact, the navigational requirements of many LBS datasets will mean that a high proportion of data will be provided by the private sector through the value adding of navigation and additional attributes to base state and local scale datasets.

The data integrator role, following current trends, is expected to be taken up by private industry. Organizations will continue to develop easy to use applications for end users. An environment that facilitates LBS application development by providing access to a repository of appropriate datasets governed by associated policies and standards would relieve developers from having to invest significant time and resources in establishing arrangements to obtain data or implementing independent data capture regimes.

Finally, the role of end users encompasses using location based services to solve common problems related to their location and either some destination or target object. For example, this category could include: people interested in getting from one location to another, learning more about a particular area, finding a central meeting place for a dispersed group of friends; managers tracking a mobile workforce; clients tracking shipments; etc. The challenge is to develop applications that anybody, regardless of their spatial information knowledge or training, can use to solve their everyday problems.

16.5 CASE STUDY – PUBLIC TRANSPORT APPLICATION

To demonstrate the proposed augmented SDI approach and determine any additional requirements for SDIs supporting wireless data dissemination, a prototype public transport information service has been developed. This service relies on both static and real-time information, and its presentation within a mobile environment.

The public transport information system aims to provide the most accurate and up to date information possible regarding the status of public transport services in metropolitan Melbourne without users having to physically participate within the public transport service. Melbourne, along with other cities around the world, is currently implementing and trialling real-time information systems for public transport patrons (Yarra Trams, 2001; Department of Infrastructure, 2002). However, these systems provide real-time information only at public transport stops or stations. This information is also of value to users who will be using the public transport system in the immediate future, or those who are contemplating using the public transport system. Existing services available to these user groups revolve around static timetable data that represents the expected, rather than the actual state of the public transport service.

The operating environment and situation in which wireless applications are used impact on the content, structure, rate of delivery, quantity and format of the information to be presented. Rainio (2001) notes that it is imperative that information provided to users on a wireless device is not contradictory, does not lead to incorrect actions and does not overload the user cognitively. In addition to these issues, Rainio (2001) also recognises that users should always be informed about the quality of the information presented and where responsibility lies for such items as location accuracy, currency, pricing and service guarantees.

16.5.1 Use Case Scenarios

Examination of example user scenarios has led to the development of a use case model and requirements specification. Three categories of system user were identified: planner, imminent traveller and active traveller. A 'planner' may be deciding between public and private transport options, or planning a future public transport journey. An 'imminent traveller' intends to use public transport immediately or in the short-term future (i.e. within the current day). An 'active traveller' is a user who is currently on board a public transport service. The requirements for these three user categories are shown in Table 16.1.

Table 16.1: Public Transport Information System User Requirements

	Requirement	Planner	Imminent Traveller	Active Traveller
Pedestrian journey	Origin of journey (may correlate with position of mobile device)	✓	✓	-
	Destination of journey	✓	✓	-
	Textual information from origin to first public transport point	✓	✓	✓
	Textual information from final public transport point to destination	✓	✓	✓
	Pedestrian route map (beginning and end of journey)	✓*	✓*	✓*
	Pedestrian journey duration	✓	✓	-
Public transport journey description	Mode identifier	✓	✓	-
	Embarkation identifier (stop number or station name, platform)	✓	✓	✓
	Disembarkation identifier (stop number or station name, platform)	✓	✓	✓
	Route identifier (number and/or name)	✓	✓	✓
	Direction of service (inbound/outbound)	✓	✓	✓
	Scheduled departure time at disembarkation point	✓	-	-
	Expected departure time at embarkation point	(✓)	✓	✓
	Scheduled arrival time at disembarkation point	✓	-	-
	Expected arrival time at disembarkation point	(✓)	✓	✓
	Indication of route/mode interchange	✓	✓	✓
	Public transport journey 'snap shot' maps (e.g. interchanges)	✓*	✓*	✓*
	Journey duration (including connection waiting times if necessary)	✓	✓	✓
	Zone/Fare information	✓	(✓)	-

Legend: ✓ necessary requirement; (✓) optional requirement (depending on mode, or time of service access); ✓* 'nice to have' requirement; - not applicable

In all cases, the detail and form of the information presented would ideally vary depending on the user's requirements and the platform on which they access the service. To accommodate this, a hierarchical information interface has been proposed. Figure 16.3 shows an example of the interface flow for a user interested in finding a route between two locations. Upon activating the service, the user would select the 'Determine route' option from the opening screen. The current position of the mobile device would then be determined and the user prompted to enter their destination. This destination, entered in the form of a street address, is reverse geocoded for later use in the routing algorithms. The user then requests a route between the origin and destination, and if a route exists between the two a summary of the journey details is provided (syntax: stop embarkation number, route number, route direction, stop disembarkation number). This summary information is the first level of the interface hierarchy and is intended to provide enough information for an experienced user of the system to be able to determine if this route suits their needs. Less experienced users, or those requiring more information can drill down through the hierarchy to gain more details. Preliminary testing of the interface hierarchy reveals that the volume of information presented is suitable for the wireless environment; users can examine two levels of the interface for three journey segments (as shown in Figure 16.3d) in under one minute, and for the cost of spare change.

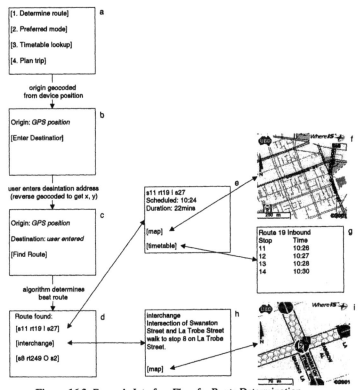

Figure 16.3: Example Interface Flow for Route Determination

Preliminary results relating to the investigation of SDI requirements for LBS have led to the representation of an augmented SDI model (Figure 16.4). This model has been based on the standard SDI components and their interconnection (as represented in Figure 16.1) and has proved to be a sound foundation from which to add the additional requirements for LBS. The augmentation lies in additional levels of detail for each of the components. At this stage, the layers relate to the three distinct roles of people that have been proposed for the LBS SDI environment, and their unique demands in relation to each of the components, as identified in Figure 16.2. It is expected that there will be additional augmentation layers describing in greater detail the interfaces between the components and between the component requirements themselves. Further development and testing of the prototype will be undertaken to verify the model and define enhancements as required.

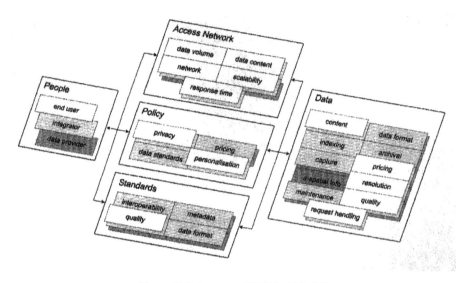

Figure 16.4: Augmented SDI Model for LBS

16.6 DISCUSSION, RECOMMENDATIONS AND FUTURE DIRECTIONS

Technological advances are presenting many new challenges for the spatial information industry. The ability to access information on a portable wireless device whilst limited in many ways, is providing new ways in which to work as well as conduct daily activities. Additionally, linking a user's position with spatial and non-spatial data in the form of LBS is increasing the user base of spatial information.

Location based services are consistently cited as a range of mobile Internet services that are likely to drive the development of the mobile Internet. Irrespective of the range of services encapsulated by the term location based services, all of them require spatial data management capabilities to link position information with other data sources. The underlying technologies required to support these services must relate to the rapidly evolving Internet and communications standards.

The potential for a location based service SDI containing common datasets exists, and could promote LBS development through the provision of high quality spatial data. However, characteristics of the delivery of spatial information via wireless communication channels needs to be investigated, along with the specific requirements of data storage and wireless access.

The interconnected nature of the five SDI components means that modifying one component will probably require modifications to be made for other components. For example the policy component will have to examine privacy in response to the concerns of the community in relation to the ability of a mobile phone to determine its location and transmit and use this information to provide additional information.

The issues presented in relation to augmenting the SDI model need to be refined to ensure that SDIs are capable of supporting the predicted range of applications and services. The mechanisms that will enable mobile users to gain access to small, appropriate sections of larger datasets (such as those stored in SDIs) also need to be determined in order to make wireless spatial information access a feasible option.

16.7 REFERENCES

Arias, P., 2001, Colombian Spatial Data Infrastructure - ICDE: Technical implementation issues as a contribution to build SDI components in the developing world. *International Symposium on Spatial Data Infrastructure*, (Melbourne: Centre for Spatial Data Infrastructure and Land Administration).

AUSLIG, 2001, Australian Spatial Data Infrastructure, AUSLIG, Online. <http://www.auslig.gov.au/asdi/index.htm> (Accessed November 2001).

Blonz, T. and McCarthy, C., 1998, *Mobile Location Services*, Ovum, Boston.

Buckingham, S., 2000, *What is WAP?*, GSM World, Online. <http://www.gsmworld.com/technology/wap/intro.shtml> (Accessed August 2001).

Cho, G., 1995, Legal dilemmas in geographic information: property, ownership and patents. *Journal of Law and Information Science*, 6:2, pp. 193-207.

Clarke, R., 2001, Person-Location and Person-Tracking: technologies, risks and policy implications. *Information Technology and People*, 14:2, pp. 206-231.

Coleman, D.J. and McLaughlin, J., 1998, Defining Global Geospatial Data Infrastructure (GGDI): components, stakeholders and interfaces. *Geomatica*, 52:2, pp. 129-143.

Department of Infrastructure, 2002, What is SmartBus?, Online. <http://www.linkingvictoria.vic.gov.au/DOI/Internet/transport.nsf/AllDocs/A6F7 49C95A15876BCA256BF1001577FC?OpenDocument> (Accessed August 2002).

Drane, C. R. and Rizos, C., 1998, *Positioning Systems in Intelligent Transportation Systems*, The Artech House ITS series, (Boston: Artech House).

ETSI, 2000, *ETSI TS 101 724, v7.3.0 (2000-02)*, European Telecommunications Standards Institute, RTS/SMG-030371Q7R2.

FCC, 1999, Third Report and Order, Federal Communications Commission, Online. <http://www.fcc.gov/Bureaus/Wireless/Orders/1999/fcc99245.pdf> (Accessed July 2000).

Guptill, S. C. and Morrison, J. L. (Eds.), 1995, *Elements of Spatial Data Quality*, (Oxford: Elsevier Science).

Hayes, H. B., 2001, Working off the same map - OMB's I-team initiative pulls together federal, state and local mapping efforts. *Federal Computer Week,* October 29, pp. 18-21.

Hayes, N., 2000, Locating your location based services provider, Wireless Developer Network, Online. <http://www.wirelessdevnet.com/channels/lbs/features/newsbite11.html> (Accessed July 2001).

Held, G., 2000, *Understanding Data Communications: From Fundamentals to Networking,* 3rd ed., (Chicester: John Wiley and Sons).

Hissong, F., 1999, Will locals lead the way to a national GIS? *American City and Country,* August, pp. 22-24.

Leiner, B. M., Cerf, V. G., Clark, D. D., Kahn, R. E., Kleinrock, L., Lynch, D. C., Postel, J., Roberts, L. G. and Wolff, S., 2000, A Brief History of the Internet, Internet Society, Online. <http://www.isoc.org/internet/history/brief.shtml> (Accessed October 2001).

McCabe, E., 1999, Location-based services offer a global opportunity for new revenue, Telecommunications Americas, Online. <http://www.telecoms-mag.com/default.asp?journalid=3&func=articles&page=location&year=1999&month=10&srchexpr=mccabe> (Accessed October 1999).

MobileInfo.com, 2001, Wireless Application Protocol, MobileInfo, Online. <http://www.mobileinfo.com/WAP/what_is.htm> (Accessed November 2001).

Nairn, A. D., 2000, Australia's Developing GIS Infrastructure - achievements and challenges from a federal perspective. *5th International Seminar on GIS,* Seoul, Korea, Online. <http://www.auslig.gov.au/corpinfo/publications/technical/docs/gis.pdf>.

Office of the Federal Privacy Commissioner, 2001, 1-2001 Overview of the Private Sector Provisions, Office of the Federal Privacy Commissioner, Online. <http://www.privacy.gov.au/publications/IS1_01.pdf> (Accessed November 2001).

Onsrud, H. J., Johnson, J. P. and Lopez, X., 1994, Protecting personal privacy in using Geographic Information Systems. *Photogrammetric Engineering and Remote Sensing,* LX:9, pp. 1083-1095.

Open GIS Consortium, 2000, Open Location Services - Call for sponsors, OGC, Online. <http://www.openls.org/docs/cfs.htm> (Accessed February 2000).

Open GIS Consortium, 2001, OGC's Role in the Spatial Standards World, Open GIS Consortium, Online. <http://www.opengis.org/pressrm/summaries/20010225.TS.Stds.htm> (Accessed August 2001).

PricewaterhouseCoopers, 2001, Technology Forecast: 2001-2003 - Mobile Internet: Unleashing the Power of Wireless, (California: PricewaterhouseCoopers Technology Centre).

Rainio, A., 2001, Location-Based Services and personal navigation in mobile information society. *The New Technology for a New Century, International*

Conference FIG Working Week 2001, (Seoul, Korea: FIG), Online. <http://www.fig.net/figtree/pub/proceedings/korea/fullpapers/plenary1/rainio.htm

Rajabifard, A., Feeney, M., and Williamson, I.P., 2002, Future Directions for the Development of Spatial Data Infrastructure. *Journal of the International Institute for Aerospace Survey and Earth Sciences*, ITC, Vol. 4, No. 1, pp. 11-22.

Rajabifard, A., Williamson, I. P., Holland, P. and Johnstone, G., 2000, From Local to Global SDI Initiatives: a pyramid of building blocks. *4th GSDI Conference*, South Africa, Online. <http://www.geom.unimelb.edu.au/research/publications/IPW/ipw_paper41.pdf>.

Somers, R., 1997, Framework Introduction and Guide, FGDC, Online. <http://www.fgdc.gov/framework/frameworkintroguide/> (Accessed October 2000).

Tosta, N., 1999, NSDI was Supposed to be a Verb. In *Integrating information infrastructures with geographical information technology*, Innovations in GIS 6, edited by Gittings, B. M., (London: Taylor and Francis), pp. 3-24.

Viergever, K., 2001, Spatial Data Infrastructure - Successful and Easy Implementation in Southern Africa? *GIM International*, April 2001, pp. 13-15.

Yarra Trams, 2001, Superstops, Online. <http://www.yarratrams.com.au/> (Accessed October 2001).

CHAPTER SEVENTEEN

Positional Frameworks for SDI

John Manning and Neil Brown

17.1 INTRODUCTION

The overall objective of Spatial Data Infrastructure (SDI) is to facilitate sharing of data that describes the position, distribution and attributes of objects in space. Data quality is of particular concern to both data providers and data users as it affects the utility and market value of the data. There are numerous aspects to data quality including completeness, logical consistency, positional accuracy, temporal accuracy and thematic accuracy. This chapter focuses on some of the broader issues concerning positional accuracy of spatial data in the context of SDI.

If spatial data are not built on a solid positional foundation, problems with data compatibility and positional uncertainty will arise. The ability of SDI to facilitate efficient collection, management, access and utilization is contingent upon having an accurate, efficient and well-maintained positional framework as the fundamental layer. This chapter examines some of the key issues regarding geodetic infrastructure and positional accuracy that are essential for a successful SDI. It will look at what constitutes a good geodetic infrastructure with regards to its definition and realization. Additionally, techniques that may be used to transform data between different geodetic systems will be reviewed. Two case studies will be used to illustrate the complexities and issues in selecting and maintaining a solid geodetic datum to support National, Regional and Global SDI.

17.2 SDI AND POSITIONAL ACCURACY

Key objectives of any SDI are to enable efficient and effective collection, management, access and utilization of spatial data (Coleman and McLaughlin, 1998; Rajabifard *et al.,* 2000). To achieve these objectives, positional accuracy, as a component of data quality, is of vital importance due to its influence on the utility of the data.

Due to the developments in SDI including the dynamic nature of digital datasets and advances in positioning technology and infrastructure, positional accuracy is becoming more of an issue than it was in the past as datasets are being integrated more. Previously data were collected with a particular use in mind and, importantly, the organization that collected the data was often the user of the data. Additionally, many datasets were used in isolation and rarely, if ever, integrated with other datasets from other agencies. With the advent of the wider SDI concept and the awakening awareness of the commercial value of spatial data, many spatial datasets are now accessible by a wide range of potential users with an almost unlimited range of applications. Given this situation there is a great potential for data to be used for purposes that were not foreseen by those who collected it and for which the data may not be suited. Furthermore, the users of the data are often removed from the collection of the original data and may have no background knowledge of the data and its quality - hence the importance of having adequate metadata (data about data).

The documentation and understanding of the positional accuracy of data is of fundamental importance if the data is to be correctly interpreted and for it to be integrated with other data without introducing anomalies. To avoid confusion about data being misplaced due to errors in its position, metadata that describes the positional quality of the data is essential. Depending on the uniformity of the data, it may be necessary to have metadata that covers:

- entire datasets,
- sub blocks of data,
- isolated datasets for integration, or
- single point objects within a dataset.

Even with the availability of sufficiently detailed metadata, communicating data quality to a user and propagating the data quality through various GIS operations (such as overlaying) is difficult and the on-going focus of research (Hunter, 1999). Online digital data is especially susceptible to misuse because it can be readily accessed and re-used at larger scales than it was collected without considering the accuracy of the original data or its relevance to the new application. It was stated previously that the objectives of SDI are to enable efficient collection, management, access and utilization of spatial data. Each of these objectives is in some way affected by positional accuracy. Table 17.1 contains a non-exhaustive list of positional accuracy issues that may influence these objectives.

It is clear looking at Table 17.1 that positional accuracy has many impacts on how spatial data is collected, managed, distributed and used. Understanding, assessing and communicating data quality are all important issues. However, central to all stages of SDI are the problems of knowing and, hence, being able to measure and define positional accuracy. When data from multiple sources or jurisdictions are involved, any assessment of positional accuracy must start with the original geodetic datum involved. This is essentially the fundamental data step in any SDI. If the underlying geodetic infrastructure and processes are flawed, then the accuracy of any further datasets will be adversely affected.

The positional accuracy of points within an SDI depends on the quality of the geodetic framework on which it is built. Classically, each area or country is covered by an observed network of Geodetic (trigonometric) points, with varying accuracy. Their locations are then calculated using a geometric model. This framework is then used to break down to a connected but less accurate network of coordinated survey points at the local level which is thus traceable to the national or regional geodetic infrastructure. Many spatial datasets are built up from secondary sources dependent on the local survey control such as digitised from maps; compiled from aerial photography; or acquired directly by terrestrial survey from the local network of ground survey points.

Table 17.1 Positional Accuracy and Its Influences on SDI Objectives

Collection	• Positional accuracy requirements of the intended data use. • Data collection procedures, equipment and cost. • Quality assurance. • Connection of data to local survey control. • Transformation of data collected using global survey control (such as GPS derived coordinates).
Management	• Data upgrading integration of data of higher (or lower) positional accuracy. Problems with distortion and non-uniform quality. • Data editing addition of data of higher (or lower) positional accuracy. Problems with discontinuities and non-homogenous quality.
Access	• Accessible and descriptive data quality information and other metadata. • Ability to assess fitness-for-purpose.
Utilization	• Transformation of datasets onto common reference frames. Problems with distortion and preservation of the accuracy of the original data. • Communication of data quality to the user. • Integration of datasets with varying levels of positional accuracy. • Assessing risk and soundness of decisions based on the data.

However, data is increasingly being captured directly from Global Navigation Satellite Systems (GNSS), especially the NAVSTAR Global Positioning System (or simply GPS), that is related to global reference frames, rather than to local or even national networks. In light of this, issues regarding the definition and maintenance of geodetic datums, datum transformations and the measurement and recording of positional accuracy will be examined in detail.

17.3 OPENING THE LID ON GEODESY

Geodesy, like many professions, is fraught with jargon that often causes confusion amongst the wider spatial community. Given this problem it is worth taking the time to clarify some terms that will be used throughout the rest of this chapter.

17.3.1 Reference Systems, Reference Frames and Geodetic Datums

All real world objects have a position in three-dimensional space. In order to measure and mathematically represent the spatial characteristics of these objects, for instance for use in a GIS, a positioning framework is required. The term reference system is used to describe a well-defined mathematical framework against which the positions of real world objects may be measured and recorded.

The important components of a reference system are the origin, the scale and the orientation. For the terrestrial reference system (Earth-fixed) the Z-axis is the

Earth's rotation axis, the X-axis passes though the Prime Meridian (Greenwich) and the Y-axis completes a right-handed system. The XY plane lies on the equator.

When a reference system is realised by a set of positions, typically Earth-centred Cartesian positions (X, Y, Z), it is known as a *reference frame*. When a subset of these positions is adopted for a local region, usually at a given instant in time, it is commonly referred to as a geodetic datum. In this case an ellipsoid is also selected to allow the positions to be expressed as Latitude, Longitude and ellipsoidal height. A single real world object will have different coordinates in different datums. The positions of a spatial dataset in one datum may be *transformed* into another datum.

In the past the origin and ellipsoid were traditionally chosen to provide a best fit to a certain geographical region, such as a single country or continent. In Australia the first national geodetic adjustment was undertaken in 1966. It used the best fitting local ellipsoid and identified it as the Australian National Spheroid (ANS). This figure minimised the local geoidal separation and simplified geodetic computation across Australia. Even though the next year this spheroid was adopted by the global community as the preferred best fitting International Spheroid for the Global Reference System 1967 (GRS67), the development of satellite based positioning technology in space geodesy has since developed better overall models.

In the last decade many countries have moved to adopt geocentric datums, so that spatial data positions may be more easily collected using satellite based positioning techniques, such as the GPS. In these geocentric datums the origin is nominally at the Earth's centre of mass and the ellipsoid to be used for subsequent calculations of positions (usually WGS84 or GRS80) provides a global best fit to the shape of the Earth.

17.3.2 Dynamic Datums

Historically, the geodetic ground marks, which constitute the realization of the reference system, were considered *fixed positions on the earth's surface*. The concept of continental drift was not widely accepted until the late 1960s when it became measurable with the increased accuracy from satellite positioning techniques.

These advances in positioning ability lead to a better understanding of the movement of the Earth's crust due to tectonic and volcanic activity, which is now readily measurable within a precise global reference frame. In regions that are tectonically active, such as those near or spaning tectonic plate boundaries, it is no longer reasonable to assume that the geodetic marks are fixed or even have a constant velocity vector.

In addition to the components of a regular (static) reference frame, a dynamic datum has a time varying component, such as a velocity model, that models how the coordinates of the ground marks change with time. As such coordinates in a dynamic frame are four-dimensional because they must include a time tag (called the epoch) to be valid. The positions of ground marks in tectonically active areas, such as Japan (at the junction of three tectonic plates) and New Zealand can change relative to each other even in the one datum set. If centimetre accuracy datasets are required, as is the case for the digital cadastral database in New Zealand, some type of dynamic or time-varying geodetic datum may need to be used so that the actual movement of the marks on the ground is reflected in changes in the coordinates. Even if a region is internally stable, absolute movements of the continental plate may need to be considered for some centimetre accurate datasets.

17.3.3 Height Datums

Ellipsoids are used to model the shape of the earth because of their sound mathematical properties for computing horizontal positions. However, due to the Earth having an irregular gravity field, ellipsoidal heights are not suitable for defining the flow of water for drainage. The geoid, a complex surface of equal gravitational potential that is approximated by mean sea level, is the most appropriate surface for a height datum where water flow is important. Practically, mean sea level is usually adopted at one or more tide gauges and optical levelling, which generally follows the geoid surface, propagates the (orthometric) heights. The difference between the ellipsoidal height and the geoid at any point is known as the geoid undulation and may be many tens of metres. Orthometric heights may be computed from ellipsoidal heights and vice versa using a geoid model but the accuracy depends on the quality of the local Geoid model. In Australia, for example, the Australian Height Datum (AHD) is based on mean sea level, but differs from the geoid by up to half a metre due to the sea level slope across the continent.

17.3.4 Coordinate Systems

The position of an object, in context of the reference frame, may be equivalently expressed by various coordinate types. For instance, the coordinates of a point may be converted to Latitude, Longitude and ellipsoidal height, or projected onto a Universal Transverse Mercator (UTM) grid to give easting, northing and zone number. Alternatively, the coordinates may be expressed as Earth-centred Cartesian coordinates (X, Y and Z) as is common practice in Space Geodesy. Coordinates may be rigorously converted from one coordinate type to another as long as the metadata on the coordinates is known.

17.3.5 Error

Regardless of the positioning technique, measurements must be taken in order to establish a position (a spatial reference) for a feature within an SDI. Errors or biases invariably contaminate all such determinations as the result of the equipment, technique, and measurement used, can cause position determinations to be offset from their true value. The numerous possible sources of error can be classified into three categories- gross, systematic, and random.

Gross errors are blunders caused by human mistakes or equipment failure. Systematic errors are the result of the influence of scientifically explainable (though not necessarily understood) physical processes on the measurements. Random errors are the result of imperfect measurement technique or equipment.

Frequently, spatial datasets are built from digitising hard copy maps produced by analogue means. Positional errors typically build up as a square root of the square of the component parts ("combination of variances"), such that:

Positional error = Geodetic network error + mapping control error + mapping compilation error + cartographic generalization error + printing and paper distortion error + digitising error + data base construction error.

For data built up from the analogue 1:100 000 national topographic mapping coverage this could have a positional error of 15 to 30 metres in absolute terms. In comparison, using GPS technology a point in a dataset can be observed using geodetic quality dual frequency equipment and processed online using the AUSPOS service (Dawson, *et al.*, 2001) thus eliminating most of the error budget described above. The positional error may then be expressed as:

Positional error = Global reference frame error + GPS observation error + AUSPOS computation error + transformation to the local datum model error

Indicatively, across Australia using this technique, this point could be determined in absolute terms within the GDA94 geodetic datum, with a positional error of 3 to 5 cm. Clearly one has nearly one thousand times the positional error (uncertainty) of the other. Complications arise when the two datasets are combined, for example as part of a data upgrade or maintenance program. Obviously it is important to preserve information about the positional error and source of the different data.

17.3.6 Positional Accuracy

Commonly data quality is understood to mean the characteristics that affect the ability to meet specified or implied needs. For positional information the quality measure most often sought is accuracy. But accuracy should not be confused with precision. Accuracy is the relationship of a measurement to the "truth", while the precision of a measurement is its repeatability, whether or not it is close to the "truth". Sometimes the quoted accuracy of positions in a spatial dataset may actually be the precision. The accuracy of a position may be a measure of the true relationship between two or more positions (the relative accuracy) or it may be the measure of the true relationship of a position to the coordinate datum ("absolute" accuracy).

In many applications the relationships between features is more important than the absolute position of the features but it is of much less value when it is integrated with a dataset from a completely different source. As the accuracy relationship between two independent datasets is rarely known, the only way to understand their accuracy relationship is through their common link to the datum – the "absolute" accuracy.

17.3.7 Precision

In practice absolute positional accuracy of data built from a number of secondary or tertiary sources may be difficult to obtain without access to robust metadata, and the quality of spatial information is often described in terms of precision, usually with a standard deviation or RMS value. Often the terms accuracy and precision are (incorrectly) used interchangeably. Precision is a statistical term that, in conjunction with a probability distribution, describes repeatability.

17.4 SPATIAL DATA REFERENCING

An SDI needs to be built on a solid positional foundation. Without a good geodetic base many of the potential problems regarding positional accuracy become more pronounced. It is desirable if the data are both compatible and homogeneous across the SDI to reduce duplication and uncertainty. This means that the geodetic infrastructure, which is essentially the base or framework dataset in any SDI, is well defined, homogenous and may be easily related to the geodetic infrastructures of other jurisdictions for building Regional and Global SDIs.

A good reference frame is one in which the coordinates of the ground marks are uniformly accurate in both an absolute and a relative sense. In practice this is difficult to achieve because of mis-modelled systematic and random errors and distance dependent errors that affect traditional surveying methods. Geodetic datums that suffer from these effects will be in error such that they present a distorted and non-uniform picture of reality. Over time fixed (static) datums will also distort in part or in total due to tectonic movements in the Earth's crust.

17.4.1 Adopting a Reference Frame to Support SDI

The concept of the SDI hierarchy, ranging from local through to global levels, has been covered in earlier chapters. In choosing a positional framework for an SDI, it obviously makes sense to select one that will facilitate easy transfer of data both between and within the different levels of the hierarchy. The ease of transfer criterion is determined by the computational effort required to transform the data from one geodetic datum to another. Computational effort is not necessarily limited to CPU time, but may also include allowances for the sophistication of the transformation software and the associated transformation parameters that may be required.

Given this criterion, it would make sense for all jurisdictions to adopt exactly the same global reference frame and geodetic datum to define positions within an SDI. However, there are a number of reasons why this is not feasible, or desirable. The prime reason is that any such reference framework would have to be dynamic in order to maintain high accuracy in unstable areas and to have a long service life. Many jurisdictions, in both developed and developing nations are not ready administratively or technically for dynamic reference frames because of the increased operational complexity they impose. In addition, a global survey campaign would be required to link the geodetic infrastructures of each jurisdiction. Participation in such campaigns will often be limited because of political, military or technological considerations. For instance only seventeen of the 55 member nations of the Permanent Committee on GIS infrastructure for Asia and the Pacific (PCGIAP) participated in the 1998 regional GPS campaign (Manning and Chen, 1999). As such it is not practical for a single datum to be adopted by all jurisdictions. Further consideration will be given to these practical issues later in the chapter with a case study of the PCGIAP.

Instead, this criterion is best met nationally, regionally and globally by realising the reference system with a uniformly accurate geodetic infrastructure based on a single, well-defined datum. Rather than calculate transformation parameters to transform between each and every reference frame and geodetic datum used by the various jurisdictions, it is more efficient to relate each individual reference frame to a common base. Consequently there are two components commonly used for referencing spatial data;

- a global reference frame,
- a geodetic datum within that global reference frame.

17.4.2 Global Reference Frames as a Basis for SDI

The best global choice for geodetic reference frame is the International Terrestrial Reference Frame (ITRF), maintained by the civilian International Earth Rotation Service (IERS) and propagated through the International GPS Service (IGS) products (IERS, 2001). ITRF is a highly accurate, dynamic global reference frame developed primarily for scientific research. Since 1992 IERS have made seven versions of the ITRF, the latest being ITRF2000. In addition IERS computes transformation parameters that allow accurate transformations between the various realizations of the ITRF.

However, the network of ground marks that define the ITRF is sparse and, to a large extent, has not been primarily linked to national geodetic networks. Traditionally geodetic marks have been located for access or visibility and are often unsuitable as stable geodynamic points for IERS purposes. Furthermore, the use of VLBI, SLR and DORIS in the computation of ITRF limits the ready access to geodetic control points except through the use of GPS.

The preferred option is then to link the local or national geodetic networks to the ITRF network using continuously operating GPS reference stations or observation campaigns. The choice between adopting a dynamic or fixed datum will depend on the local circumstances and requirements of the users of the dependent spatial data. In an area where a dynamic datum or constantly changing coordinates is not acceptable to users, and the application of velocities on all spatial data is not popular, the pragmatic decision is to use a static datum or semi-dynamic datum. However, this will need revision in time depending on the magnitude or complexities of the moving plates.

If the geodetic infrastructure is linked to the IERS network through the use of GPS and the products of the International GPS Service (IGS), then the link between the local reference frame and ITRF may be readily maintained. By maintaining this link, many of the concerns about selecting a static datum may be negated. However, maintenance of a highly accurate geodetic infrastructure in an unstable region is always going to be problematic requiring a very good understanding of the tectonic or volcanic based motions. Such issues and their impacts will be further clarified later in the chapter with a case study of the Geocentric Datum of Australia.

Adopting a datum that is directly related to ITRF has another advantage in that collecting and transforming data using GPS is greatly simplified. The WGS84 datum defined by NIMA and used in GPS is now maintained to be within a few centimetres of ITRF (NIMA, 2001) (Figure 17.1). Therefore, if a geodetic datum that is compatible with ITRF is used, there is no need to transform to and from WGS84 when collecting data using GPS.

17. 5 TRANSFORMING BETWEEN DIFFERENT REFERENCE SYSTEMS

Transformations between different ITRF versions are well documented and straightforward below the decimetre level of difference. For transformations between global and local datums a number of approaches can be taken.

The preferred and most rigorous method is to actually re-calculate the positions using the original survey measurements and the new coordinates for the common geodetic control. As very few spatial databases store the raw observations, preferring instead to store only coordinate information, the use of less rigorous transformation methods is necessary.

The simplest method is the block shift, where the coordinates of the dataset are corrected using average difference between the two datums. The block shift is very efficient, but is generally only valid for small geographic regions.

Alternatively, a 3-dimensional Earth-centred origin shift, combined with the differences between the two ellipsoids used, can be used to simply transform positions from one datum to another (Molodensky transformation). This is the method that is commonly used in inexpensive GPS receivers.

A Helmert (seven-parameter similarity) transformation is commonly used in GIS and survey software packages. This method uses a conformal transformation employing three Earth-centred origin shifts, three rotations and a scale change. Provided the two datums are homogenous, the Helmert transformation is suitable for transforming data across a wide geographic region. However, when the datums are not of uniform quality, transforming the data using a conformal technique will propagate the error from one datum into the other. Instead a non-conformal technique should be used.

Non-conformal transformation methods are able to correct for the distortions in the datum, thereby preserving the relative accuracy of the original data. Examples of this type of transformation are the Minimum Curvature method used in the USA; the Multiple Regression method used in Canada; and the Collocation method used in Australia.

The Collocation method used in Australia caters for both the datum transformation and also the distortion in the local survey network, providing improved transformation for derived datasets. Where enough data of sufficient quality has been used, transformation with an accuracy of better than 10 cm may be achieved. Although initially mathematically complex, the results from this method are presented to users as a simple file of shifts in terms of Latitude and Longitude on a regular grid, which can be used to interpolate the shifts for any other point.

This grid is provided in the Canadian NTv2 format that is supported by many GIS packages.

Obviously it is preferable to use the simplest transformation technique that provides the desired level of accuracy, but whatever method is used, the appropriate metadata must be included to allow the history of the positions to be traced.

17.6 MEASURING AND RECORDING POSITIONAL ACCURACY

The International Standards Organization Technical Committee 211 (TC211) on geographic information has released a draft international standard on quality principles, known as ISO/DIS 19113. Another standard, ISO/DIS 19114, covers quality evaluation procedures. The standard on quality principles identifies five elements of data quality: completeness, logical consistency, positional accuracy, temporal accuracy, and thematic accuracy. Positional accuracy is further subdivided into absolute or global accuracy; relative or internal accuracy and gridded data position accuracy.

For many years the Australian Inter-governmental Committee on Surveying and Mapping (ICSM) has maintained through its Geodesy Technical Sub-Committee (GTSC) its Special Publication 1 (SP1) - "Standards and Practices for Control Surveys" (ICSM, 2002). Since its inception this document has used the concepts of *Class* and *Order* to define the quality and relative accuracy of control surveys coordinates. Although SP1 is widely used in Australia to report the quality of survey control, the concepts of *Class* and *Order* are difficult to understand, particularly for non-specialists. In addition, *Order* does not cater for positions that have been obtained independent of the local datum. Examples of such independent systems include real-time Wide Area Differential GPS services (WADGPS) and their imminent successors that will produce positions with an accuracy of a few decimetres. In addition, online geodetic GPS processing is also now available and with suitable observations will provide ITRF & GDA94 positions with an accuracy of a few centimetres, with about 15 minutes turnaround (Dawson, *et al.*, 2000).

In accordance with TC211 ICSM has now included *Positional Uncertainty and Local Uncertainty* in SP1 (ICSM, 2002). *Positional Uncertainty* provides a simple means of specifying the accuracy of positions obtained independent of the local survey control. *Local Uncertainty* is a simpler way of specifying relative accuracy, and will by 2005 replace *Order*.

Positional Uncertainty is the uncertainty of the coordinates or height of a point, in metres, at the 95% confidence level, with respect to the defined reference frame. this value is the total uncertainty propagated from the datum, which in Australia is realised by the Australian Fiducial network (AFN) or, in case of heights, the Australian Height Datum.

Local Uncertainty is the average measure, in metres at the 95% confidence level, of the relative uncertainty of the coordinates, or height, of a point(s), with respect to adjacent points in the defined frame.

Calculation of Positional Uncertainty is possible in geodesy, surveying and photogrammetry when sufficient redundant measurements are available to check the measurements. The types of tests that may be used to evaluate data quality during measurement, after measurement and for dynamic datasets are covered by ISO/DIS 19114. Importantly, the technique(s) used to evaluate the data quality should be included as part of the metadata, as they are important in assessing the meaningfulness of the quality estimates. Further details on how this should be done may be found in the previously mentioned standards, plus ISO/DIS 19115, which covers metadata. This standard is currently being implemented through the MetaData Working Group of Australian and New Zealand Land Information Council (ANZLIC – the Spatial Information Council). Currently, ANZLIC's MetaData Guidelines for geographic data include a mandatory field for positional accuracy, to allow a "brief assessment of the closeness of the location of spatial objects in the dataset in relation to their true position on the Earth" (ANZLIC, 2001). However, this is a large text field, which invites a range of verbose possibilities that would be very difficult, if not impossible, to use in an automated matching of diverse datasets.

The Positional Uncertainty of data that is collected autonomously, using geodetic GPS receivers and computed in a global framework or through the online AUSPOS, can be readily determined at about ± 5 centimetres. However for a hand held GPS receiver where the data is not recorded, a traceable accuracy value is not as easily obtained. Accuracy estimates may be determined from evaluating samples against known positions and the identified status of the GPS system at that time or from expectation based on experience or empirical tests. For example before Selective Availability was turned off, the criteria was ± 100 metre at 95% confidence level. This has improved dramatically to ± 13 metres global average and in practice ± 5 metres or better is common.

Generally speaking the errors that affect survey measurements and the appropriate modelling and reduction techniques are well understood. Where a number of data points are tested both the Positional and Local Uncertainty can be determined for the SDI. As such it is possible to calculate or Estimate Positional Uncertainty in most situations. However, there are other sources of error and concerns regarding positional uncertainty than those originating from survey measurements.

Well known cartographic examples are when data is abstracted through the use of symbols, or the data is adjusted to make sure the data is 'graphically accurate'. For instance, a road that runs along side a railway may be shifted to ensure the two features are distinct on the map when it is produced at a small scale. Often when new data is incorporated into a dataset, the GIS operator will use 'rubber sheeting' to fit the new data to the old. Propagation of error during GIS operations such as overlays, area and volume calculation and the like is difficult, especially when data classes such as lines and polygons are involved. The Positional Uncertainty of data resulting from these edits and operations is very difficult to determine and is often not managed by the software. Such problems must be overcome by establishing careful procedures and quality checks.

In an ideal case (from a geodetic point of view), all vector-based spatial data would be stored, not as categorical data, but as measurements. Data updating and upgrading would then simply entail a re-adjustment and would permit rigorous quality checking of the data. Also the data could be easily adjusted onto a new reference frame, avoiding the problems associated with transformations altogether. Unfortunately, such a strategy has not been widely employed and few software packages exist that are able to manage the data in this way, especially when large datasets are involved. The development of *Dynamic Network Adjustment (DNA)*, University of Melbourne (Leahy and Collier, 1998), *SNAP* in New Zealand or *ArcSurvey* though a partnership between Leica Geosystems and ESRI (Nix and Hill, 2001) shows promise for future developments in this regard. Land Information New Zealand (LINZ) have taken the approach of creating a survey-accurate digital cadastral data base (DCDB) by entering all the bearing and distance information from existing paper plans and the information on the survey control (Spaziani, 2002). Cadastral surveyors are required to digitally lodge their measurements rather than the traditional survey plans, allowing LINZ to perform automatic quality checking and integration of the data (Haanen, *et al.,* 2002). The flow-on benefits of having a survey accurate DCDB, which like the geodetic infrastructure is a fundamental dataset for many applications, are considerable.

17.7 CASE STUDIES

In order to clarify and reinforce the points raised in the previous sections, two case studies will be examined. The first case study looks at the operational issues regarding the definition, maintenance and life span of the Geodetic Datum of Australia. The second case study draws on the experience of the PCGIAP and examines the complexities of establishing uniform geodetic infrastructure for Regional SDI.

17.7.1 Case Study 1: The Geocentric Datum of Australia

On 1 January 2000 Australia implemented a new geocentric datum known as the Geocentric Datum of Australia 1994 (GDA94). There were numerous reasons for the adoption of the new datum, including (Collier and Steed, 2001):

- Compatibility with GPS and future global satellite navigation systems within the ITRF.
- Standardised reference frame for the whole of Australia, removing the complications from a mixture of the AGD66 and AGD84 datums being used by the states and territories as well as the use of WGS84.
- Support to underpin SDI with a defined and internationally accepted geodetic datum identified and connected to the global reference frame.

Several GPS campaigns were used to connect the Australian Fiducial Network (AFN), and the Australian National Network (ANN) to the global IGS network, a subset of which is used by the IERS in the definition of ITRF. Using data collected during these campaigns, coordinates for the control marks were produced (Morgan *et al*, 1996) at the 1994.0 epoch (1 January 1994) of ITRF92. These positions for the AFN were adopted and gazetted as the definition of GDA94. The rest of the geodetic network was adjusted on to GDA94 using the AFN and ANN as control.

As mentioned previously, ITRF and WGS84 are closely related. Therefore, for most practical purposes it is safe to assume that ITRF, and by association GDA94, and WGS84 are equivalent. However, because GDA94 was fixed at the

1994.0 epoch of ITRF and the Australian continent is moving, there is a widening gap between the GDA94 positions (ITRF92 @ 1 January 1994) and their current positions (ie ITRF2000 @ 1 December 2002). This relationship between the datums is represented diagrammatically in Figure 17.1. Australia is moving more or less uniformly north north-east (NNE) at about seven centimetres per year (Manning and Steed, 2001) and on 1 January 2003 the difference between ITRF and GDA94 was in the order of 60cm. Whilst the difference is, and will continue to be, insignificant for many spatial data applications, already it may be detected using advanced positioning techniques. To facilitate the on-going use of GDA94 datum rather than moving to a dynamic datum in Australia, the AUSPOS online GPS processing system, producing positions at the centimetre level of accuracy, has been established utilising IGS products for post processing. AUSPOS accepts data from users' computers and rapidly processes the data, initially in ITRF at epoch time of observation, before applying a transformation including a continental plate motion model to produce GDA values in an emailed response. This tool is being widely used also by the Government, Australian Maritime Safety authority, Airservices Australia and private sector companies to verify their DGPS base station positions.

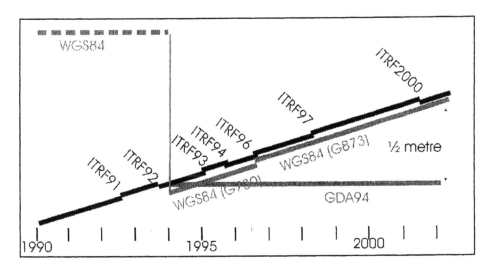

Figure 17.1: Indicative Relationship Between GDA94, WGS84 and ITRF

For other users the transformation model from current observation epoch in ITRF back to GDA94 is readily available. The National Mapping Division of Geoscience Australia (formerly AUSLIG) has for some time produced transformation parameters for advanced user to transform between ITRF and GDA94 using a Helmert transformation (Dawson and Steed, 2002). These transformation parameters are updated quarterly and are accurate to about one centimetre across Australia and are monitored by the inclusion of the AFN stations in the IGS network on a daily basis.

Other high accuracy DGPS providers such as FUGRO and Starfire are promising to supply real time 10cm level accuracy positions using satellite communications in the near future. When these services become available their compatibility with GDA94 will need carefully consideration.

Transformation from the older national datums, ADG66 and AGD84 is more complicated. Due to some systematic and random error not being fully modelled the AGD datums are not uniform, with the distortions relative to GDA94 reaching up to several metres in some areas (Collier and Steed, 2001). In order to correct for this, national grids of Latitude and Longitude shifts were calculated using least squares collocation based on approximately 30,000 survey marks. These grids enable transformation accuracies of a few centimetres or better for much of the country, however transformation errors do exceed one metre in some areas when extended offshore outside of the common point networks (Collier and Steed, 2001). Such errors illustrate that even a sophisticated, intensive effort to develop a transformation will be limited by the available data and will not produce perfect results. It also highlights the dangers of using a simple transformation technique, such as a block shift or Helmert transformation, in the presence of distortions.

17.7.2 Case Study 2: PCGIAP

In this case study the geodetic aspects regarding the PCGIAP, which was introduced in Chapter 4, will be discussed. The members of the PCGIAP are the heads of the national survey and mapping organizations in the 55 counties comprising the United Nations region of the Asia Pacific. The PCGIAP now meets annually and is coordinated by an elected executive board and Working Groups established to undertake required activities.

The PCGIAP was initially tasked by the UNRCC-AP with development of a Regional SDI infrastructure:

- To Maximise the economic, social and environmental benefit of geographic information in accordance with environmental Agenda 21
- To cooperate in development of Asia Pacific SDI for local GIS, GSDI, and for the Global Map
- To develop an Asia Pacific SDI (APSDI)

To meet these objectives required that a regional geodetic framework to be established to be able to integrate all reliant spatial data together in an orderly manner as a Regional SDI. A regional geodesy working group (Working Group 1) was established to assist in this task. Following an investigative process it was decided not to establish a unique geodetic datum for the region but to adopt the ITRF series of reference frames to provide the global setting, and for countries to work within that frame when suitable. A series of annual observation campaigns was arranged, aimed at extending the ITRF geodynamic observation sites and to link them to the national geodetic networks in individual countries. After a series of five regional campaigns using SLR, VLBI as well as GPS, these campaigns have developed links with existing datums in most of the 55 countries. This has allowed not only the densification of ITRF primary positions in the region with connections to national geodetic networks, but it has produced movement velocities for sites with repeated epoch observations. It has produced horizontal block shifts or seven parameter transformations required for countries to be able to bring spatial data from local existing datums to the regional reference frame.

Some countries such as Japan and Australia have recomputed their networks in terms of ITRF at different epochs and others, such as China, are moving on producing specific transformation parameters for spatial data. It is noted that although Australia only moved coordinates of its geodetic framework some 200 metres from its old datum to the new GDA94 geocentric datum some nations have to move positions kilometres onto ITRF geocentric values. However there are many issues yet to be resolved including:

- Many countries do not have a contiguous horizontal network on a single datum, others have non homogeneous networks, where a single set of parameters cannot be used;
- Others, such as New Zealand, have markedly different site velocities across their county;
- Some have a number of isolated islands within their jurisdiction. Of the 20 pacific island nations very few have access to GPS technology.

So in many cases the ability to produce data at the one metre positional accuracy level is a significant step towards an SDI, without being immediately concerned over centimetre or millimetre distortion of the network arising from tectonic movement. However, ultimately when the use of the SDI becomes widely adopted for economic or other reasons such as cadastral boundary determination, as in New Zealand, then will need to be a progressive strategy to greater accuracy.

In November 2002 the fifth annual PCGIAP campaign was observed and it was noted that the approach of precisely fixing just a few points on each geodetic network with a simultaneous one-week or ten days occupation campaign for millimetre level accuracy now has shortcomings with improvement in GPS technology. For example the use of receivers at any time to observe one or two days of data, and to calculate a number of points to centimetre accuracy positions using the AUSPOS online system, offers much better returns to upgrade individual networks, in a difficult country, or an isolated island archipelago than an extended observation campaign.

Working Group 1 of the PCGIAP, is endeavouring to provide a uniform framework for the many countries in the Asia Pacific area. Currently most use local datums varying considerably from ITRF (e.g. Japan differs by 400 metres and Korea differs by some 3½ km). The goal is to have enough accurate ITRF positions established in each country, to enable their local coordinate system to be related to the global system either by transformation or readjustment, thus allowing them to take advantage of global spatial data and contribute to it.

The identification of local datum transformation parameters to the regional datum through PCGIAP activity is continuing with details listed on the PCGIAP web site (PCGIAP, 2003). In the cooperative framework of PCGIAP each country has responsibility for determination and publication of its own transformation parameters to permit spatial data to be readily incorporated into the Regional SDI. Pacific nations have a particular problem because of the many islands involved, often each one with a separate datum. For example Kiribati has many islands with 15 different datums (Table 17.2).

The activities of the PCGIAP illustrate the complexities in establishing a solid geodetic foundation for a region. Vast differences in local tectonic activity, existing infrastructure and access to technology all play significant roles in addition to the political and administrative sensitivities discussed in Chapter 4.

Table 17.2: Local Data used for Mapping in Kiribati (Adopted from Llewellyn, 2000)

Island	Ellipsoid	Datum
Butaritari	International	Makin Astro 1965
Marakei	International	HMS Cook Astro 'H' 1962
Abaiang	International	HMS Cook Astro 'H' 1962
Tarawa	WGS84	WGS84
Tarawa	International	Secor Astro 1966
Maiana	International	Maiana Astro 1965
Abemama	International	HMS Cook Astro 1959
Kuria	International	HMS Cook Astro, Kuria 1962
Aranuka	International	HMS Cook Astro, Kuria 1962
Nonouti	International	Nonuiti Astro 1965
Tabiteuea	International	TBZ1 Astro 1965
Beru	International	BRZ1 Astro 1965
Nikunau	International	Nikunau Astro 1965
Onotoa	International	ONZ 7 Astro 1970
Tamana	International	HMS Cook Astro 1962
Arorae	International	Arorae Astro 1965
Kiritimati	International	Christmas 1967 Astro

17.8 CONCLUSIONS

SDIs depend on a base of homogeneous geodetic frameworks. In order for maximum benefit of the data to be realised there should be no confusion about its positional accuracy.

Positional Uncertainty can be both considered in absolute terms within a specified datum and in terms of its Local Uncertainty to other data within an individual dataset. Where isolated datasets are used Local Uncertainty may be more important than Positional Uncertainty, however where datasets are to be integrated, knowledge of the Positional Uncertainty will enable the data to be merged and used in perspective without confusion.

Today ITRF provides the state of the art global reference frame producing global coordinates for some two hundred primary points, together with site velocities. These are published in a series of epoch determinations such as IRTF2000. It is produced and maintained by IERS and implemented at the user level by GPS through IGS products. WGS84 is now being kept compatible with ITRF at the centimetre level.

Within this global reference frame determination, all geodetic points are subject to horizontal tectonic plate motion and the primary ITRF points need to be densified through connections to national geodetic infrastructures. In turn, contributions to the ITRF determinations from permanent observation points in each country will improve ITRF determinations. In many cases it will be practical for national datums with observation campaigns undertaken at specific epochs to be declared as fixed or semi fixed locations as the basis for a National SDI and for determination of transformation parameters to regional and global applications. However, whilst ongoing centimetre variations in the positions of national spatial data are unwarranted it is important that the Positional Uncertainty of the data is clearly documented so that it is traceable and able to be recalculated as required.

Where national data is observed at an ITRF epoch and given a local datum name for the defining primary data points, there will be a need to be able to transform from the original datum to the new IRTF-based datum. It will also be necessary to be able to calculate ahead to a future ITRF epoch.

Positions, through spatial data, are the utility of the 21st century and will be used in an increasing number of applications, with diverse data. However, a position is only fully and comprehensively described when it includes the coordinate value, the datum, the epoch and the accuracy relative to the datum.

In summary, the following are a number of recommendations made regarding the establishment of a geodetic framework to support SDIs:

a) That ITRF be used as the basis for SDIs.
b) That geodetic networks be connected to and contribute to ITRF determinations.
c) That geodetic networks be upgraded with GPS observations at key sites or be completely re-observed and recalculated in ITRF at a defined and documented epoch.
d) That the positional accuracy of all points in an infrastructure be determined in the geodetic computation process to provide absolute and relative accuracy information embedded in the metadata.
e) That permanent GPS observatories be established at key geodetic datum points and contribute data daily into the IGS for inclusion in ongoing ITRF determinations.
f) That where possible vertical datums be clearly connected to sea level and the geoid, and that observed and not derived heights are stored (whether ellipsoidal or local datum) to allow future refinement.

17.9 REFERENCES

ANZLIC, 2001, ANZLIC MetaData Guidelines: Core Metadata Elements for Geographic Data in Australia, Version 2, ANZLIC Spatial Information Council, Online. <http:// www.anzlic.org.au/asdi/metaelem.htm> (Accessed February 2001).

Coleman, D.J. and McLaughlin, J., 1998, Defining Global Geospatial Data Infrastructure (GGDI): Components, Stakeholders and Interfaces, *Geomatica*, 52, 2, 129-143.

17.9 REFERENCES

ANZLIC, 2001, ANZLIC MetaData Guidelines: Core Metadata Elements for Geographic Data in Australia, Version 2, ANZLIC Spatial Information Council, Online. <http:// www.anzlic.org.au/asdi/metaelem.htm> (Accessed February 2001).

Coleman, D.J. and McLaughlin, J., 1998, Defining Global Geospatial Data Infrastructure (GGDI): Components, Stakeholders and Interfaces, *Geomatica*, 52, 2, 129-143.

PART SIX

Future Directions

CHAPTER EIGHTEEN

Future Directions for SDI Development

Ian Williamson, Abbas Rajabifard, and Mary-Ellen F. Feeney

18.1 INTRODUCTION

The objective of this book is to promote a better understanding of the nature of SDIs and to assist practitioners to design more efficient and effective SDIs. It is hoped that it will also contribute to the growing body of knowledge concerned with SDIs and as such should also be useful for researchers and students of SDI. In writing and editing the book we had hoped to cover many if not most of the key issues in the evolving SDI concept. While we are satisfied that the book addresses many of the current SDI issues we also recognise that there are some current and evolving issues which have not been addressed. We have tried to identify these both throughout the book and in this chapter.

This chapter summarises the context and achievements of other chapters. From these chapters several important issues have been raised and will be discussed based on the five SDI components (access, people, data, standards, policies) identified in Chapter 2, in order to simplify the direction and the approach needed to deal with these issues. There are also a number of other issues influenced largely by global drivers such as economic, environmental, social, technology drivers and the growth of the private sector. The chapter concludes by examining the relationships between all these issues in the context of the SDI Hierarchy by exploring different models for SDI development to enable practitioners to plan and develop an appropriate SDI development strategy.

18.2 COVERING THE SDI LANDSCAPE

The book comprises 18 chapters divided into 6 parts. The Preface provides the background and the motivation to write this book based on the International Symposium on SDI. Part One (Chapters 1-2) begins in Chapter 1 with a brief review of the need for spatial data, the growing awareness of SDIs and concentrates on broad ranging issues concerned with building SDIs. It then discusses some global challenges facing SDI development including the role of education, research and capacity building. With this in mind, Chapter 2 recognises that SDI is understood and described differently by stakeholders from different disciplines and different political and administrative levels. It is argued that current SDI definitions are individually insufficient to describe the dynamic and multi-dimensional nature of SDI. Then it discusses the concept and nature of SDIs in detail in order to facilitate their development and progressive uptake and utilization by different communities (diffusion). It proposes that an SDI comprises not only the four basic components of institutional framework, technical standards, fundamental datasets and access networks, but also an important additional component, namely, people (human resources).

Based on the nature and concept of SDIs, Chapter 2 also reviews the model of SDI hierarchy. An SDI hierarchy is made up of inter-connected SDIs at corporate, local, state/provincial, national, regional (multi-national) and global levels. Each SDI at the local level or above is primarily formed by the integration of spatial datasets originally developed for use in organizations operating at that level and below. The hierarchy provides two viewpoints of SDI structure. The first view is an umbrella view (top down) in which the SDI at a higher level, say the global level, encompasses all the components of SDIs at levels below. This suggests that ideally at a global level, the necessary institutional framework, technical standards, access network and people are in place to support sharing of fundamental spatial datasets kept at lower levels, such as the regional and national levels. The second view is the building block view (bottom up) in which any level of SDI, say the state level, serves as the building block supporting the provision of spatial data needed by SDIs at higher levels in the hierarchy, such as the national or regional levels. Each level in the SDI hierarchy model is discussed in detail through the relevant chapters of the book with a variety of chapters being devoted to specific case studies.

Part Two (Chapters 3-7) take up the challenges identified in Part One. With this in mind, in order to commence the examination of the SDI hierarchy, Chapter 3 demonstrates current global initiatives such as Global SDI (GSDI) and the Global Map of the World and their relationships with each other as well as with other SDI initiatives. The chapter explores the growing role of the Regional SDI level, the 'pivot' between the State and National SDI's and the possibilities to attain the GSDI vision. It evaluates how the organizational models of global initiatives can maintain and strengthen current relationships with other national and multi-national SDI initiatives. Based on this evaluation, recommendations and future directions are suggested to facilitate SDI development at a multi-national level and its influence on other levels in the SDI hierarchy.

Regional SDIs are examined in detail in Chapters 4 and 5. Chapter 4 takes up where Chapter 3 leaves off performing a comparative analysis of the strategies, organizational models, progress, issues and relationships of Regional SDI initiatives from Europe, Asia and the Pacific and the Americas regions, with initiatives at other levels in the SDI hierarchy. The focus on Regional SDIs is extended in Chapter 5 by exploring diffusion theory as a theoretical framework to facilitate Regional SDI development. This chapter considers the low rate of participation in Regional SDI development by identifying key factors that facilitate development through better understanding the complexity of the interaction between social, economic and political issues, as well as reviewing SDI development models. The chapter argues that the deliverables expected from SDI initiatives have frequently had more to do with aligning the access networks, policies and standards for particular stakeholders or databases, than establishing spatial data-people relationships. It draws on the experiences of SDI development in Asia and the Pacific region, as a selected case study region, and complements the comparative Regional SDI exploration completed in the previous chapter.

The nature and development of National SDIs over the last two decades and their influence on the emergence of first and second generation SDI development features is the subject of Chapter 6. The chapter explores National SDIs, their similarities and differences, their relationships between Local, State and International SDIs, and in particular discusses the organizational models supporting SDI development in each jurisdiction. The chapter concludes with a discussion on recommendations and future directions to facilitate SDI development at a national level and its influence on other levels in the SDI hierarchy.

There is currently a diversity of SDI initiatives within state (province or county) jurisdictions. However there is similarity between the strategies, organizational models, issues and relationships between these initiatives. Chapter 7 discusses State SDIs and the components of National SDIs which focus on land administration or cadastral data and support medium to large scale data. These SDIs have a particular role in providing a link between Local Government SDIs (large scale data) and National SDIs (small scale data) in the SDI hierarchy. This chapter discusses the importance of partnerships to support SDI development in each jurisdiction. Based on this evaluation, issues are identified to facilitate future SDI development at a state level and its influence on the other levels in the SDI hierarchy.

A multi-levelled examination of the SDI development occurring across Australia as a case study is explored in Part Three of the book (Chapters 8-10). The case study emphasises in-depth operational examples of existing initiatives within the SDI hierarchy in Australia. Chapter 8 demonstrates the current status of National SDI development in Australia and its relationships to Local, State and International SDI initiatives as well as providing an in-depth examination of its coordination strategies. The latter in particular explores the establishment of a distributed network of public, private and academic associations participating in SDI and the development of the spatial information industry. Chapter 9 explores the role of the State of Victoria SDI initiative in this context and presents the current status of the state's SDI, with a discussion of, amongst other things, its online initiatives and the growing role of partnerships in inter- and intra-jurisdictional SDI development. Chapter 10 documents a case of Local SDI development and how it contributes to higher level SDI development. It especially considers the development of corporate SDIs within organizations, through internal coordination and cooperation, and the development of external linkages among themselves through partnerships. From this review, the generic elements and processes of hierarchical SDI development are identified.

The importance of understanding the operation of SDI within a hierarchy of initiatives is found in the expression of SDI supporting economic, environmental and social objectives. In Part Four (Chapters 11-13) of the book aims to introduce the value of SDIs within these broader contexts of society. In this regard Chapter 11 discusses the role of SDIs in delivering the kind of information needed for sustainable development, the relationship between information and good governance and concludes by discussing the potential role of E-governance in enhancing the reach of SDIs and in progressing good governance through wider participatory processes. Chapter 12 takes up where Chapter 11 leaves off discussing the challenges and features of SDIs supporting the diversity of decision-making environments, decision complexity and the variety of participants in the decision process. It explores the role of enabling tools and mechanisms as part of a decision-infrastructure supported by SDIs. The chapter concludes with a discussion on how evaluating the decision-support capacity of different institutional frameworks may guide future SDI development. Chapter 13 encourages additional interest in the economic issues associated with SDI implementation, especially the area of funding. It reviews the economic issues associated with SDI implementation, and presents an in-depth study of funding models for SDI development in different implementation environments.

These methods of comparison and evaluation of SDI initiatives, in terms of their institutional, administrative, decision support and financial frameworks can help to better understand the issues, to find best practice for certain tasks, and to improve the system as a whole. Chapter 14, which concludes Part 4, demonstrates how evaluating and comparing public and private administration systems can be significant in terms of improvements to processes and institutional structures, through the identification and application of measurable indicators. The application of these principles to the development of SDIs will play a crucial role in the management of our spatial information.

Chapters 15 through 17 (Part Five) discuss the technical dimension of SDI initiatives. Chapter 15 demonstrates the significance of administrative boundaries within the SDI framework and highlights one of the most prevalent problems currently limiting the use of data within a number of GIS applications - the spatial hierarchy problem. The chapter demonstrates that the reorganization of boundaries into a coordinated spatial hierarchy is possible, due to the number of technical developments that have been made in surface modelling, interpolation, derived boundaries, and the re-aggregation of point and polygon data. However, it concludes by emphasising that developing a technical solution alone cannot ensure the development of a hierarchy of administrative boundaries until a coordinated SDI is in place. Chapter 16 takes this up by demonstrating the need for the development of a framework to enable SDI to incorporate new applications for data use and access. It reviews and evaluates existing access models, polices and standards and proposes new opportunities and the infrastructure requirement for the key components of SDIs in relation to wireless applications that use spatial information.

Chapter 17 focuses on some of the broader issues concerning positional accuracy of spatial data in the context of SDI. The ability of SDI to facilitate efficient collection, management, access and utilization is contingent upon having an accurate, efficient and well-maintained positional framework as the fundamental layer. This chapter examines some of the key issues regarding geodetic infrastructure and positional accuracy that are essential for a successful SDI. It then looks at what constitutes a good geodetic infrastructure with regards to its definition and realization, and concludes by employing two case studies to illustrate the complexities and issues in selecting and maintaining a solid geodetic datum to support National, Regional and Global SDIs.

Part Six (this chapter) reviews all the chapters of the book, highlights many issues facing the development of SDIs and concludes with a discussion of the applicability of different SDI models.

18.3 SDI DEVELOPMENT ISSUES

As individual chapters have identified, there are a number of important issues related to SDI development from conceptual, technical, socio-technical, political, institutional and financial perspectives. Following are some general issues that practitioners may need to consider when establishing SDIs. These are discussed broadly using the SDI component classification discussed in Chapter 2 (access, data, standards, policies and people). A number of other issues are also identified which do not fall into these categories. Finally these issues are discussed in relation to options for jurisdictional and institutional frameworks of SDI development. It is note-worthy that often the issues discussed pertain to more than one category, or complement issues in other categories, or may be classified differently depending on the priorities of different jurisdictions. What is important is that these issues should be considered in the long-term in order to achieve sustainable and ongoing development of SDIs.

18.3.1 Access Networks

The communications infrastructure, including communication technologies such as the Internet and wireless applications, are revolutionising methods of maintaining, disseminating and accessing spatial data. To fully utilise these technologies there must be a clear understanding of how they impact on and assist in implementation of an SDI that supports the delivery of accurate real and near-real time data.

For instance, one of the most effective uses of information transmission capabilities is to access data from decentralized, geographically dispersed business locations. Unfortunately Internet access, analysis on remote computers, real time data transmissions and distributed decision modelling often require more than the currently available bandwidth, node and hub capacities. Bandwidth must be available to facilitate access through the full extent of telecommunication land-lines, line-of sight microwave transmissions, wireless and broadcast technologies, or satellites (Kelmelis, 1999).

While different applications will have varying spatial data usage requirements, the role of common infrastructure elements (such as query and delivery mechanisms) for a range of applications, will enhance the accessibility and usability of these communication and information technologies and their potential role in the development of decision support products and services.

The fact that information and communication technology capacity (and infrastructure) are not evenly distributed around the world, limits access to information for many and leads to an unequal distribution of information, tools, and models among countries and continents (NAS, 2002). This is very evident between developed and developing countries. To promote access to these decision support capabilities it may be more desirable to share resources than to develop or install duplicate resources in all locations or jurisdictions. Thus inefficient and costly duplication of effort could be avoided and scarce resources could be applied to build better alternatives (models, data, hardware, software, communications) which could be shared through SDIs to achieve distributed decision support.

18.3.2 People

SDI development requires coordination, cooperation and awareness across different disciplines, borders, and between developed and developing initiatives. There is also a great need for increased awareness of the benefits of spatial information (social, environmental and economic) as well capacity building, the existence of training/courses and educational resources to strengthen awareness and skills to participate and be involved in SDI development activities. Therefore, human capital shortage can be marked as one of the potentially big issues facing SDI development for the future. It impacts on those tasked with developing SDIs as well as engaging participation in the wider community.

Sustainable development requires that good governance supports the discovery of and access to information to support decision-making processes. For SDIs this will require addressing issues of equity - of access to the information and tools a SDI can provide to support a decision-making process; and support for participation in the process by relevant stakeholders and by people with different levels of experience with spatial information.

Nevertheless, despite the optimism, SDI development will be heavily dependent on increased awareness of its markets and its users. For instance, Chapters 11 and 12 point out that to assist stakeholders towards visualizing one another's points of view and reaching agreement with regard to resource use will continue to require initiatives that aim to achieve interactive decision-making in physical and virtual meeting rooms and to adopt specific project tools, such as GIS and decision support systems.

18.3.3 Data and Standards

While the data issues in SDI development are as diverse as the initiatives and their jurisdictional distinctions, a number of issues stand out such as the need for better coordination of and access to commonly needed spatial data - Keen (1987) referred to this as the data 'energy crisis' as far back as the 1980s. We have the computing capacity, just not enough good data. New data are needed and must meet interoperability, quality and content standards, while existing data must be maintained and updated as necessary in compatible forms. The three issues of interoperability, quality and content standards are considered below:

Interoperability - to optimise existing data integration - existing data that can be used in decision-making needs to be converted into structures and formats which can be exchanged, integrated and fully utilized by all users. As discussed in Chapter 15, one of the most fundamental problems restricting the integration, comparison and transfer of data between agencies is the fragmentation of data between non-coterminous boundary systems. Datasets that can be easily integrated to meet a variety of user requirements and business needs, have the benefit of increasing confidence in data use, consistency of presentation and consistency and comparability of results. This is because all parties operate using common data and standardised methods of data aggregation. The benefits of interoperability between different information and business systems are equally numerous.

Quality - Improved methods of analyzing, communicating and interpreting data and model uncertainty must be developed so that decision-makers understand the possible ramifications of their decision-making (Kelmelis, 1999). For example, the measures of GPS quality given by common software packages are either very optimistic, or conversely, are overly conservative and therefore have low fidelity, which may result in inappropriate use or reduced utility of the data. There is a need to develop methods for the estimation, documentation and visualization of true errors at GPS base stations and the relationships between the observing conditions and the stochastic behaviour of the observable in the Geodetic framework – Chapter 17 highlights the importance of these issues to an effective SDI.

Content – the role of metadata in communicating the quality, currency, custodianship and other information about the content of datasets is an area of underdevelopment in spatial data documentation, management, custodianship and user understanding. There is a need for metadata at common (or at least compatible) standards - data must include robust metadata enabling users to effectively mine and use existing data bases. Tosta (1999) believes the value of metadata to infrastructure and systems development will continue to grow and will become of increasing emphasis in data maintenance, despite the complexity of current metadata standards, the expense of their development and implementation. Metadata is evolving to be of critical importance to supporting spatial data transactions, business activity and resource organization within the spatial data community, especially in supporting and utilising the SDI hierarchy – relationships acknowledged and explored in the ICA's Forthcoming book on metadata standards (Moellering *et al.*, 2003).

18.3.4 Policies

There will be an ongoing requirement to develop policies to support SDI development – from access and pricing issues grappling with security and data sensitivity, privacy, copyright, through to institutional issues like defining the roles of government and the private sector, as well as the institutional issues concerned with developing SDIs to meet the decision-making requirements of sustainable development. Some of these issues are identified below.

One of the biggest challenges facing SDI development is balancing privacy and security with utility of data. Copyright, licensing and other rights to the use of data and information must be addressed to ensure the appropriate data and information are accessible to all, especially from a good governance perspective.

Some data are considered to be sensitive, proprietary, or require cost-recovery, which raises issues of access and pricing. Who will pay and maintain the SDI is a question which is being faced by all levels of government. Will a public good model or a commercial model be adopted, or will a model drawing from both models be adopted to cover the costs? Who will decide? Many National SDI coordinating agencies can produce national policies. They can develop models, standards and protocols, but have difficulty mandating their acceptance or use unless adoption is mutually beneficial to all parties.

It is therefore important to engage in open and collaborative dialogue about ideas, information and technologies to advance tools and systems that will enable the use of natural and social science in decision-making, and ultimately decision support to achieve sustainable development objectives as discussed in Chapter 11. To fulfil the potential of such dialogue, an 'organizational' or 'discipline' mandate for decision support must be transformed into a broad-scale 'whole of government' mandate coupled with an appropriate supportive infrastructure, for both information and spatial technologies as discussed in Chapter 12.

18.3.5 External Developments

They are six other developments which are reflected in global trends and which are impacting on the development of SDI models and which will continue to influence the evolution of the SDI concept. They are:

a. *Expanding technologies* – the rapid development of the key SDI enabling technologies of the Internet, distributed database systems, open system architectures and the emergence of powerful application technologies such as hand-held GPS, desk-top GIS and high-resolution satellite imagery, have all challenged conventional technical models of SDIs.

b. *Market demand* - the spatial information industry and market for spatial data appears poised for rapid growth.   For example in Australia the Spatial Information Industry Action Agenda referred to in Chapter 8 reported global expenditure in the sector as \$34 billion per year and growing at 20%, with Australian expenditure over \$1 billion and expected to double by 2005. This increase market demand will inevitably affect the demand for greater SDI functionality and services.

c. *Changing business models* – the public sector is rapidly moving from an in-house mode of program delivery, to a position where all non-core functions are outsourced to the private sector. This raises the question of how much a National SDI-related activity is really core business, and whether public-private partnership models for program delivery may be more appropriate.

d. *E-government and participatory democracy* – there is a world trend to decentralise government decision-making and to make decision-making more participative by involving the wider community at the local level. This trend is being facilitated by the information and communications technology revolution and relies significantly on having access to appropriate spatial data which in turn relies on an SDI.

e. *Sustainable development* – the commitment by many governments to the objectives of sustainable development or 'Triple bottom line' is requiring more complex decision-making as a consequence of considering the economic, environmental and social dimensions of any decision. SDI s are a key component in providing the basis for this complex decision-making.

The above list of challenges, issues and external developments crystallise the following challenges for SDI coordinating agencies: accelerated development of SDI-compliant fundamental datasets; defining the role of dataset sponsors and custodians in a distributed SDI model; defining the role of the private sector in designing, developing and using an SDI; jurisdictionally consistent data access and pricing policies; as well as monitoring and evaluating the objectives of SDI development and the relativity of their outcomes. These must all be taken into account as part of the interrelationship of SDI development at any level, from local to global, as these can no longer be considered in isolation. They have to be considered as part of the SDI hierarchy. The recognition that local events can aggregate to impact at a global level, passing across state and national boundaries, highlights the importance of having infrastructures that facilitate information being exchanged within the SDI hierarchy.

18.4 RELATIONSHIPS BETWEEN SDI HIERARCHY, ISSUES AND DIFFERENT MODELS OF SDI DEVELOPMENT

All the issues discussed in the previous section together influence the strategy chosen for SDI development and its effectiveness. How then does a state, country or region decide on the best SDI model to accommodate these issues recognising that they differ between jurisdictions and levels in the SDI hierarchy, as well as between developed and developing countries. For example, as was demonstrated by Rajabifard *et al.* (2002), the predominant model adopted for SDI development by first generation SDI initiatives was product-based. This was due in part to a lack of alternative options and awareness of the use and advantages of alternative models characterising the development-coordinating SDI community. However it was also a result of a lack of understanding of the relationship between different jurisdictional initiatives and the issues affecting SDI development models including funding models.

As a result, some SDI development initiatives exhibit characteristics of different SDI development models, or of being in a transitional stage - developing a more process-based approach while having product-based origins. This has begun a process of looking beyond a single focus for strategic SDI development to the broader issues contributing to the context of any SDI initiative. Therefore, understanding of the relationships between different SDI jurisdictions, knowing more about SDI development issues and knowing about the potential and applicability of each SDI development model are important for effective SDI development and driving the flexibility required in the second generation of SDI development.

In order to facilitate understanding the relationship between development models for SDIs, an organizational classification of hierarchy levels enables the characterization of their dominant organizational structure. The relevance of this approach is that each layer of the organizational structure has distinct information requirements and hence demands support from a specific SDI level. It is thus possible to classify different levels of an SDI hierarchy (which is made up of inter-connected SDIs at corporate, local, state/provincial, national, regional and global levels), according to the roles played within different political and administrative levels and their similarities to the organizational structure (Figure 18.1).

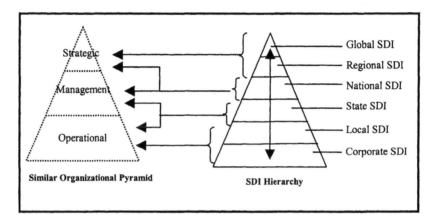

Figure 18.1: Relationships Between SDI Hierarchy and Their
Similarities to the Organizational Structure

According to the above figure and based on the nature of SDIs, any multi-national SDI (regional or global), can be considered similar to the strategic tier of an organizational structure. Due to the particularity of the National SDI level (as discussed in Chapter 6) an SDI at a national level has characteristics of both managerial and strategic tiers. Similarly, state-level SDIs can emulate management or operational organizational tiers, or both due to the wide ranging responsibilities many have in a nation which is a federation of states. The local and corporate levels of an SDI hierarchy are similar to the operational tier of an organizational structure.

Both management and operational tiers tend to adopt product-based models due to their key roles in data development. Only the strategic tier and nations with federal systems tend to adopt the process-based model of SDI development. The main reason multi-national and federated nations can benefit more from using a process-based model is that SDI participation at these levels of SDI hierarchy is voluntary.

Whatever SDI model is adopted, whether it is product or process based or is somewhere on the continuum between, the model which is best for the individual jurisdiction will need to consider all the complex issues which are discussed above if an infrastructure is to be established which facilitates the ongoing development of the SDI concept.

18.5 CONCLUSION

This chapter has summarised the main conclusions of the book, has identified some key issues which need to be considered in developing SDIs and has explored some different models to facilitate SDI development.

But what does the book contribute to our understanding of SDIs? We believe this contribution is highlighted by three themes that are reflected throughout the book and which have influenced all chapters. These are firstly the discussions surrounding the concept of SDI, secondly the investigations associated with the inter- and intra-jurisdictional relationships between SDIs in an SDI hierarchy and lastly an investigation on how SDIs can contribute to sustainable development objectives and good governance through technical, institutional, administrative and financial innovation.

The book adopts a broad view of SDI to encompass the five dimensions of access, data, standards, policies and people while recognising that the SDI concept is evolving. It has identified many of the global drivers which are currently influencing the development of the SDI concept and explores in-depth the importance of a user focus and a recognition of what constitutes an infrastructure. While this evolving concept is challenging to SDI developers, the book has attempted to provide some practical advice on how to build and maintain an SDI.

The introduction of the SDI hierarchy is an easier concept to understand and to use in practical applications. The book explores how the hierarchy influences all the dimensions of an SDI and recognises the importance of dynamic inter- and intra-jurisdictional partnerships. Importantly the case studies in the book are based around the levels in the SDI hierarchy and provide examples for analysing the concepts and theories explored in the book.

A discussion about the last theme is an appropriate place to conclude the book. We have endeavoured to show that SDIs can contribute to decision-making that supports sustainable development and good governance – and thus hopefully enables SDIs to contribute to making the world a better place.

18.6 REFERENCES

Keen, P.W.G., 1987, Decision Support Systems: The Next Decade, *Decision Support System,s* 3:253-265.

Kelmelis, J., 1999, The Changing Decision Environment. In Report of Decision Support Systems Workshop, Edited by Lessard, G. and Gunther, T., February 18-20, 1998, Denver, Colorado, U.S. Department of the Interior and U.S. Geological Survey. USGS Open-File Report 99-351, 1999: 5-16.

Moellering, H., Aalders, H.J.G.L. and Crane, A. (Eds), 2003, *World Spatial Metadata Standards*, International Cartographic Association, ISBN-008439497, Forthcoming August 2003.

NAS, 2002, Down to Earth – Geographic Information for Sustainable Development in Africa, National Academy of Sciences. National Academy Press, Washington, USA, Online. <http://www.nap.edu> (Accessed December 2002).

Rajabifard, A. Feeney, M. and Williamson, I.P., 2002, Future Directions for SDI Development. *International Journal of Applied Earth Observation and Geoinformation*, Vol. 4, No. 1, pp. 11-22.

Tosta, N., 1999, NSDI was supposed to be a verb, In *Integrating Information Infrastructures with GI Technology*, Edited by Gittings, B., (London: Taylor & Francis).

Index

Access Infrastructure 152
Access Networks 69, 188, 254, 259, 266, 305
Access Policy 118–19, 125
Accessibility 47, 123, 142, 200, 205
Accuracy 149–50, 153, 242, 251, 253, 269–70, 273
  Positional 269, 281–82, 286
  Relative 289–90
Administration systems, public and private 21, 61, 63, 131, 135, 192, 214, 220, 235, 273
Administrative Boundary 249, 254, 257
Agenda 21 48–9, 65, 74, 80, 195–97, 294
ANZLIC 132, 135
Applicability 47, 201, 205
Assessment cycle 237
Augmenting SDI Model 265
Australia 3, 10, 100–101, 103, 113, 117, 131, 138, 143–44, 147, 159, 167, 189, 219, 252–53, 268, 285
Australian SDI (ASDI) 26, 131, 133, 138–39, 142, 144
Australian Spatial Data Directory (ASDD) 133, 140–41
Availability 33, 44, 47, 50, 56, 60, 69, 87, 139, 148, 200, 205, 265
Awareness 8, 10, 18, 45–6, 48, 61–2, 75, 87, 91, 97–8, 101, 105, 108, 118, 124, 154, 172, 176, 184, 188, 214, 268

Brazil 185
Business Models 24, 144, 308

Cadastre 17, 111, 115, 136, 157, 167–68, 189, 244, 257
Canada 102, 111, 188–89, 191, 220, 289
Capacity Building 45, 50–1, 62, 67, 87, 106, 126, 204, 206, 270
Capacity Factors 87
Classic Infrastructure 217, 219–20
Clearinghouse 26, 69, 87–8, 142, 189, 200, 206, 223
Colombia 71, 75, 222
Cooperative Research Centre for Spatial Information (CRC-SI) 138
Coordinate Systems 285, 295
Corporate GIS 165–66, 176
Corporate SDI 25, 29, 165, 167, 172–73, 177

Data aggregation 306
Data custodians 178, 259, 269
Data Environment 25, 199, 202, 206, 216
Data Interpolation 250–51
Data re-aggregation 252
Data Quality 269, 281–83
DCDB (Digital Cadastral Database) 292
Decision Environment 198–99
Decision Process 20, 195–96, 199, 206–207
Decision support 45, 47, 195, 197, 200, 203
Decision support capability 196–97
Decision Support System (DSS) 19, 47–8
Decision-making 3, 18–20, 48, 51, 59, 79, 95, 105, 117, 135, 183–85, 195–97, 208, 255, 270

Developing nations/countries
    5, 45, 49, 190, 287
Diffusion   79, 82, 85
Digital Earth   3, 14, 102, 145

Economic Drivers   120
Economic Issues   212–213
E-governance   183, 191
E-government   73, 157, 191, 308
Emerging nations   213–14, 225
Environmental Factors   86, 89
EUROGI   52, 59, 75
Evaluation   56, 74, 101, 124, 235,
    237, 241, 263
  Aspects   237, 239, 241
  Elements   237–38, 241
    Objectives   237
    Outcomes   237
    Strategies   237
  Framework   237
External developments   308

Financing models   212–13, 222
First generation SDIs   215, 309
Fundamental Datasets   53, 68, 139
Funding Models   213, 215–16, 219

Geodesy   283, 290
Geodetic infrastructure   281–82,
    287–88, 292
Geodetic network   286, 288,
    293–95
Global drivers   301, 311
Global Map   51, 67–8
Global SDI (GSDI)   14, 26, 43
Good Governance   14, 60, 115,
    183–84, 306
Good Practice   4, 177, 239, 241
Governance   14, 60, 115,
    183–84, 306
GPS (Global Positioning System)
    10, 13, 122, 143, 156–57,
    159–60, 283, 286, 288–91

Great Britain   113, 215, 222

Hierarchical Reasoning   30
Hierarchical Spatial Reasoning
    (HSR)   30, 256

Influencing Factors   91, 85, 103,
    239–40, 242
Infrastructure Economics   212
Institutional Arrangement   5, 12,
    25, 52, 115, 144, 158, 211
Institutional Framework   26, 65–6,
    108, 135
Indonesia   102
Interoperability   14, 48, 62, 105,
    107, 112, 118, 144, 180, 202,
    223, 268, 306
Iran   102
ISCGM (Intergovernmental Steering
    Committee on Global
    Mapping)   52
ISO TC 211 (International Standards
    Organisation Technical
    Committee on International
    Geographic Information
    Standards)
    52, 56, 67–9, 156, 290

Japan   51–2, 55, 64, 284, 295

Kiribati   296
Korea   102, 295

Land Administration   6–7, 11, 69,
    73, 111, 147, 235–36
Land administration system
    132, 235–36, 244
Licensing   26, 122, 142, 150, 153,
    213, 254, 307
Local Councils   174–75, 177
Location Based Services   160, 236,
    263–72, 276

Macro-economic issues 212
Metadata 63, 68, 124,
140, 152, 281
Micro-economic issues 212

National Mapping Agencies 52,
56, 61, 71, 74, 107, 113, 211
National SDI 8, 25, 34–5, 43, 95,
97, 101–02
Network Infrastructure 221, 224
New Zealand 25, 80, 111, 135

Open GIS Consortium (OGC)
62, 68, 223, 268
Organizational levels 238–39, 241
Management level 240, 242
Operational level 240, 242
Policy level 240, 241
Organizational Models 52, 133
Organizational Pyramid 238, 241
Organizational Structure 46, 71,
81, 87–8, 228, 309

Partnerships 121, 123, 188
Partnerships, intra- and inter-
jurisdictional 7, 121
PCGIAP (Permanent Committee on
GIS Infrastructure for Asia and
the Pacific) 64, 81, 294
PC IDEA 70
People 26, 254, 259, 270, 306
Data providers 9, 96, 107, 241,
266, 272, 281
End users
9, 96, 266, 270, 272, 281
Integrators
96, 104, 142, 242, 266
People Environment 204, 272
Performance Assessment 241, 243,
244–45
Performance Indicators 114, 144,
235, 239, 245
Policies 187, 256, 259, 267, 307
Pricing 65, 142, 153

Privacy 11, 14, 65, 112, 142,
172, 187, 254, 267–68, 307
Portugal 102, 215
Positioning Framework 281, 283
Privacy 11, 14, 65, 112, 142, 172,
187, 254, 267–68, 307
Private Sector 5–8, 12–3, 19, 21,
73, 114, 118–20, 131, 137,
155
Project Management 177, 179

Reference Frame
65, 68, 283, 292, 294
Reference System 70, 84, 159,
256, 268–69, 283, 288–89
Regional SDI
7, 33, 59, 79, 85, 292
Regions
Asia and the Pacific 64, 67,
75, 79, 102
Europe 7, 52, 59–60
the Americas (PC IDEA) 70

SDI Challenges 10, 195
SDI Components 10, 23–4, 26–7,
45, 65, 211, 241, 254, 275
Access Networks 69, 254, 266,
305
Data 189, 255, 259, 269, 306
People 254, 270, 306
Policies 187, 256, 267, 307
Standards 256, 268, 306
SDI Concept 3–4, 7, 10, 13, 17,
27, 88
SDI Cookbook 4, 21, 47
SDI Coordinating Agencies 82, 85,
90–1, 102, 206, 212, 307
SDI Design 97, 105, 107, 190,
263
SDI development 13, 53, 176, 299
building block view 32, 166
first generation 95, 101–02, 215,
241, 309

inter- and intra-jurisdictional 7,
    21, 34
macro-scale 177–78
micro-scale 177–78
organizational models 45, 52,
    65, 131, 133
productional perspective 166
second generation 95, 102–03,
    105, 214, 309
success factors 85, 87, 89, 97,
    107, 120, 144, 177, 216
umbrella view 32, 166
SDI development models 88, 97,
    309
  Process-based 7, 88, 91, 103
  Product-based 7, 88, 91, 97, 103,
    309
SDI Diffusion 79, 82, 85
SDI Hierarchy 28, 98
SDI Issues 309
SDI political or administrative Level
    34, 37
SDI, Corporate 165, 173
SDI, Global 14, 26, 43–4
SDI, Local 99, 165, 167
SDI, National 8, 25, 34–5, 43, 95,
    97, 101–02
SDI, Nature 24
SDI, Regional 59
SDI, State 35, 111, 114, 116, 119,
    121, 124–25, 147, 156
SDI, User 9
SDI, Vision 10
SDI, Sustainable Development
    4, 6, 13, 20, 43, 45, 67, 96,
    183, 186, 195

Sharing spatial data 18
Singapore 191
Social system 14, 83, 85, 90
Spatial Accuracy 149–50, 153,
    242, 251, 253, 269–70, 273
Spatial data 4–5, 17–9
Spatial data and DSS 19
Spatial hierarchy 257–58
  problem 250
Standards 67, 188, 256, 259, 268,
    306
State SDIs 35, 111, 114, 116, 119,
    121, 124–25, 147, 156
Sustainable Development 4, 6, 13,
    20, 43, 45, 67, 96, 183, 186,
    195

Technical Standards 67, 188, 256,
    259, 268, 306
Technological Environment 6, 199,
    202
Technological Frameworks 186,
    192
Technology Drivers 9, 301
The Netherlands 61, 111, 223
Triple Bottom Line 6, 192, 308

United States of America 100, 103,
    222
User Environment 23, 208, 272

Wireless Application Protocol
    (WAP) 266
Wireless Applications 3, 9, 118,
    160, 263, 271

9 781138 372504